Olfa Kanoun (Ed.)

Energy Harvesting for Wireless Sensor Networks

Also of Interest

Impedance Spectroscopy
O. Kanoun (Ed.), 2018
ISBN 978-3-11-055712-1, e-ISBN (PDF) 978-3-11-055892-0,
e-ISBN (EPUB) 978-3-11-055716-9

Progress Reports on Impedance Spectroscopy
O. Kanoun (Ed.), 2016
ISBN 978-3-11-044756-9, e-ISBN 978-3-11-044982-2,
e-ISBN (EPUB) 978-3-11-044767-5

Low Power VLSI Design
A. Sarkar, S. Dep, M. Chanda, C. K. Sarkar, 2016
ISBN 978-3-11-045526-7, e-ISBN 978-3-11-045529-8,
e-ISBN (EPUB) 978-3-11-045545-8, Set-ISBN 978-3-11-045555-7

Nano Devices and Sensors
J. J. Liou, S.-K. Liaw, Y.-H. Chung (Eds.), 2016
ISBN 978-1-5015-1050-2, e-ISBN 978-1-5015-0153-1,
e-ISBN (EPUB) 978-1-5015-0155-5, Set-ISBN 978-1-5015-0154-8

Energy Harvesting for Wireless Sensor Networks

Technology, Components and System Design

Edited by
Olfa Kanoun

DE GRUYTER
OLDENBOURG

Editor
Prof. Olfa Kanoun
Technical University of Chemnitz
Faculty of Electrical Engineering and Information Technology
Reichenhainer Str. 70
09126 Chemnitz, Germany
Olfa.Kanoun@etit.tu-chemnitz.de

ISBN 978-3-11-044368-4
e-ISBN (PDF) 978-3-11-044505-3
e-ISBN (EPUB) 978-3-11-043611-2

Library of Congress Control Number: 2018954539

Bibliographic information published by the Deutsche Nationalbibliothek
The Deutsche Nationalbibliothek lists this publication in the Deutsche Nationalbibliografie; detailed bibliographic data are available on the Internet at http://dnb.dnb.de.

Cover image: Sandipkumar Patel/DigitalVision Vectors/Getty Images
Typesetting: le-tex publishing services GmbH, Leipzig
Printing and binding: CPI books GmbH, Leck

www.degruyter.com

Preface

Wireless sensors and wireless sensor networks (WSNs) are nowadays becoming increasingly important due to their decisive advantages. They enable mobility and flexibility, and significantly reduce installation and maintenance costs. Different on-going trends towards the Internet of Things (IoT), Industry 4.0 and 5G Networks are based on WSNs. All these trends address massive sensing and admit to having wireless sensors deliver measurement data directly to the World Wide Web in a reliable and easy manner. They can only be supported, if sufficient energy efficiency and flexible solutions are developed for energy-aware wireless sensor nodes (WINO). Within system design of wireless sensors, several aspects related to technology, electronics and algorithms need to be considered to reach a high energy efficiency on the node and network level. In the last years, different possibilities for energy harvesting have been investigated and several works have been reported on theoretical and practical applications using these technologies. Generally, special aspects are treated and the overview of these methods remains on the level of personal effort.

Writing a book on energy harvesting is a huge work and can be only done in an aim-oriented way. In a previous book *Energy Harvesting*, we described different technologies and concepts more from the perspective of feasibility to introduce this technology to applications. At the time, the potential of ambient energy was not as clear and evident as it is today. This success is due to the developments towards low power consumption and also towards efficient and suitable converters based on manifold principles and able to generate enough energy. In this book, in addition to these topics, we focus on advanced and novel developments in vibration converters, wireless energy transfer and energy saving methods on the network level. This is no longer from the point of view of feasibility, but with the aim of giving an overview of novel developments that make technologies for using ambient energy and transferred energy to supply sensors realistic. These methods, systems and technologies are enabling technologies for the wide spread use of sensors in general and, especially, in the IoT era.

The specific motivation for writing this book is based on the convincement that an interlocked interaction between suitable methods and technologies is necessary to realize optimized solutions for energy aware WSNs. In the last years, this has become more visible with the increasing number of converters, hybrid solutions, energy transfer and energy management methods acting at node and network level.

The system design of an energy aware WINO begins with the choice of the suitable available ambient energy, such as solar radiation, vibration and electromagnetic fields and its characterization. Therefore, suitable converters need to be designed and improved in their performance to meet necessary requirements and to reach enough energy to guarantee the function of the WSNs. It is thereby not possible every time to supply a sensor node from just one ambient source. In order to enhance the energy income, a combination of similar or diverse converters can be envisaged. Energy

https://doi.org/10.1515/9783110445053-201

transfer also plays an important role, as it supports wireless sensor nodes by additional energy in cases where ambient energy sources are not sufficient or the behavior of the converters is affected by environmental conditions or aging. Additionally, different techniques for enhancing the efficiency of energy converters and reducing energy consumption on node and network level are developed and need to be considered together to realize optimal solutions.

This book gives an overview on fundamentals and techniques for energy harvesting and energy transfer from different points of view. A distinct focus is dedicated to vibration converters as an often available ambient energy source where special optimization of the converter is needed. Different techniques and methods for energy transfer, management and energy saving on network level are reported together with interesting applications.

The first part of the book introduces essential fundamentals and methods. The first chapter describes techniques for finite element modeling of energy harvester at the example of vibrational devices. The focus is on the investigation of active materials, which have an intrinsic ability for coupling several branches of physics and, consequently, are commonly used for manufacturing harvesters. In particular, an investigation is carried out for the energy production from mechanical vibrations in high-speed railway bridges. The second chapter focuses on solar energy harvesting as one of the most predictable ambient energy sources with a good availability and energy level. The third chapter describes methods for the determination of incoming power and the most used methods for maximum power point tracking together with an efficiency analysis. The fourth chapter deals with energy management concepts, including maximum power point tracking, voltage and current conversion, energy income forecast and modeling and tuning the energy consumption profile.

The second part of the book also deals with vibration converters and hybridization. It begins with the fifth chapter, which deals with magnetoelectric vibration converters, which are based on promising conversion mechanisms for kinetic energy harvesting combining magnetostrictive and piezoelectric materials. The chapter begins with the basics of magnetoelectricity and an overview of the recent developments in the state of the art of magnetoelectric energy harvesting. After that, a model-based design method for magnetoelectric converters is presented.

In the sixth chapter, a nonlinear electromagnetic vibration converter based on the use of magnetic spring principle is discussed. It enables energy harvesting for a large bandwidth compared to the usual linear vibration converters. In order to improve efficiency, bistable random mechanical switching harvesting on inductors (RMSHI) is proposed to realize synchronization between mechanical movements and energy management. In the seventh chapter, an oscillating vertical piezoelectric cantilever with clearance is proposed for vibration energy harvesting. A mechanical resonator is realized as an inverted elastic pendulum excited by kinematic excitation and then characterized.

One of the most important possibilities to maximize the energy income from ambient vibration is to design a hybrid vibration converter. This is the subject of the eighth chapter that deals with hybridization of electromagnetic vibration converters dedicated to low vibration frequencies. Several macro hybrid vibration converters based on the combination with the electromagnetic principle are reviewed and compared. Thereby, electromagnetic and magnetoelectric energy harvesting techniques deliver energy with relative high efficiencies and are, therefore, very promising principles. The ninth chapter presents a combination of a piezoelectric and a magnetostrictive coil transducer. Thereby, the mechanical resonator moves in two degrees of freedom, which have been monitored in two channels to show the broadening of the total frequency transduction.

The third part of the book is dedicated to wireless energy transfer, including both RF and inductive energy transfer as an interesting solution to extend the lifetime of wireless nodes and to overcome the lack of ambient energy sources. Inductive energy transfer is contactless and reaches a high efficiency over short distances. Thereby, data can be transferred at radio frequencies concurrently to the energy transfer. This is the subject of the tenth chapter, where a secure simultaneous wireless information and power transfer (SWIPT) system is presented. The focus is on the use of multiple antennas to improve the efficiency of wireless power transfer (WPT) and secure information transmission. In particular, the objective is to maximize the secrecy rate via beamforming by a model-based approach to achieve a significant gain in the secrecy rate compared to conventional methods.

Over a relative high distance, radio frequency (RF) energy transmission and harvesting techniques enable proactive energy replenishment for wireless sensors. With RF power transfer, long distances can be overbridged. The eleventh chapter focuses on the use of RF energy harvesting into usable electrical form, providing a certain voltage or enabling delivery of a desired current to load.

The finite element method simulation of inductive wireless power transmission is the subject of the twelfth chapter, where a wireless power transfer system is analyzed, including important parts such as coil, core and driver, showing good agreement with experimental results.

Efficiency improvement of wireless power transmission via inductive links is the focus of the thirteenth chapter. It investigates the use of multi-coil inductive systems, in which the sending and/or the receiving sides have multiple coils in the case of misalignment, to provide energy to a movable receiver and to increase the system flexibility. Suitable energy management is also indispensable for wireless inductive power transmission via inductive links to reduce losses and compensate coil position perturbations. The fourteenth chapter examines energy management circuits applied to the sending and receiving sides of inductive power transmission systems.

The fourth part of the book treats energy saving and management strategies. It begins with Chapter 15, in which energy-saving concepts for WSNs based on scheduling, dynamic voltage, frequency scaling and dynamic power management are reported. Chapter 16 emphasizes the optimal energy allocation in WSNs by introducing a system engineering approach modeling and optimizing the system in order to derive policies on when and how to use the available energy in order to achieve the best possible performance. In Chapter 17, energy management concepts in WSNs are presented for large networks. The chapter presents a survey of energy conservation and energy optimization techniques for WSNs.

In the eighteenth chapter, the design of wake-up receivers for energy harvesting is discussed. The main idea thereby is that a wake-up receiver consumes much less power than the WINO. During non-active time, the WINO can, therefore, be turned off while only the wake-up receiver is switched on to detect possible signals and wake up the node. Hence, a high energy efficiency is realized.

Applications for energy harvesting and energy transmission are nowadays manifold. In the final part of the book, in the last part about system design and applications, a selection of specific solutions related to agriculture, structural health monitoring of bridges and power grids is given.

In total, the book reports about new advances and approaches in energy-aware wireless sensors and sensor networks including theory, methods and applications. The book is interesting for researchers, developers and students in the field of sensors, wireless sensors, WSNs, IoT and manifold application fields using related technologies.

The author would like to thank all experts for the interesting contributions and the reviewers who supported the decision about publication with their valuable comments.

Chemnitz, June 2018 Prof. Dr.-Ing. Olfa Kanoun

Contents

Preface — V

Part I: **Fundamentals and methods**

Roberto Palma, José L. Pérez-Aparicio, and Pedro Museros
Finite element modeling of energy harvesters: application to vibrational devices — 3
1 Introduction — **3**
2 Active materials — **5**
3 Continuum physics — **9**
3.1 Outline of continuum mechanics — **10**
3.2 Outline of continuum electrodynamics — **13**
3.3 Outline of thermodynamics — **16**
3.4 Summary of governing equations — **18**
4 Finite element method — **20**
4.1 Outline of the finite element method — **20**
4.2 Finite element modeling of active materials — **22**
5 Application: Energy production in high-speed railway bridges — **24**
5.1 Overview of high-speed railway bridges — **24**
5.2 Numerical modeling of harvesters — **30**
6 Concluding remarks — **33**

Christian Viehweger
Solar energy harvesting for wireless sensor systems — 37
1 Introduction — **37**
2 Assessment of the incoming energy — **38**
3 Maximum power point tracking — **40**
3.1 The most common methods — **41**
4 Conclusion — **44**

Zdenek Hadas and Jan Smilek
Efficiency of vibration energy harvesting systems — 45
1 Introduction — **45**
2 Vibration energy harvesting systems — **46**
2.1 Mechanical resonators of vibration energy harvesting systems — **47**
2.2 Complex model of a vibrating structure with an energy harvesting system — **48**
2.3 Efficiency of vibration energy harvesting systems — **49**

3 Model of an electromagnetic energy transducer — 50
4 Efficiency analysis based on vibration energy harvesting simulation — 51
4.1 Mechanical model of the vibration energy harvesting system — 51
4.2 Equivalent RLC circuits of the vibration energy harvesting system — 53
5 Analyses and simulation results of equivalent RLC circuits — 54
5.1 Mechanical and electrical parameters for analyses — 54
5.2 Analyses of efficiency and output power for mechanical parameters — 55
5.3 Analyses of efficiency and output power for coupling parameters — 58
6 Comparing the performance of vibration energy harvesters — 59
7 Conclusions — 61

Martin Götz
Energy management concepts for wireless sensor nodes — 65
1 Introduction — 65
2 Energy harvesting from ambient power sources — 67
3 Wireless power transfer — 68
4 Energy storage — 70
5 Loads — 70
6 Energy management approaches — 71
6.1 Active and passive energy management — 71
6.2 Microcontroller-based energy management — 72
6.3 Multi-level energy management — 72
7 Extraction power management strategies — 73
7.1 Maximum power point — 73
7.2 Maximum power point tracking — 74
7.3 Maximum energy extraction — 75
7.4 Hybrid harvesting — 78
7.5 Non-linear energy management — 80
8 Core system energy management strategies — 80
8.1 Energy storage management — 80
8.2 Multi-storage management — 86
8.3 Voltage supervision — 89
8.4 System optimization — 91
9 Wireless energy management strategies — 92
9.1 Wireless communication/energy conservation in link and network layers — 92
9.2 Energy aware routing, multi-hop deployments and clustering — 94
10 Conclusion — 95

Part II: **Vibration converters and hybridization**

Slim Naifar, Sonia Bradai, Slim Choura, and Olfa Kanoun
Magnetoelectric vibration energy harvesting — 103
1 Introduction — 103
2 Magnetoelectric composites — 104
2.1 Magnetoelectric effect — 104
2.2 Magnetoelectric materials — 105
2.3 Magnetoelectric laminate composites — 106
3 Overview of magnetoelectric energy harvesters — 106
3.1 Rectangular magnets — 107
3.2 Hollow magnets — 108
4 Finite element model of magnetoelectricity — 110
4.1 Modeling of magnetostriction — 110
4.2 Modeling of adhesively bonded joints — 111
4.3 Simulation model — 111
5 Results and discussion — 112
6 Conclusion — 112

Carlo Trigona, Sonia Bradai, Slim Naifar, Olfa Kanoun, and Salvatore Baglio
Nonlinear electromagnetic vibration converter with bistable RMSHI for power harvesting from ambient vibration — 117
1 Introduction — 117
2 The proposed nonlinear vibration converter — 118
2.1 Principle — 118
2.2 Experimental setup — 119
3 Results — 120
4 Conclusion — 123

Krystian Łygas, Piotr Wolszczak, Paweł Stączek, and Grzegorz Litak
Energy harvesting from an oscillating vertical piezoelectric cantilever with clearance — 125
1 Introduction — 125
2 Methods and experimental setup — 126
3 Results on power output — 128
4 Solution identification — 128
5 Frequency sweep: Evolution of the solution — 131
6 Conclusions — 134

Sonia Bradai, Slim Naifar, Riadh Elleuch, and Olfa Kanoun
On hybridization of electromagnetic vibration converters — 137
1 Introduction — 137
2 Hybrid converters — 138
2.1 Electromagnetic-piezoelectric vibration converter — 138
2.2 Electromagnetic-magnetoelectric vibration converter — 142
2.3 Electromagnetic-magnetostrictive vibration converter — 143
2.4 Comparison of hybrid systems — 144
3 Hybrid magnetoelectric-electromagnetic converters — 146
3.1 Design of an EM/ME vibration converter — 147
3.2 Effect of combined EM-ME on the energy outcome — 147
4 Conclusion — 150

Andrzej Rysak, Marek Borowiec, Arkadiusz Syta, and Grzegorz Litak
Hybrid vibrational energy harvesting using piezoelectric and magnetostrictive transducers — 153
1 Introduction — 153
2 Experimental setup and measurement results — 154
3 Conclusions — 157

Part III: **Wireless energy transfer**

Elena Boshkovska, Nikola Zlatanov, Xiaoming Chen, Derrick Wing Kwan Ng, and Robert Schober
Beamforming design for secure SWIPT systems under a non-linear energy harvesting model — 161
1 Introduction — 161
2 Channel model — 164
2.1 Nonlinear energy harvesting model — 165
2.2 Achievable secrecy rate — 167
3 Problem formulation and solution — 168
4 Results — 171
5 Conclusions — 174
6 Appendix — 174
6.1 Proof of proposition 1 — 174
6.2 Proof of theorem 1 — 175

Issam Chaour, Ahmed Fakhfakh, and Olfa Kanoun
Radio frequency power transfer for wireless sensors in indoor applications — 181
1 Introduction — 181
2 RF to DC conversion and storage — 182
3 Reflection coefficient and matching circuit — 185

4 RF-DC rectifier circuits for RF power transfer — **186**
4.1 Schottky diode — **186**
4.2 Topologies for RF to DC rectifiers — **188**
4.3 RF to DC rectifier based on a diode-connected CMOS transistor — **192**
5 Conclusion — **194**

Mohammad Haerinia
Modeling and simulation of inductive-based wireless power transmission systems — 197
1 Introduction — **197**
2 Fundamental principles of the inductive coupling approach — **198**
3 Modeling and simulation of inductive power transfer (IPT) systems — **200**
3.1 Resonant inductive coupling as a potential means for WPT to printed spiral coils — **200**
3.2 Investigation of the receiving pot core effect on the magnetic flux density in inductive coupling-based wireless power transfer — **206**
3.3 Electromagnetic analysis of different geometries of transmitting coils for wireless power transmission applications — **211**
4 Conclusion — **217**

Bilel Kallel, Ghada Bouattour, Olfa Kanoun, and Hafedh Trabelsi
Wireless power transmission via a multi-coil inductive system — 221
1 Introduction — **221**
2 Effect of vertical and lateral distances on Inductive Power Transfer (IPT) systems — **222**
3 Classification of the multi-coil inductive systems — **223**
4 Examples of the use of multi-coil inductive systems — **224**
4.1 Three-coil systems — **224**
4.2 Four-coil systems — **225**
4.3 Multi-coil systems — **226**
5 Modeling of multi-coil inductive systems — **227**
5.1 Basic laws — **227**
5.2 Analytical modeling — **229**
5.3 Analytical expression for equivalent inductance L_{eq} — **229**
5.4 Analytic expression for equivalent mutual inductance M_{eq} — **230**
5.5 Analytical expression for induced voltage and current — **231**
6 Optimization of multi-coil inductive systems — **231**
6.1 Received power optimization — **231**
6.2 Topology optimization — **233**
6.3 Detection of the receiver — **234**
7 Conclusion — **234**

Ghada Bouattour, Bilel Kallel, Olfa Kanoun, and Houda Derbel
Energy management for inductive power transmission — 237
1 Introduction — **237**
2 Coil shape and orientation — **239**
2.1 Coil shape — **239**
2.2 Coil orientation — **239**
3 Energy management at the sending side — **240**
3.1 DC/AC inverters — **240**
3.2 Resonance topologies — **243**
4 Energy management on the receiving side — **246**
4.1 Rectification stage — **246**
4.2 DC to DC converter — **248**
4.3 Charge storage — **249**
5 Control system on inductive power transmission systems — **251**
6 Conclusion — **252**

Part IV: **Energy saving and management strategies**

Rym Chéour, Mohamed Wassim Jmal, Olfa Kanoun, and Mohamed Abid
Towards energy-efficient power management for wireless sensors networks — 257
1 Introduction — **257**
2 Energy saving techniques in wireless sensor networks — **259**
3 HEEPS: Hybrid energy-efficient power manager scheduling — **261**
4 Performance evaluation — **264**
5 Conclusion — **266**

Steffi Knorn and Daniel E. Quevedo
Optimal energy allocation in energy harvesting and sharing wireless sensor networks — 269
1 Introduction — **269**
2 Modeling — **270**
2.1 Energy harvesting model — **271**
2.2 Energy sharing model — **272**
2.3 Battery model — **273**
2.4 Channel gain model — **274**
3 Application: Multi-sensor estimation network with energy harvesting and energy sharing — **275**
3.1 Source model and sensor measurements — **275**
3.2 Energy harvester, energy sharing and battery dynamics — **276**
3.3 Transmission model — **276**

3.4 Distortion measure at the FC — 277
3.5 Dynamic energy control — 277
4 Optimization techniques — 278
4.1 Finite time horizon optimization — 278
4.2 Infinite time horizon optimization — 279
4.3 Suboptimal solutions — 280
5 Simulation study — 281
5.1 Varying energy transfer efficiency — 282
5.2 Varying battery capacity — 284
6 Conclusions — 285

Sabrine Khriji, Dhouha El Houssaini, Ines Kammoun, and Olfa Kanoun
Energy-efficient techniques in wireless sensor networks — 287
1 Introduction — 287
2 Energy consumption in wireless sensors networks — 289
3 Energy supply sources in wireless sensors network — 291
4 Energy conservation in wireless sensors networks — 292
4.1 Data driven approaches — 293
4.2 Duty cycling — 295
5 Energy-efficient routing — 298
5.1 Flat routing — 298
5.2 Hierarchical routing — 299
5.3 Mobility-based energy conservation schemes — 301
6 Conclusion — 301

Sadok Bdiri, Faouzi Derbel, and Olfa Kanoun
A wake-up receiver for online energy harvesting enabled wireless sensor networks — 305
1 Introduction — 305
2 System implementation — 306
2.1 Envelope detector — 307
2.2 Baseband amplifier — 309
2.3 BBamp controller — 309
2.4 The decoder — 310
2.5 Wireless module — 311
3 System evaluation — 313
3.1 Measurements — 313
4 Comparison with the literature — 317
5 Conclusion — 319

Part V: **System design and applications**

Dhouha El Houssaini, Sabrine Khriji, Kamel Besbes, and Olfa Kanoun
Wireless sensor networks in agricultural applications — 323
1 Introduction — **323**
2 Applications of wireless sensor networks — **324**
3 Wireless sensor network deployment scenarios — **325**
3.1 Terrestrial wireless sensor networks — **326**
3.2 Underground wireless sensor networks — **326**
4 Wireless technologies and standards — **327**
5 Wireless sensor platforms for agricultural applications — **329**
6 Taxonomy of agricultural monitoring applications — **329**
6.1 Soil preparation system — **330**
6.2 Crop yield system — **331**
6.3 Livestock monitoring — **331**
6.4 Greenhouse monitoring — **332**
6.5 Horticulture monitoring — **333**
6.6 Forest monitoring — **333**
7 Application-specific sensors in agriculture — **333**
7.1 Soil analysis — **334**
7.2 Detection and classification of crops, weeds and fruits — **334**
7.3 Weather and climate-related sensors — **335**
8 Solutions for irrigation systems based on WAN in the measurement and sensor technology laboratory — **335**
8.1 The nodes developed — **337**
8.2 Mote deployment and network architecture — **338**
8.3 Software application — **339**
9 Conclusion — **340**

Paul Cahill and Vikram Pakrashi
Piezoelectric energy harvesting for monitoring of rail bridge infrastructure — 343
1 Introduction — **343**
2 Piezoelectric energy harvesting — **345**
2.1 Piezoelectric cantilever energy harvester — **346**
2.2 Piezoelectric patch energy harvester — **346**
3 Bridge–vehicle interaction model — **347**
3.1 Modeling of bridge and train loadings — **347**
3.2 Attributes of an international train fleet — **348**
3.3 Train-induced vibrations in the bridge model — **349**

4 Energy output from a bridge structure using a cantilever harvester — **350**
4.1 Properties of the cantilever energy harvester — **350**
4.2 Energy harvesting output from the cantilever energy harvester — **352**
5 Energy harvesting from a bridge structure using a patch energy harvester — **354**
5.1 Properties of the patch energy harvester — **354**
5.2 Energy harvesting output from the patch energy harvester — **354**
6 Conclusions — **357**

Feng Yang and Lin Du
Hybrid energy harvesting methodologies for energizing sensors towards power grid applications — 359
1 Introduction — **359**
2 Energy profiles and methodologies — **360**
2.1 Overview — **360**
2.2 Magnetic energy harvesting — **360**
2.3 Thermoelectric energy harvesting — **363**
2.4 Vibrational energy harvesting — **364**
2.5 Hybrid energy management system — **364**
3 Envisioned prototype — **366**
4 Conclusions — **367**

Xinming Zhao, Thomas Keutel, and Olfa Kanoun
Energy harvesting for a wireless monitoring system of overhead high-voltage power lines — 369
1 Introduction — **369**
2 Theoretical consideration and simulation — **371**
2.1 Theoretical analysis — **371**
2.2 Analytical simulation — **372**
2.3 Finite element simulation — **375**
3 Experiment in the high-voltage laboratory — **376**
3.1 Measurement in the high-voltage laboratory using the test harvester and test power module — **376**
3.2 Measurement in the high-voltage laboratory using the prototype of harvester housing and power module — **378**
4 Performance test in the field test — **380**
4.1 Test strategy for the field test — **380**
4.2 Field test results — **381**
5 Conclusion — **382**

Part I: **Fundamentals and methods**

Roberto Palma, José L. Pérez-Aparicio, and Pedro Museros

Finite element modeling of energy harvesters: application to vibrational devices

Abstract: This chapter presents the set of governing equations to study the behavior of active materials that have an intrinsic ability for coupling several branches of physics and, consequently, are commonly used for manufacturing harvesters. Once the equations are defined, a numerical formulation based on the finite element method is developed in order to model these materials. In particular, this chapter studies the energy production from the mechanical vibrations present in high-speed railway bridges. For this purpose, a review of the basic parameters of these bridges, their vibrations, frequencies and the dynamic characteristics are highlighted. Then, cantilever harvesters made out of piezoelectric and piezomagnetic materials are simulated under typical mechanical vibrations, and several conclusions are highlighted.

Keywords: Active materials, finite element method, piezoelectric devices, piezomagnetic devices, high-speed bridges, vibrating harvesters

1 Introduction

As is well known, the energy obtained from residual sources is called renewable or clean energy. Depending on the level of power produced, this renewable energy is commonly divided into two groups:

- Macro-energy harvesting plants, generating in the order of [kW] or [MW], for example: watermill, geothermal and solar energies. These plants are designed as alternatives to traditional fossil-fuel-based and nuclear plants.
- Micro-energy harvesting technologies, which produce in the order [mW] or [μW], and are based on residual sources such as mechanical vibrations, heat, sunlight, chemical or biological sources, etc. In contrast to macro-energy, this technology is conceived as an alternative to conventional electro-chemical batteries.

The present chapter is focused on micro-energy technology using harvesters. In general, harvesters are manufactured with modern materials, denominated active or smart materials that have the intrinsic ability to couple up to four physic energies, such as mechanical, thermal and electromagnetism. According to [1], the global market for smart materials was 26 USD billion in 2014 and will be approximately 42 USD billion in 2019.

Roberto Palma, Department of Mechanical Engineering and Construction, Universitat Jaume I, Spain
José L. Pérez-Aparicio, Pedro Museros, Department of Continuum Mechanics and Theory of Structures, Universitat Politècnica de València, Spain

https://doi.org/10.1515/9783110445053-001

There are many examples of energy harvesting applications depending on the residual sources. For instance, thermoelectric devices are used to generate energy from residual heat and piezoelectric/piezomagnetic materials from residual mechanical vibrations.

In connection with thermoelectric materials, they are commonly used to reduce the weight in aeronautics and astronautics. In particular, the radioisotope thermoelectric generator (RTG) is an electrical generator that draws its energy from the heat released by radioactive disintegration of the fuel – usually plutonium. The assembly of an RTG can be seen in Figure 1, consisting of a container of nuclear fuel that generates the necessary heat through a nuclear fission. This heat is converted into electrical energy by thermocouples in the radioisotope heat unit, powered by a thermal flux between the core at up to 1200 [K] and the hot spot, usually a liquid metal cooling system connected to the space at up to 300 [K]. In spite of this very large temperature increment, the efficiency usually lies between 4–7%, with a maximum of 10%.

Fig. 1: Radioisotope thermoelectric generator for space rover, [2].

Concerning vibrating harvesters, they are used for wearing shoes and backpacks, for computer hard disks, for medical micro-robots and, especially, for energy recovering from environmental vibrations, see [3]. In this context, the possibility of energy harvesting from vibrations of bridges is a very new field. Recently, [4] and others have been investigating the feasibility of recovering electric energy for high-speed trains passing over a bridge. Small amounts of electric energy can be recovered by the induced vibrations, and this energy be used for the WiFi signals of structural health monitoring (SHM) systems.

In short, the energy generation from residual sources – such as heating and mechanical vibrations – is a challenge for both engineering and scientific communities.

Therefore, both the theoretical understanding and numerical modeling are important challenges in order to design sophisticated harvesters that generate as much energy as possible. From a modeling point of view, some of the authors of the present chapter have published several numerical formulations based on the finite element (FE) method to study active materials, see [5] to [6].

From a practical point of view, it is also very important to know the energy orders of magnitude that the residual sources can yield, for instance, temperature gradients in buildings, mechanical vibration amplitudes in bridges, etc. In this way, one author of the present chapter has wide experience in the design of high-speed railway bridges, see [7] to [8].

On these grounds, this chapter defines the active materials and their advantages in comparison with classical materials, see Section 2. Then, an outline of the governing equations that theoretically describe the behavior of active materials is reported in Section 3. In particular, the main equations of continuum mechanics, electromagnetism and thermodynamics, are briefly described. Section 4 introduces the FE formulation to numerically model the active materials. First, for the sake of clarity, a brief outline of the FE method is introduced. Regarding practical applications, Section 5 presents a summary of railway bridges and their mechanical vibrations produced by passing vehicles. Finally, numerical simulations of harvesters that produce energy from mechanical vibrations are reported. In particular, basic cantilever beams made of both piezoelectric and piezomagnetic materials are simulated, and a comparison between both is highlighted.

Through the chapter, several simplifications are introduced. From a mechanical point of view, small strain is considered: it is a good approximation for most of the harvester applications, since classical active materials are made out of ceramics. Electrodynamically, free electromagnetic sources such as free electric charges and currents are not considered, since these active materials are, in general, non-conductors. Furthermore, the high frequencies produced by the electromagnetic field (speed of light) are neglected in comparison with the mechanical frequencies; this is a reasonable approximation, since this chapter deals with the energy production from the residual mechanical vibrations. Finally, and from a thermodynamical point of view, conservative materials are considered, namely, dissipations generated by the heating of harvesters are neglected.

2 Active materials

In classical physics, every cause has an effect. For instance, applying forces to an elastic body (cause) produces strains (effect); applying electric fields to a dielectric medium (cause) polarizes the material (effect), etc.

Historical note

Early work on active materials was carried out by the French brothers Jacques and Pierre Curie; in particular, piezoelectricity was discovered in 1880 by both. Jacques Curie (1855–1941) was Professor of Mineralogy at the Université de Montpellier. His younger brother Pierre Curie (1859–1906) received the Nobel Prize in Physics with his wife, Marie Curie, and Henri Becquerel. Signature taken from *Wikipedia*.

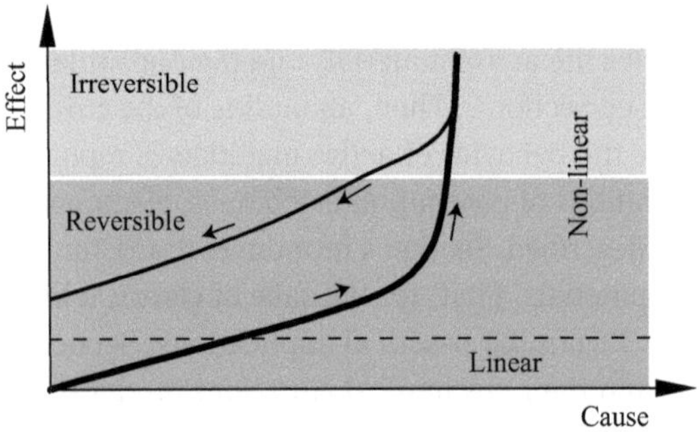

Fig. 2: Cause-effect curve. Two regions are observed: linear/non-linear and reversible/irreversible.

Causes and effects are related by constitutive equations that describe the behavior of the material and are obtained by experimental techniques. For example, in mechanics, the stress-strain curve is calculated by applying equal and opposite forces at the ends of a bar and measuring its relative deformation. From this curve, several important material properties, such as the modulus of elasticity and ductility, are obtained. Figure 2 shows a general cause-effect curve in which two regions are observed:

- Linear and non-linear. The linear region is mathematically represented by first-order material properties in a Taylor series expansion. On the contrary, the non-linear region requires high-order material properties that, in general, depend on the high-order expressions of the causes and/or effects.
- Reversible and irreversible regions. The former represents a hypothetical behavior for which the entropy of the system is conserved. For this reason, the material properties inside this region are denominated conservative. In contrast, the irreversible region is characterized by the increasing of the entropy and, consequently, by the production of heat. For example, plasticity is an irreversible response for which the internal structure of the material is irreversibly transformed and heat is dissipated. Therefore, the initial state is not recovered when the cause ceases; this process is path-dependent.

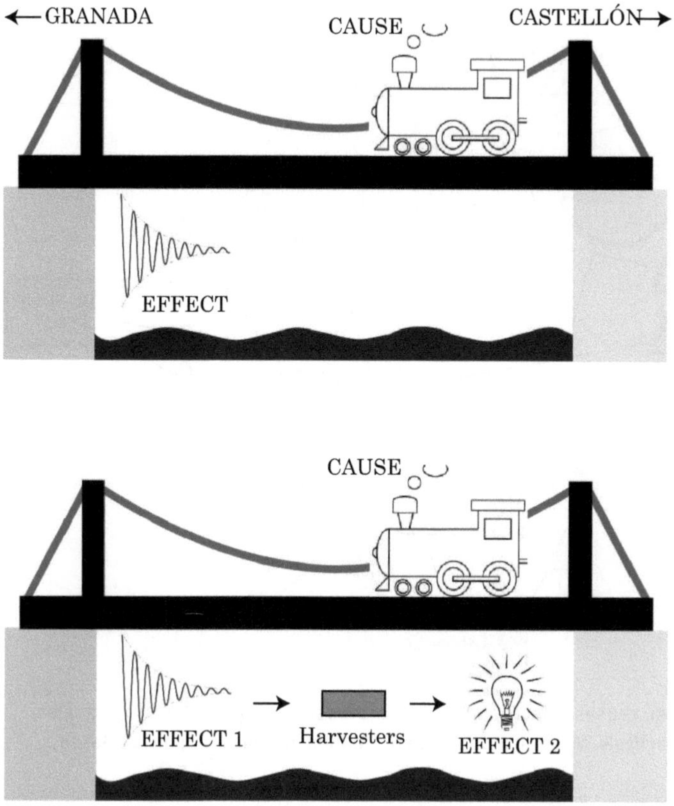

Fig. 3: Passage of a train over a bridge. Top: the train (cause) produces mechanical vibrations (effect). Bottom: the bridge incorporates harvesters made of active materials that produce electrical energy from mechanical vibrations and, consequently, the train generates two effects.

Traditional materials – denominated *passive* materials in the remainder of this chapter –, are characterized by a one-to-one function between cause and effect, namely, stresses generate strains and vice versa. For example, consider a railway bridge made of passive materials such as concrete and steel. The passage of a train over the bridge (cause) generates mechanical vibrations (effect), as observed in Figure 3 (top). On the contrary, active materials are characterized for their intrinsic ability to couple several branches of physics; consequently, a cause produces one or more effects. Continuing with the bridge example, consider that this bridge also incorporates harvesters made out of active materials. Now, the mechanical vibrations generate electrical energy, see Figure 3 (bottom). Therefore, the passage of a train generates two sequential effects: mechanical vibrations and electrical energy.

As commented, in this work, an active material can couple up to four fields: thermal, mechanical, electrical and magnetic. Heckmann's diagram [9] allows us

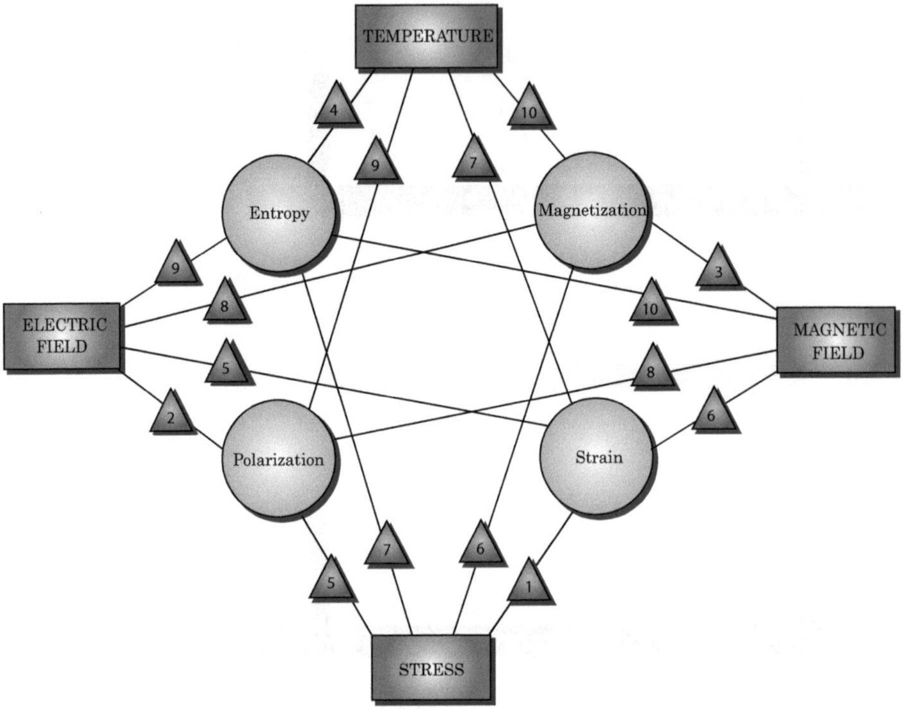

Fig. 4: Heckmann's diagram: causes, effects and first-order properties represented by rectangles, circles and triangles, respectively. Material properties are clarified in Tables 1, 2; picture taken from [9].

to schematically visualize these couplings, as shown in Figure 4. The causes are represented by rectangles and the effects by circles. Then, the first-order materials properties that relate causes and effects are represented by triangles, and they are listed in Tables 1 and 2. Table 1 refers to the main interactions, namely, the *passive* properties; Table 2 shows the coupling, the *active* properties.

Tab. 1: First-order *passive* properties. Triangles in first column refer to Figure 4. Table taken from [9].

Symbol	Conjugate variables	Passive property
1	Stress – Strain	Elastic tensor
2	Electric field – Polarization	Electric susceptibility
3	Magnetic field – Magnetization	Magnetic susceptibility
4	Temperature – Entropy	Heat capacity

Tab. 2: First-order *active* properties. Triangles in first column refer to Figure 4. Table taken from [9].

Symbol	Conjugate variables	Active property
5	Mechanic – Electric	Direct, converse piezoelectric
6	Mechanic – Magnetic	Direct, converse piezomagnetic
7	Mechanic – Thermal	Thermal expansion, piezocaloric
8	Electric – Magnetic	Direct, converse magnetoelectric
9	Electric – Thermal	Direct, converse pyroelectric
10	Magnetic – Thermal	Direct, converse pyromagnetic

In short, the active materials allow the production of energy from residual sources such as heat, mechanical vibrations, etc. These materials will be studied in the remainder of this chapter. Furthermore, since these materials couple several fields of physics, the following section introduces the basic governing equations of continuum physics.

3 Continuum physics

Consider a continuum solid of domain Ω, boundary Γ and its outward normal $\mathbf{n}$. This solid is subjected to mechanical, electromagnetism and thermal energies and, consequently, the governing equations of the four fields must be considered.

For the sake of convenience, the present chapter uses both tensor and matrix notations. For the first, the following tensor operators are used:

- Scalar product or single contraction denoted by the symbol $(\cdot)$: $\mathbf{a} \cdot \mathbf{b} = a_i b_i$.
- Double contraction denoted by the symbol $(:)$: $\mathbf{A} : \mathbf{B} = A_{ij} B_{ij}$.
- Dyadic or outer product denoted by the symbol $(\otimes)$: $(\mathbf{a} \otimes \mathbf{b})_{ij} = a_i b_j$.
- Cross product denoted by the symbol $(\times)$: $(\mathbf{a} \times \mathbf{b})_i = \epsilon_{ijk} \, a_j b_k$, where ϵ is the Levi–Civita tensor.
- The transposition is denoted by $(\,)^{\mathsf{T}}$.
- The symbol ∇ denotes the *Del* operator.

3.1 Outline of continuum mechanics

Classical continuum mechanics is a branch of continuum physics that studies the kinematics and the mechanical response of deformable bodies subjected to the action of forces. Firstly, the word deformable refers to the capacity to change the medium shape – conversely to the rigid body mechanics that studies non-deformable solids. Secondly, the term continuum indicates that matter completely fills the medium, that is, the medium is assumed to be continuous. This assumption is known as the continuum hypothesis [9]. Finally, and considering this hypothesis, each point of the body (commonly called material point) can be identified by a position vector and, consequently, the tensor algebra can be applied to study the motion and deformation of solids.

Historical note

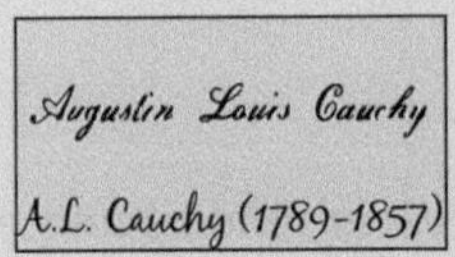

The French mathematician Augustin Louis Cauchy (1789–1857) pioneered the use of continuum instead of discrete models. In spite of his initial 3 years of working as an engineer, he was more attracted to the abstract beauty of mathematics. Cauchy was a prolific scientist: in continuum mechanics he holds 16 theorems. Signature taken from *Wikipedia*.

3.1.1 Kinematics

Consider a continuum body before (at time $t = 0$) and after deformation (at time t; Figure 5). The position of a material point P_0 inside a domain Ω_0 with border Γ_0 "before"

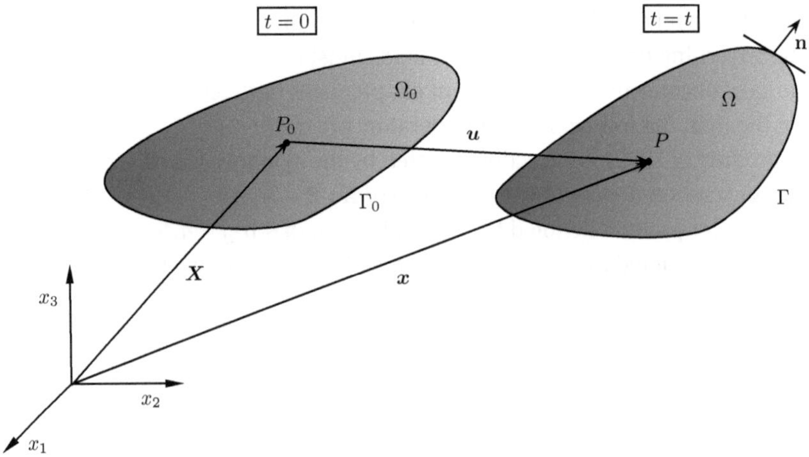

Fig. 5: Continuum body before and after deformation.

is located by $\boldsymbol{X}$; "after" the point becomes P inside another domain Ω of boundary Γ and located by $\boldsymbol{x}$. From vector calculus, the displacement $\boldsymbol{u}$ of any material point can be represented by:

$$\boldsymbol{u}(\boldsymbol{X}, t) = \boldsymbol{x}(\boldsymbol{X}, t) - \boldsymbol{X}(t)\,. \tag{1}$$

The dependency of $\boldsymbol{u}$ and $\boldsymbol{x}$ on $\boldsymbol{X}$ is commonly denominated as "Lagrange coordinates" in the continuum mechanics community. However, under the assumption of small displacements $\boldsymbol{X} \approx \boldsymbol{x}$, the notation $\boldsymbol{u}(\boldsymbol{x}, t)$ is adopted in the remainder of the present work.

3.1.2 Strain measure

An objective strain measure can be defined as the change of length and rotation of an elementary differential of displacement. Mathematically, this definition can be expressed as:

$$\mathrm{d}\boldsymbol{u}(\boldsymbol{x}, t) = \frac{\partial \boldsymbol{u}}{\partial \boldsymbol{x}} \cdot \mathrm{d}\boldsymbol{x} = \boldsymbol{u} \otimes \nabla \cdot \mathrm{d}\boldsymbol{x}\,, \tag{2}$$

where the non-symmetric tensor $\boldsymbol{u} \otimes \nabla$ is called the displacement gradient. As commented, this tensor contains information on the change of lengths and on the rotations of $\mathrm{d}\boldsymbol{u}$. In particular, it is observed in Figure 6 that the information relative to the change of lengths is stored in the symmetric part of $(\boldsymbol{u} \otimes \nabla)^{\mathrm{sy}}$ and the rotation in the skew-symmetric $(\boldsymbol{u} \otimes \nabla)^{\mathrm{sk}}$.

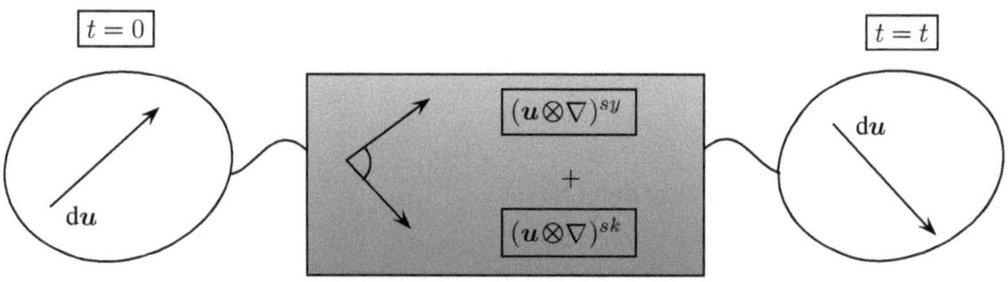

Fig. 6: Box representing the deformation process: elongation $(\boldsymbol{u} \otimes \nabla)^{\mathrm{sy}}$ and rotation $(\boldsymbol{u} \otimes \nabla)^{\mathrm{sk}}$.

According to the theory of tensor algebra, any tensor can be decomposed in symmetric and skew-symmetric parts. Therefore, the displacement gradient can be expressed as:

$$\boldsymbol{u} \otimes \nabla = \frac{1}{2}\left(\boldsymbol{u} \otimes \nabla + \nabla \otimes \boldsymbol{u}\right) + \frac{1}{2}\left(\boldsymbol{u} \otimes \nabla - \nabla \otimes \boldsymbol{u}\right), \tag{3}$$

Since the rotations do not produce deformation, the strain measure is given by the symmetric part of (3); under the assumption of small strains, the symmetric displacement gradient is finally given by:

$$\boldsymbol{S} = \frac{1}{2}\left(\boldsymbol{u} \otimes \nabla + \nabla \otimes \boldsymbol{u}\right) = \nabla^{\mathrm{sy}}\boldsymbol{u}\,, \tag{4}$$

where the symbol ∇^{sy} is used for simplicity and $\boldsymbol{S}$ denotes the second-order small strain tensor.

For the sake of convenience, $\boldsymbol{S}$ is expressed in Voigt's notation with indexes $11 = 1$, $22 = 2$, $33 = 3$, $12 = 4$, $23 = 5$, $13 = 6$ to obtain:

$$\{\boldsymbol{S}\} = \{S_1,\ S_2,\ S_3,\ S_4,\ S_5,\ S_6\}^\top \tag{5}$$

As observed, Voigt's notation allows us to represent a symmetric second-order tensor by a vector of six coefficients.

3.1.3 Linear momentum balance

From the second Newton's law, the linear momentum balance states that the rate of momentum $\boldsymbol{p} = \rho_\mathrm{m}\dot{\boldsymbol{u}}$ is equal to the total forces acting on the body of Figure 7 (ρ_m is the mass density). Notice that the domain is drawn after deformation, commonly denominated current configuration.

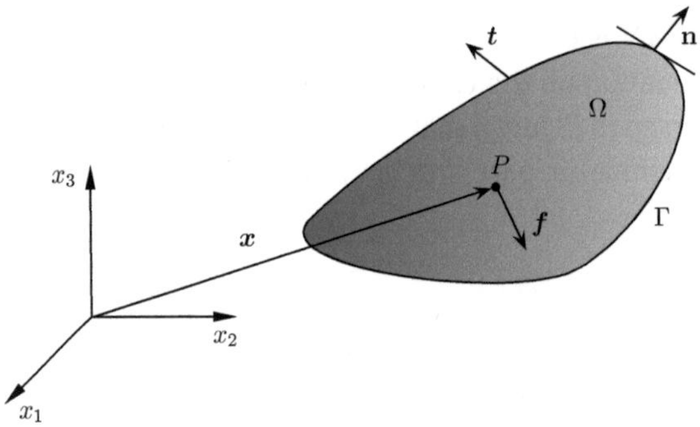

Fig. 7: Forces acting on continuum body after deformation.

The forces acting on Ω can be classified into long-range $\boldsymbol{f}$ and short-range $\boldsymbol{t}$ forces. The former represents the forces exerted by external fields such as gravity, and the latter the forces that interact among material points.

On this basis, the linear momentum balance is mathematically stated as:

$$\frac{\mathrm{d}}{\mathrm{d}t}\int_\Omega \boldsymbol{p}\ \mathrm{d}\Omega = \int_\Gamma \boldsymbol{t}\ \mathrm{d}\Gamma + \int_\Omega \boldsymbol{f}\ \mathrm{d}\Omega\,. \tag{6}$$

Considering the notation $\boldsymbol{T}^\mathrm{c}$ as the second-order Cauchy stress tensor, the mass conservation $\mathrm{d}\rho_\mathrm{m}/\,\mathrm{d}t = 0$, Cauchy's lemma $\boldsymbol{t} = (\boldsymbol{T}^\mathrm{c})^\top \cdot \mathbf{n}$ and applying the divergence theorem to the first term on the right-hand side (6), the linear momentum balance in local form becomes:

$$\rho_\mathrm{m}\ddot{\boldsymbol{u}} = \nabla\cdot(\boldsymbol{T}^\mathrm{c})^\top + \boldsymbol{f}\,. \tag{7}$$

3.1.4 Angular momentum balance

The angular momentum balance states that in any inertial frame the rate of change of the torque is equal to the total momentum generated by the forces acting on the body. Considering the Figure 7 and (6), this equality is mathematically expressed as:

$$\frac{\mathrm{d}}{\mathrm{d}t}\int_{\Omega} \boldsymbol{x} \times \boldsymbol{p} \,\mathrm{d}\Omega = \int_{\Gamma} \boldsymbol{x} \times \boldsymbol{t} \,\mathrm{d}\Gamma + \int_{\Omega} \boldsymbol{x} \times \boldsymbol{f} \,\mathrm{d}\Omega \,. \tag{8}$$

Again, taking into account mass conservation, Cauchy's lemma, the divergence theorem to the first term of (8) on the right-hand side and using (6), the angular momentum balance reads:

$$\epsilon : \boldsymbol{T}^{\mathrm{c}} = \boldsymbol{0} \quad \Rightarrow \quad \boldsymbol{T}^{\mathrm{c}} = (\boldsymbol{T}^{\mathrm{c}})^{\mathsf{T}} \,. \tag{9}$$

In conclusion, the symmetry of the Cauchy stress automatically guarantees the balance of angular momentum. As for $\boldsymbol{S}$, $\boldsymbol{T}^{\mathrm{c}}$, it can be expressed in Voigt's notation as:

$$\{\boldsymbol{T}^{\mathrm{c}}\} = \{T_1^{\mathrm{c}},\ T_2^{\mathrm{c}},\ T_3^{\mathrm{c}},\ T_4^{\mathrm{c}},\ T_5^{\mathrm{c}},\ T_6^{\mathrm{c}}\}^{\mathsf{T}} \,. \tag{10}$$

3.2 Outline of continuum electrodynamics

Classical electrodynamics (also called classic electromagnetism) is a branch of continuum physics that deals with the interactions between matter and electric charges and currents. This formalism is founded on the Maxwell's equations and on the Lorentz force.

Historical note

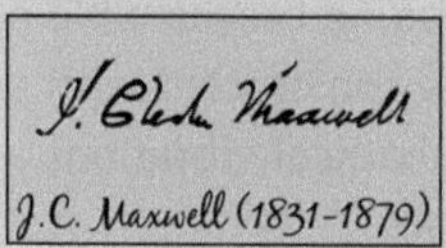

Although many scientists worked on electromagnetism, the Scottish mathematician James Clerk Maxwell (1831–1879) achieved the formulation of the classical theory of electromagnetism, denominated "second great unification in physics" after the first was realized by Isaac Newton. Signature taken from *Wikipedia*.

3.2.1 Maxwell equations

Maxwell's equations are a set of four coupled and empirical equations that relate the macroscopic variables of electromagnetism: electric field $\boldsymbol{E}$, electric displacement $\boldsymbol{D}$, magnetic field $\boldsymbol{H}$ and magnetic induction $\boldsymbol{B}$. Furthermore, the sources of electromagnetism, free electric charges ρ_q^{f} and free electric currents $\boldsymbol{j}^{\mathrm{f}}$ also appear in Maxwell's equations:

$$\begin{aligned} \nabla \cdot \boldsymbol{D} &= \rho_q^{\mathrm{f}} \,, \\ \nabla \cdot \boldsymbol{B} &= 0 \,, \\ \nabla \times \boldsymbol{E} + \frac{\partial \boldsymbol{B}}{\partial t} &= \boldsymbol{0} \,, \\ \nabla \times \boldsymbol{H} - \frac{\partial \boldsymbol{D}}{\partial t} - \boldsymbol{j}^{\mathrm{f}} &= \boldsymbol{0} \,. \end{aligned} \tag{11}$$

The first equation is denominated electric Gauss' law and states that the scalar sources of the electric field are the free electric charges. The second is called magnetic Gauss' law and establishes the absence of magnetic monopoles in nature, namely, the magnetic field is solenoidal. The third law is the Maxwel–Faraday law and states that the rate of change of a magnetic field generates an electric field. Therefore, this law couples both electric and magnetic fields. Finally, the fourth equation is Ampère's law, which asserts that the free electric currents and the rate of change of the electric displacement produce magnetic fields. Consequently, this law also couples both fields.

3.2.2 Electromagnetic potentials

In accordance with the Helmholtz theorem – also known as the fundamental theorem of vector calculus [9] – and according to Maxwell's laws of (11), there are four electromagnetic potentials:
- Two scalar potentials obtained from the Maxwel–Faraday and Ampère laws.
- Two vector potentials deduced from the electric and magnetic Gauss' laws.

Nevertheless, in the present chapter, the free sources ρ_q^{f} and $\boldsymbol{j}^{\mathrm{f}}$ and the partial derivatives $\partial \boldsymbol{B}/\partial t$ and $\partial \boldsymbol{D}/\partial t$ are assumed to be zero. This is a good and reasonable approximation for most applications of harvesters, since they are constructed with polarizable/magnetizable media and the frequencies of the mechanical vibrations are several orders of magnitude lower than the electromagnetic ones.

Assuming $\rho_q^{\mathrm{f}} = 0$, $\boldsymbol{j}^{\mathrm{f}} = \boldsymbol{0}$, the electric V and magnetic φ scalar potentials can be obtained from the Maxwel–Faraday and Ampère laws, respectively [9]:

$$\begin{aligned} \nabla \times \boldsymbol{E} = \boldsymbol{0} \quad &\Rightarrow \quad \boldsymbol{E} = -\nabla V \,, \\ \nabla \times \boldsymbol{H} = \boldsymbol{0} \quad &\Rightarrow \quad \boldsymbol{H} = -\nabla \varphi \,. \end{aligned} \tag{12}$$

Furthermore, these scalar potentials are more amenable for an FE formulation, as was reported in [9]. Finally and for convenience, the electromagnetic constitutive equations that relate polarization $\boldsymbol{P}$ and magnetization $\boldsymbol{M}$ vectors with $\boldsymbol{E}$, $\boldsymbol{D}$ and $\boldsymbol{H}$, $\boldsymbol{B}$, respectively, are introduced in the same reference:

$$\begin{aligned} \boldsymbol{D} &= \boldsymbol{P} + \epsilon_0 \boldsymbol{E} , \\ \boldsymbol{B} &= \mu_0(\boldsymbol{H} + \boldsymbol{M}) . \end{aligned} \tag{13}$$

3.2.3 Linear momentum balance

The linear momentum balance of electromagnetism is obtained by combining Maxwell's laws and Lorentz forces. There exist four representations of the electromagnetic linear momentum, see [9], closely related to the choice of the Poynting vector. Despite the fact that the best choice of this vector has generated controversy in the literature (for instance, the famous Abraham–Minkowski debate), in the present chapter, the Minkowski representation is assumed. Taking into account this representation, the linear momentum balance in local form is given by [10]:

$$\frac{\partial}{\partial t}(\boldsymbol{D} \times \boldsymbol{B}) = \nabla \cdot (\boldsymbol{T}^{\mathrm{EM}})^{\mathrm{T}} - \boldsymbol{f}^{\mathrm{EM}} , \tag{14}$$

where $\boldsymbol{T}^{\mathrm{EM}}$ and $\boldsymbol{f}^{\mathrm{EM}}$ denote the Maxwell stress tensor and the Lorentz forces, respectively. Obviously, both terms depend on the electromagnetic variables and, in the absence of free sources, are given by:

$$\begin{aligned} \boldsymbol{T}^{\mathrm{EM}} &= \boldsymbol{D} \otimes \boldsymbol{E} + \boldsymbol{B} \otimes \boldsymbol{H} - \frac{1}{2}(\boldsymbol{D} \cdot \boldsymbol{E} + \boldsymbol{B} \cdot \boldsymbol{H})\boldsymbol{I} , \\ \boldsymbol{f}^{\mathrm{EM}} &= \frac{1}{2}(\nabla \otimes \boldsymbol{E} \cdot \boldsymbol{D} - \nabla \otimes \boldsymbol{D} \cdot \boldsymbol{E} + \nabla \otimes \boldsymbol{H} \cdot \boldsymbol{B} - \nabla \otimes \boldsymbol{B} \cdot \boldsymbol{H}) . \end{aligned} \tag{15}$$

3.2.4 Angular momentum balance

As deduced in (9), the mechanical angular momentum balance is guaranteed by the symmetry of the Cauchy stress tensor. Similarly, the electromagnetic angular momentum balance requires the symmetry of $\boldsymbol{T}^{\mathrm{EM}}$.

In vacuum (absence of matter), $\boldsymbol{T}^{\mathrm{EM}}$ is symmetric; however, its symmetry in ponderable media (media with matter) has generated controversy in the literature [9]. Under classical continuum physics, the definition of a total stress tensor composed of two non-symmetric tensors (Cauchy and Maxwell) is the best approach.

The use of a total stress implies the modification of the classical traction vector $\boldsymbol{t}$ (Figure 7), by the mechatronic vector $\boldsymbol{t}_{\mathrm{EM}}$, which contains information of both mechanic and electromagnetic fields. For more details on the non-symmetry of the Maxwell tensor, see the recently published scientific article by the present authors [6].

3.3 Outline of thermodynamics

Thermodynamics is a branch of continuum physics that deals with the conservation and conversion of the energy, among several fields such as mechanical, thermal, electromagnetic fields, etc. Therefore, this formalism is the cornerstone to theoretically study harvesters.

Historical note

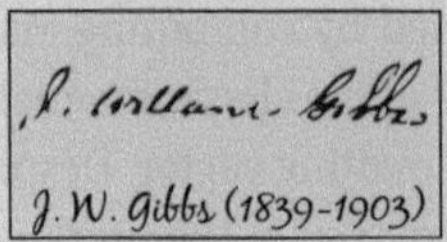

The American scientist Josiah Willard Gibbs (1839–1903) introduced the idea of incorporating the entropy into the internal energy of a system. Therefore, he was a pioneer in combining the first and second laws of thermodynamics and, consequently, he is the father of modern thermodynamics. It was declared that "Gibbs's name not only in America but in the whole world will ever be reckoned among the most renowned theoretical physicists of all times". Signature taken from *Wikipedia*.

The first law of thermodynamics states the conservation of the total energy in a closed system as: the total energy U is equal to the sum of heat Q and work W performed on the system, mathematically:

$$\mathrm{d}U = \delta Q + \mathrm{d}W\,, \tag{16}$$

where the symbols -d- and -δ- denote exact and inexact differentials, respectively. As observed, the heat is an inexact differential since it is path-dependent, namely, part of the total energy is irreversibly converted into heat. An example of an irreversible process is the mechanical plastic deformation for which part of the mechanical energy is employed in the transformation of the internal structure of the material. Then, overheating is produced – heat is dissipated – and the initial state of the material is never recovered.

As commented, this chapter deals with conservative processes for which the effects of the temperature are neglected. Consequently, $\delta Q = 0$ and $\mathrm{d}U = \mathrm{d}W$.

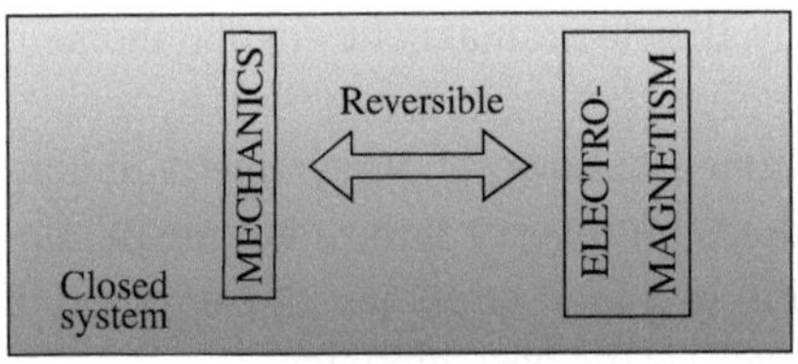

Fig. 8: Reversible exchanges of energies between mechanical and electromagnetic fields with thermal interaction neglected.

Consider the closed thermodynamical system shown in Figure 8. The system is composed of mechanical and electromagnetic energies that can be reversibly exchanged with each other given that $\delta Q = 0$.

As reported in Section 2, causes and effects play an important role in the study of active materials and, consequently, in harvesters. From a thermodynamical point of view, causes and effects are represented by intensive and extensive variables, respectively, and their products represent the total work performed by the system:

$$\mathrm{d}U(\boldsymbol{S},\boldsymbol{P},\boldsymbol{M}) = \mathrm{d}W(\boldsymbol{S},\boldsymbol{P},\boldsymbol{M}) = \boldsymbol{T}^{\mathrm{c}} : \mathrm{d}\boldsymbol{S} + \boldsymbol{E}\cdot\mathrm{d}\boldsymbol{P} + \boldsymbol{H}\cdot\mathrm{d}\boldsymbol{M}\,. \tag{17}$$

Table 3 lists the intensive and extensive variables used in the present work. Simple and double contractions product are used, and the fact that the energy is a scalar variable facilitates the following analyses.

Tab. 3: Intensive (cause) and extensive (effect) variables used in this chapter.

Intensive variables			**Extensive variables**		
Stress	–	$\boldsymbol{T}^{\mathrm{c}}$	Strain	–	$\boldsymbol{S}$
Electric field	–	$\boldsymbol{E}$	Polarization	–	$\boldsymbol{P}$
Magnetic field	–	$\boldsymbol{H}$	Magnetization	–	$\boldsymbol{M}$

There are several thermodynamic potentials to represent the total internal energy of the system. In this work, and for an amenable numerical implementation, it is convenient to use the electromagnetic enthalpy Π, which results from a Legendre transformation of (17) to exchange the pair **P**, **M** by the corresponding **E**, **H**:

$$\mathrm{d}\Pi(\boldsymbol{S},\boldsymbol{E},\boldsymbol{H}) = \boldsymbol{T}^{\mathrm{c}} : \mathrm{d}\boldsymbol{S} - \boldsymbol{P}\cdot\mathrm{d}\boldsymbol{E} - \boldsymbol{M}\cdot\mathrm{d}\boldsymbol{H}\,. \tag{18}$$

Since the problem is conservative, Π can be expressed as an exact differential:

$$\mathrm{d}\Pi(\boldsymbol{S},\boldsymbol{E},\boldsymbol{H}) = \left.\frac{\partial\Pi}{\partial\boldsymbol{S}}\right|_{\boldsymbol{E},\boldsymbol{H}} : \mathrm{d}\boldsymbol{S} + \left.\frac{\partial\Pi}{\partial\boldsymbol{E}}\right|_{\boldsymbol{S},\boldsymbol{H}}\cdot\mathrm{d}\boldsymbol{E} + \left.\frac{\partial\Pi}{\partial\boldsymbol{H}}\right|_{\boldsymbol{S},\boldsymbol{E}}\cdot\mathrm{d}\boldsymbol{H}\,, \tag{19}$$

Finally, comparing (18) and (19), the constitutive equations can be obtained from Π as:

$$\begin{aligned} \boldsymbol{T}^{\mathrm{c}} &= \left.\frac{\partial\Pi}{\partial\boldsymbol{S}}\right|_{\boldsymbol{E},\boldsymbol{H}}\,, \\ \boldsymbol{P} &= -\left.\frac{\partial\Pi}{\partial\boldsymbol{E}}\right|_{\boldsymbol{S},\boldsymbol{H}}\,, \\ \boldsymbol{M} &= -\left.\frac{\partial\Pi}{\partial\boldsymbol{H}}\right|_{\boldsymbol{S},\boldsymbol{E}}\,. \end{aligned} \tag{20}$$

These three expressions can be arranged to obtain a Hessian matrix to describe the behavior of the material:

$$\begin{Bmatrix} \mathrm{d}\boldsymbol{T}^{\mathrm{c}} \\ \mathrm{d}\boldsymbol{P} \\ \mathrm{d}\boldsymbol{M} \end{Bmatrix} = \begin{bmatrix} \dfrac{\partial\boldsymbol{T}^{\mathrm{c}}}{\partial\boldsymbol{S}} & \dfrac{\partial\boldsymbol{T}^{\mathrm{c}}}{\partial\boldsymbol{E}} & \dfrac{\partial\boldsymbol{T}^{\mathrm{c}}}{\partial\boldsymbol{H}} \\ \dfrac{\partial\boldsymbol{P}}{\partial\boldsymbol{S}} & \dfrac{\partial\boldsymbol{P}}{\partial\boldsymbol{E}} & \dfrac{\partial\boldsymbol{P}}{\partial\boldsymbol{H}} \\ \dfrac{\partial\boldsymbol{M}}{\partial\boldsymbol{S}} & \dfrac{\partial\boldsymbol{M}}{\partial\boldsymbol{E}} & \dfrac{\partial\boldsymbol{M}}{\partial\boldsymbol{H}} \end{bmatrix} \begin{Bmatrix} \mathrm{d}\boldsymbol{S} \\ \mathrm{d}\boldsymbol{E} \\ \mathrm{d}\boldsymbol{H} \end{Bmatrix}\,. \tag{21}$$

In order to obtain the constitutive equations and considering (20), an explicit form of the potential Π must be calculated. In a first and reasonable approximation, this potential can be found by a Taylor expansion in the vicinity of a natural state $\Pi(\boldsymbol{S} = \boldsymbol{E} = \boldsymbol{H} = \mathbf{0}) = 0$ and by keeping in mind the linearity of the problem, to give:

$$\begin{aligned}\Pi(\boldsymbol{S}, \boldsymbol{E}, \boldsymbol{H}) &= \frac{1}{2}\left(\boldsymbol{S} : \mathbf{C} : \boldsymbol{S} - \boldsymbol{E} \cdot \boldsymbol{\epsilon} \cdot \boldsymbol{E} - \boldsymbol{H} \cdot \boldsymbol{\mu} \cdot \boldsymbol{H}\right) \\ &\quad - \boldsymbol{e}^{V} : \boldsymbol{S} \cdot \boldsymbol{E} - \boldsymbol{e}^{\varphi} : \boldsymbol{S} \cdot \boldsymbol{H} - \boldsymbol{H} \cdot \boldsymbol{\nu} \cdot \boldsymbol{E}\,, \end{aligned} \tag{22}$$

where $\mathbf{C}$, $\boldsymbol{\epsilon}$, $\boldsymbol{\mu}$, $\boldsymbol{\nu}$, $\boldsymbol{e}^V$ and $\boldsymbol{e}^{\varphi}$ denote elastic, permittivity, permeability, electromagnetic, piezoelectric and piezomagnetic material tensors, respectively. Finally, considering (13), (22) and (20), the Hessian (21) becomes:

$$\begin{Bmatrix} \boldsymbol{T}^{\mathrm{C}} \\ \boldsymbol{D} \\ \boldsymbol{B} \end{Bmatrix} = \begin{bmatrix} \mathbf{C} & -\boldsymbol{e}^V & -\boldsymbol{e}^{\varphi} \\ \boldsymbol{e}^V & \boldsymbol{\epsilon} & \boldsymbol{\nu} \\ \boldsymbol{e}^{\varphi} & \boldsymbol{\nu} & \boldsymbol{\mu} \end{bmatrix} \begin{Bmatrix} \boldsymbol{S} \\ \boldsymbol{E} \\ \boldsymbol{H} \end{Bmatrix}. \tag{23}$$

3.4 Summary of governing equations

This section summarizes the governing equations to model conservative active materials under small strains and displacements, for low electromagnetic frequencies and assuming material linearity. The equations are equilibrium and constitutive complemented by the boundary conditions.

3.4.1 Equilibrium equations

For an amenable FE implementation and considering $\boldsymbol{T}^{\mathrm{EM}}$, the equilibrium equations are the linear momentum balance (7) and the electric and magnetic Gauss laws (11):

$$\boxed{\begin{aligned} \rho_{\mathrm{m}}\ddot{\boldsymbol{u}} &= \nabla \cdot \boldsymbol{T}^{\mathrm{T}} + \boldsymbol{f}\,, \\ \nabla \cdot \boldsymbol{D} &= 0\,, \\ \nabla \cdot \boldsymbol{B} &= 0\,, \end{aligned}} \tag{24}$$

where $\boldsymbol{T}^{\mathrm{T}} = \boldsymbol{T}^{\mathrm{C}} + (\boldsymbol{T}^{\mathrm{EM}})^{\mathrm{sy}}$ is the total stress tensor, composed of both symmetric Cauchy and Maxwell stress tensors.

3.4.2 Material constitution

For the sake of clarity, the multi-coupled constitutive (23) is expressed in matrix form using Voigt's notation (5) and (10):

Piezoelectrics

$$\begin{Bmatrix} T_1 \\ T_2 \\ T_3 \\ T_4 \\ T_5 \\ T_6 \\ D_1 \\ D_2 \\ D_3 \end{Bmatrix} = \begin{bmatrix} C_{11} & C_{12} & C_{13} & 0 & 0 & 0 & 0 & 0 & -e^V_{13} \\ C_{12} & C_{11} & C_{13} & 0 & 0 & 0 & 0 & 0 & -e^V_{13} \\ C_{13} & C_{13} & C_{33} & 0 & 0 & 0 & 0 & 0 & -e^V_{33} \\ 0 & 0 & 0 & C_{66} & 0 & 0 & 0 & 0 & 0 \\ 0 & 0 & 0 & 0 & C_{44} & 0 & 0 & -e^V_{15} & 0 \\ 0 & 0 & 0 & 0 & 0 & C_{44} & -e^V_{15} & 0 & 0 \\ 0 & 0 & 0 & 0 & 0 & e^V_{15} & \epsilon_{11} & 0 & 0 \\ 0 & 0 & 0 & 0 & e^V_{15} & 0 & 0 & \epsilon_{11} & 0 \\ e^V_{13} & e^V_{13} & e^V_{33} & 0 & 0 & 0 & 0 & 0 & \epsilon_{33} \end{bmatrix} \begin{Bmatrix} S_1 \\ S_2 \\ S_3 \\ S_4 \\ S_5 \\ S_6 \\ E_1 \\ E_2 \\ E_3 \end{Bmatrix},$$

Piezomagnetics

$$\begin{Bmatrix} T_1 \\ T_2 \\ T_3 \\ T_4 \\ T_5 \\ T_6 \\ B_1 \\ B_2 \\ B_3 \end{Bmatrix} = \begin{bmatrix} C_{11} & C_{12} & C_{13} & 0 & 0 & 0 & 0 & 0 & -e^\varphi_{13} \\ C_{12} & C_{11} & C_{13} & 0 & 0 & 0 & 0 & 0 & -e^\varphi_{13} \\ C_{13} & C_{13} & C_{33} & 0 & 0 & 0 & 0 & 0 & -e^\varphi_{33} \\ 0 & 0 & 0 & C_{66} & 0 & 0 & 0 & 0 & 0 \\ 0 & 0 & 0 & 0 & C_{44} & 0 & 0 & -e^\varphi_{15} & 0 \\ 0 & 0 & 0 & 0 & 0 & C_{44} & -e^\varphi_{15} & 0 & 0 \\ 0 & 0 & 0 & 0 & 0 & e^\varphi_{15} & \epsilon_{11} & 0 & 0 \\ 0 & 0 & 0 & 0 & e^\varphi_{15} & 0 & 0 & \epsilon_{11} & 0 \\ e^\varphi_{13} & e^\varphi_{13} & e^\varphi_{33} & 0 & 0 & 0 & 0 & 0 & \epsilon_{33} \end{bmatrix} \begin{Bmatrix} S_1 \\ S_2 \\ S_3 \\ S_4 \\ S_5 \\ S_6 \\ H_1 \\ H_2 \\ H_3 \end{Bmatrix}. \quad (25)$$

As observed, both constitutive matrices are transversely isotropic, since the present active materials are polarized along the x_3 direction.

3.4.3 Boundary conditions

The set of multi-coupled governing equations is mathematically closed by including the boundary conditions. As is common, these equations are composed of natural and essential boundary conditions; the former are called Neumann and the latter Dirichlet equations. Both are given by:

$$\begin{array}{ll} \underline{\text{Dirichlet type}} & \underline{\text{Neumann type}} \\ \boldsymbol{u} = \overline{\boldsymbol{u}}, & \boldsymbol{T}^{\mathrm{T}} \cdot \mathbf{n} = \overline{\boldsymbol{t}}_{EM}, \\ V = \overline{V}, & \boldsymbol{D} \cdot \mathbf{n} = \overline{q}_{\Gamma}, \\ \varphi = \overline{\varphi}, & \boldsymbol{B} \cdot \mathbf{n} = 0, \end{array} \quad (26)$$

where $\overline{\boldsymbol{u}}$, $\overline{V}$, $\overline{\varphi}$, $\overline{\boldsymbol{t}}_{EM}$ and $\overline{q}_{\Gamma}$ denote prescribed displacements, voltage, scalar magnetic potential, mechatronic vector and electric charges on Γ, respectively.

4 Finite element method

The finite element method (FE) is a numerical technique used to model many problems in science and engineering. Sophisticated situations, such as the governing equations for the active materials reported in Section 3, result in complex systems of partial differential equations for which there are no analytical solutions; but the FE method allows us to approximate these sets of partial differential equations in an algebraic system that can be solved using numerical algorithms. In fact, at present, FE is the more widespread method in technological applications, and a wide number of commercial codes exist. For more details, the reader is referred to the classical books [11].

Historical note

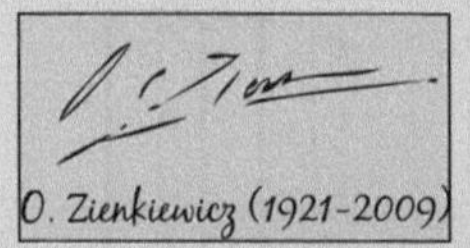

The FE method was developed in the 1960s by, among others, the Greek John Argyris (1913–2004) at the University of Stuttgart, the American Ray William Clough (1920–2016) at the University of California (Berkeley) and the Anglo-Polish Olgierd Zienkiewicz (1921–2009) at the University of Swansea. The main contribution of the latter was to recognize the general potential of FE to resolve problems in areas outside solid mechanics. Signature taken from *www.nap.edu*.

4.1 Outline of the finite element method

As mentioned, FE is probably the most advanced method for the solution of multi-coupled problems, however, for these applications the method involves complex mathematical concepts. Consider the continuum system shown in Figure 9; the FE method is constructed from the following steps:

i) The continuum domain Ω is divided into subdomains or finite elements Ω_e, interconnected at the nodal points.
ii) The nodal values of the degrees of freedom are assumed to be the unknown parameters of the problem.
iii) A set of functions denominated "shape functions" are chosen to interpolate the solution within each finite element in terms of their nodal values.
iv) The principle of virtual work is applied to the governing equations to obtain "weakened" forms of the problem.
v) The solution is calculated by solving a set of linear or non-linear equations.

Basically, the resolution of non-linear transient problems implies three steps:

a) The time interval is divided into small time increments Δt.
b) The analytical time derivatives are replaced by discrete forms using, for instance, the Newmark-β scheme.

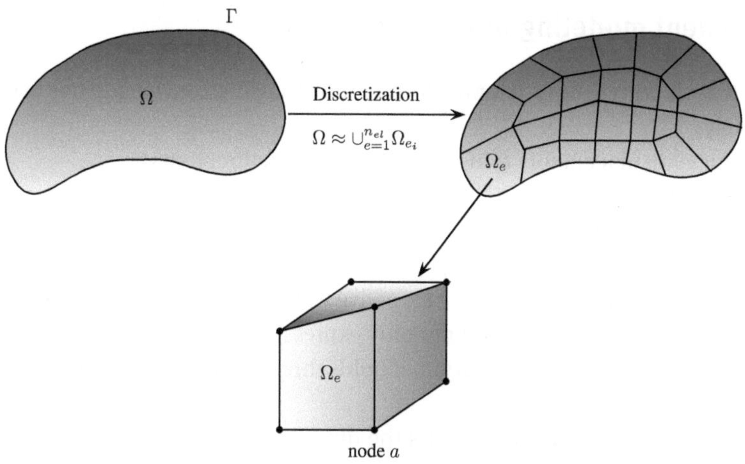

Fig. 9: Finite element discretization of a continuum domain.

c) The non-linear algebraic problem for each time increment is solved using the Newton–Rhapson algorithm.

The assembled non-linear FE equations are written in a residual form $\mathcal{R}$ and are linearized by the derivative:

$$\mathcal{R}_a^k = -\left.\frac{\partial \mathcal{R}_a}{\partial \mathbf{g}_b}\right|^k \mathrm{d}\mathbf{g}_a^k \tag{27}$$

where a, b are the local numbering of two generic nodes, k the Newton–Raphson iteration counter and $\mathrm{d}\mathbf{g}_a^k$ the derivatives of the degrees of freedom at node a. The algorithm for time integration is written as:

$$-\left.\frac{\partial \mathcal{R}_a}{\partial \mathbf{g}_b}\right|^k = c_1\, \mathcal{K}_{ab} + c_2\, \mathcal{C}_{ab} + c_3\, \mathcal{M}_{ab}\,, \tag{28}$$

where the parameters c_1, c_2 and c_3 are given in [6].

The consistent tangent $\mathcal{K}$, capacity $\mathcal{C}$ and mass $\mathcal{M}$ matrices are derived for each iteration as:

$$\mathcal{K}_{ab} = -\frac{\partial \mathcal{R}_a}{\partial \mathbf{U}_b}\,, \qquad \mathcal{C}_{ab} = -\frac{\partial \mathcal{R}_a}{\partial \dot{\mathbf{U}}_b}\,, \qquad \mathcal{M}_{ab} = -\frac{\partial \mathcal{R}_a}{\partial \ddot{\mathbf{U}}_b}\,, \tag{29}$$

where $\mathbf{g}_b^t = \{\mathbf{U}_b, \dot{\mathbf{U}}_b, \ddot{\mathbf{U}}_b\}$ represents the zero, first and second derivatives, respectively. The degrees of freedom at the generic node b are denoted by $\mathbf{U}_b$.

Finally, the solution is updated using $\mathbf{g}_b^{k+1} = \mathbf{g}_b^k + \mathrm{d}\mathbf{g}_b^k$. Notice that the Newton–Raphson iteration counter should exhibit a quadratic asymptotic rate of convergence if the tangent matrices are correctly calculated.

4.2 Finite element modeling of active materials

The present work deals with a multi-coupled formulation for which each node has five unknowns, namely:

- three mechanical displacements $\boldsymbol{u}$ in the three Cartesian directions
- voltage potential V
- scalar magnetic potential φ

Consequently, the FE formulation contains a set of five residuals. Notice that the residual for the mechanical displacements is commonly expressed in compact notation by a single equation; however, in matrix notation, it holds three equations due to the fact that $\boldsymbol{u}$ is a vector.

For the sake of clarity, this section reports the discretizations, residuals and the final assembled FE matrices. For more details on the FE formulation, the reader is referred to [9].

4.2.1 Discretizations

As commented, the continuum domain must be discretized. In this work, this is done by n three-dimensional eight-noded brick elements such that $\Omega \approx \sum_i^n \Omega_e$. Furthermore, standard shape functions $\mathcal{N}$ of Lagrangian type are used to interpolate the unknowns through the element. For simplicity, an isoparametric interpolation is adopted, namely, global coordinates $\boldsymbol{x}$ and degrees of freedom are approximated by the same $\mathcal{N}$:

$$\begin{aligned} &\boldsymbol{x} \approx \mathcal{N}_a\,\tilde{\boldsymbol{x}}_a\,, \\ &\boldsymbol{u} \approx \mathcal{N}_a\,\tilde{\boldsymbol{u}}_a\,, \quad \ddot{\boldsymbol{u}} \approx \mathcal{N}_a\,\tilde{\ddot{\boldsymbol{u}}}_a\,, \\ &V \approx \mathcal{N}_a\,\tilde{V}_a\,, \quad \varphi \approx \mathcal{N}_a\,\tilde{\varphi}_a\,, \end{aligned} \tag{30}$$

where $\tilde{\boldsymbol{x}}_a$, $\tilde{\boldsymbol{u}}_a$, $\tilde{V}_a$ and $\tilde{\varphi}_a$ refer to the nodal value at the local node a, and the Einstein summation convention is applied. Furthermore, (4) and (12) are approximated by:

$$\begin{aligned} \boldsymbol{S} &\approx \nabla^{\mathrm{sy}}\mathcal{N}_a\,\tilde{\boldsymbol{u}}_a = \mathcal{B}_a^{\mathrm{sy}}\,\tilde{\boldsymbol{u}}_a\,, \\ \boldsymbol{E} &\approx -\nabla\mathcal{N}_a\,\tilde{V}_a = -\mathcal{B}_a\,\tilde{V}_a\,, \\ \boldsymbol{H} &\approx -\nabla\mathcal{N}_a\,\tilde{\varphi}_a = -\mathcal{B}_a\,\tilde{\varphi}_a\,, \end{aligned} \tag{31}$$

where the derivation operators $\mathcal{B}$ are expressed in matrix notation as:

$$\mathcal{B}_a^{\mathrm{sy}} = \begin{bmatrix} \mathcal{N}_{a,1} & 0 & 0 \\ 0 & \mathcal{N}_{a,2} & 0 \\ 0 & 0 & \mathcal{N}_{a,3} \\ \mathcal{N}_{a,2} & \mathcal{N}_{a,1} & 0 \\ 0 & \mathcal{N}_{a,3} & \mathcal{N}_{a,2} \\ \mathcal{N}_{a,3} & 0 & \mathcal{N}_{a,1} \end{bmatrix}, \quad \mathcal{B}_a = \begin{Bmatrix} \mathcal{N}_{a,1} \\ \mathcal{N}_{a,2} \\ \mathcal{N}_{a,3} \end{Bmatrix}, \tag{32}$$

and $(\cdot_{,i})$ denotes differentiation with respect to the i-th Cartesian coordinate.

4.2.2 Residuals

The governing equations reported in Section 3.4 are commonly called strong forms, since they are second-order differential functions of the degrees of freedom $\boldsymbol{u}$, V and φ. The term "strong" refers to the higher continuity on the dependent variables. In contrast, the weak forms are often integral expressions that require weaker continuity field variables, namely, they are first-order equations and, consequently, allow an FE discretization. In order to obtain "weakened" forms two procedures exist:

I. energy principles such as Washizu or Hamilton
II. weighted residual methods

The first procedure is particularly suited for solid mechanics and structures. On the contrary, the second procedure is more general and can be applied to solve all kinds of partial differential equations. In this sense, the second approach is used in the present work, and it consists of the following steps:

i) Equilibrium equations (24) are multiplied by arbitrary test functions.
ii) The divergence theorem is applied to the gradient terms of these equations.
iii) The Neumann boundary conditions (26) are enforced (the Dirichlet type are automatic).

After the application of the discretizations of (30) and (31) to the weak forms, the three multi-coupled residuals read:

$$\begin{aligned}
\mathcal{R}_a^u &= \int\limits_{\Omega_e} \left[\mathcal{B}_a^{\mathrm{sy}\mathrm{T}}\, \boldsymbol{T}^{\mathrm{T}} + \mathcal{N}_a (\boldsymbol{f} - \rho_{\mathrm{m}}\, \mathcal{N}_b\, \tilde{\ddot{\boldsymbol{u}}}_b) \right] \mathrm{d}\Omega_e + \oint\limits_{\Gamma_e} \mathcal{N}_a\, \bar{\boldsymbol{t}}^{\mathrm{EM}}\, \mathrm{d}\Gamma_e\,, \\
\mathcal{R}_a^V &= \int\limits_{\Omega_e} \boldsymbol{B}_a^{\mathrm{T}}\, \boldsymbol{D}\, \mathrm{d}\Omega_e - \oint\limits_{\Gamma_e} \mathcal{N}_a\, \bar{q}_{\Gamma}\, \mathrm{d}\Gamma_e\,, \\
\mathcal{R}_a^\varphi &= \int\limits_{\Omega_e} \boldsymbol{B}_a^{\mathrm{T}}\, \boldsymbol{B}\, \mathrm{d}\Omega_e\,.
\end{aligned} \tag{33}$$

4.2.3 Assembled matrix

From the residuals of (33) and using (29), the tangent matrices are directly calculated by simple derivations. The final assembled matrix becomes:

$$\begin{bmatrix} \mathcal{K}_{ab}^{uu} + c_3\, \mathcal{M}_{ab}^{uu} & \mathcal{K}_{ab}^{uV} & \mathcal{K}_{ab}^{u\varphi} \\ \mathcal{K}_{ab}^{Vu} & \mathcal{K}_{ab}^{VV} & \mathcal{K}_{ab}^{\varphi V} \\ \mathcal{K}_{ab}^{\varphi u} & \mathcal{K}_{ab}^{\varphi V} & \mathcal{K}_{ab}^{\varphi\varphi} \end{bmatrix}^k \begin{Bmatrix} \mathrm{d}\tilde{\boldsymbol{u}}_b^{n+1} \\ \mathrm{d}\tilde{V}_b^{n+1} \\ \mathrm{d}\tilde{\varphi}_b^{n+1} \end{Bmatrix}^k = \begin{Bmatrix} \mathcal{R}_a^u \\ \mathcal{R}_a^V \\ \mathcal{R}_a^\varphi \end{Bmatrix}^k , \tag{34}$$

where the parameter c_3 contains information on the time integration algorithm. As observed, the assembled matrix is a set of five fully coupled equations.

First, no capacity matrices are present, since the current problem is conservative and, consequently, there are no dissipative terms. Second, only the mechanical mass matrices are included, given that the only hyperbolic equation is the linear momentum balance, (7). Third, the main diagonal represents direct (or *passive*) interactions, namely, elasticity, permittivity and permeability, see Table 1. In contrast, the off-diagonal terms are due to the couplings (or *active* interactions) of Table 2. Fourth, the explicit forms of the tangent matrices are reported in [9]. Finally, the set of algebraical equations of (34) is implemented into the research FE code FEAP [12], which belongs to the University of California at Berkeley (USA).

5 Application: Energy production in high-speed railway bridges

This section introduces an overview of high-speed railway bridges with the objective to obtain their main vibration variables. Then, harvesters are numerically simulated by the previous FE formulation, and their energy production is calculated.

5.1 Overview of high-speed railway bridges

Structures have evolved over the years, and this evolution has been driven both by the increasing demands of society and the advances of technology; bridges in general and particularly railway bridges are no exception. Every possibility of reducing traveling times has always been explored with the greatest interest, producing a progressive increase in the speed of these railways. In turn, such increases have led to more demanding requirements for rail infrastructure. Tracks have become heavier and stiffer, a greater radius of curvature has been adopted for bents, cants have also been increased, etc. Like general infrastructure, bridges have to withstand the dynamic effects induced by vehicles traveling at speeds of over 300 [km/h]. Typically, one speaks of high-speed bridges when the design of the line withstands more than 200–250 [km/h].

A wide variety of bridges has been designed to meet the requirements of the new high-speed lines. Because high-speed tracks must avoid sharp curves, long viaducts are very often required to cross over valleys, riverbeds and hollows; additionally, wide streams and estuaries also require long structures. In countries such as Italy, Germany, Spain, USA, China and Japan, prestressed concrete is usually preferred for such long bridges, often (but of course not always) resorting to simply supported, prefabricated decks on top of piers. In China and Japan, a large part of the railway lines are constructed on top of – concrete – viaducts to avoid excessive occupation of land. In addition, this strategy avoids interference between existing and new high-speed lines.

Typical span lengths range from some 30 [m] for simply supported prefabricated decks up to 70 [m] for continuous decks, with some singular constructions featuring more than 150 [m].

Continuous bridges are also present in both steel or mixed construction. Metallic viaducts are typical in France, but some modern designs have also earned a reputation in Spain (spans from 50 [m] to more than 200 [m]). Metallic arches are used in France, China and other countries, with main spans up to some 200 [m]. In Spain, steel arches have also been used in certain singular locations, as is shown in Figure 10 (top), while concrete arches have reached world-record spans in the high-speed lines Madrid-Valencia and Madrid-Portugal. Figures 10 (middle and bottom) and 11 show other representative examples.

In any case, shorter bridges are also required to cross over local roads and small rivers. Simply supported concrete slabs, pseudo-slabs or twin girder bridges are very well suited for such cases, with spans ranging from some 15 up to 40–50 [m]. Twin metallic girder or modern truss bridges are also advisable for single or multiple spans. For shorter spans below some 15 [m], rigid frames are often preferred in order to avoid excessive vibration (Figure 12).

High-speed bridges should to satisfy certain structural requirements in order to guarantee adequate performance. The limits of deflection required for good behavior are rather strict, since passenger comfort is a demanding *serviceability limit state* (SLS) in high-speed transportation. As a rule-of-thumb, in Europe the maximum permitted deflection (or vibration amplitude) under the pass of a high-speed convoy is about $L/2000$, L being the span length. Lower $L/1500$ or higher $L/2500$ values may, of course, be found in practice.

The number of cycles of vibration that the bridge undergoes is variable, depending on the type of train and speed. If resonance or near resonance occurs, vibration will take place during more cycles and with greatest amplitudes. Typically, one cycle of strong oscillation is observed for each bogie when the speed of the train is far from resonance; for a convoy with an axle load pattern such as the one in Figure 13, one would expect around 12 cycles, given that two adjacent bogies are very close in the link between power and passenger cars (also referred to as *coaches*). The motion of the mid-span section would be similar to that shown in Figure 14, in which oscillations of large amplitude arise when the four loads corresponding to the connection between power and passenger cars act at the same time on the bridge ($t = 0.76$ and 5.5 [s]). The approximate maximum amplitudes of the 12 cycles are marked with diamonds in the figure.

Figure 14 corresponds to a simply supported bridge of 22 [m] span, with 5.5 [Hz] of fundamental frequency, a linear mass equal to 22,000 [kg/m] and a 1% damping ratio. The response is computed resorting to the fundamental mode only, which plays a very predominant role in simply supported beams. The total wheelbase of the passenger cars is 18.7 [m], corresponding to a typical Eurostar vehicle. The bogie wheel-

Fig. 10: Top: continuous concrete deck with arch over the motorway Granada–Seville (Spain). Image courtesy of José Lavado Rodríguez (http://hlestructuras.com). Middle: double viaduct over the Rhône river in Avignon (France). Image courtesy of Philip Bourret by way of structurae.net. Bottom: simply supported spans on the Madrid-Barcelona high-speed line (Spain) with prefabricated concrete beams.

Fig. 11: Archidona (Spain) viaduct: continuous composite twin-girder bridge with upper and lower concrete decks. Image courtesy of Alejandro Castillo Linares (http://www.acl-estructuras.com).

Fig. 12: Skewed (left) and straight (right) rigid portal frames. Madrid-Valencia line (Spain).

Fig. 13: Scheme of an articulated train with two power cars and nine passenger cars.

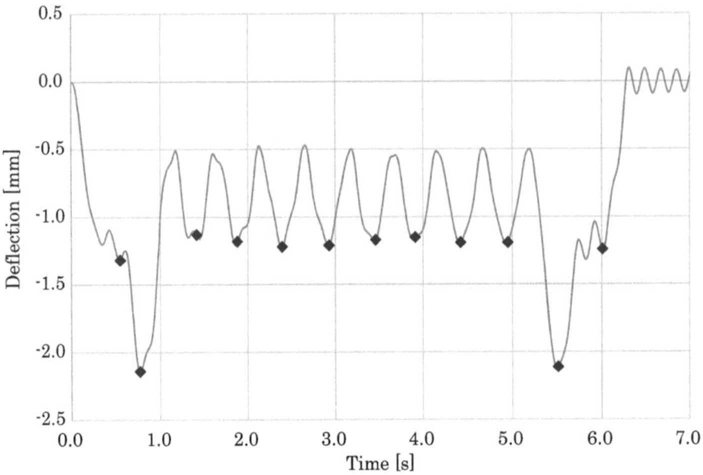

Fig. 14: Typical response in an out-of-resonance situation (v = 133.2 [km/h]).

base is 3 [m], with loads per axle of 170 [kN]. A rough number is useful to verify that the analysis yields reasonable results: the static deflection under two point loads concentrated at mid-span, computed solely for the first bending mode is 1.18 [mm]. This is approximately the maximum amplitude of the deflection observed during the passage of the coaches. For the interested reader, the main aspects of the dynamics of railway bridges have been explained well in a number of publications, for instance [7, 13, 14] and [8].

When the vehicle circulates at a speed of resonance (or sub-resonance), more cycles of strong oscillation will appear, because the free vibration induced after the train passage will be larger, and usually the decay of such free vibration is slow; this decay is due to the usual low values of damping of high-speed bridges. Two examples of resonance/sub-resonance are shown in Figures 15 and 16. This resonance occurs when the passage of a coach takes a time lapse equal to one period of the structure; in this example, the vehicle travels at 18.7 [m] in 1/5.5 [s] to reach the first resonance speed. If the time lapse is two, three or more (integer) times the period of the structure, the phenomenon is called sub-resonance (also *resonance of higher order*). Resonance and sub-resonance are jointly referred to simply as *resonances*. In some particular cases, resonances may disappear unexpectedly if the theoretical resonance speed is equal to one of the so-called *cancellation speeds*. This is a particular behavior of simply supported bridges, which is treated in great detail in [7] and [8].

As can be seen, the second resonance is not too strong in this case, and the amplitudes are not much larger in Figure 16 than in Figure 14, except for the free vibration. The reasons beyond maximum or minimum amplitudes of resonances in railway bridges are dealt with in depth in [7]. It should be emphasized that in resonance situations the amplitudes of vibration may be large in general, and the vertical accelera-

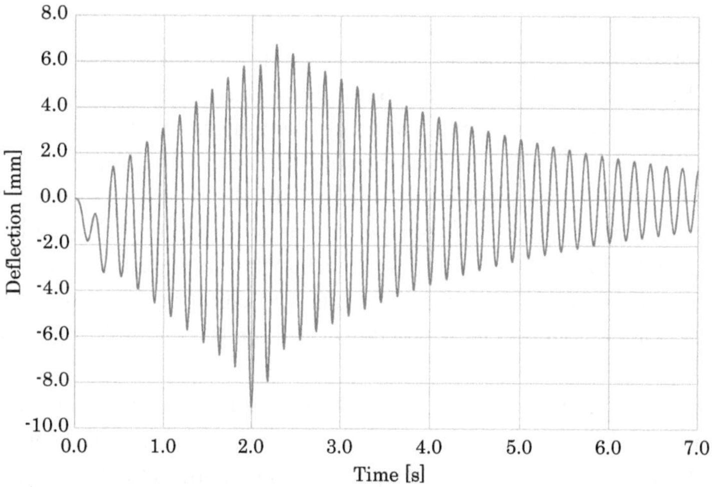

Fig. 15: Typical response in the (first) resonance situation: one oscillation cycle related with the passage of a succession of coaches (v = 370.3 [km/h]).

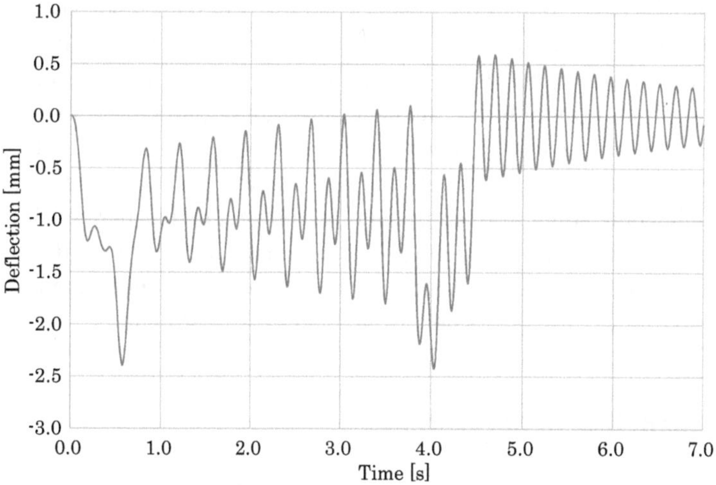

Fig. 16: Typical response in the (second) resonance situation: two cycles of oscillation during the passage of a succession of coaches (v = 185.15 [km/h]). This phenomenon is also known as sub-resonance.

tion of the deck could reach unacceptable values. This is also a very demanding SLS for high-speed bridges; usually the vertical acceleration is limited to 0.35 g in bridges with ballasted track, while 0.5 g is permitted for slab tracks. These limit values are linked to the ballast instability phenomenon and to the loss of the contact phenomenon, respectively; they include an overall security coefficient of 2.0 because of the extreme importance of those phenomena for the running safety, [15].

Other key properties for understanding the dynamic behavior are frequency and damping. Regarding the latter, it should be pointed out that high-speed bridges essentially remain in linear elastic behavior throughout their service life due to their stiffness, vibration and fatigue requirements. Therefore, damping is very low because the structure is virtually undamaged, and values of the critical viscous damping ratio as low as 1% can be expected for prestressed concrete and 0.5% for steel or mixed bridges. These are conservative numbers prescribed by the building codes. It should not be surprising that real values are somewhat higher, a fact that should be duly taken into account for energy harvesting assessments because the vibration levels will be slightly reduced. The reader is referred to, for instance, [13] and [15] for additional information on damping.

As regards the ranges of natural frequencies, a reference is provided in [13]: some values in simply supported bridges would be 10 [Hz] for a 10 [m] span, 6 [Hz] for a 20 [m] span, 4 [Hz] for a 40 [m] span and 2.5 [Hz] for a 70 [m] span; the margins are by no means narrow given the variety of bridge types. The article [16] provides useful ranges for total mass.

The behavior of continuous bridges, frame bridges and arches is more complex than that of the previous cases, due to the greater complexity of their mode shapes and also due to the significant contribution of more modes to the response. The variety of bridge and train types makes it very difficult to give concrete orders of magnitude regarding the vibration amplitudes, frequency levels, etc., particularly in arch bridges due to their singularity. For continuous decks and frames, the limits of deflection are similar (approx. $L/2000$) to the ones with simply supported structures, the frame designs usually being stiffer and, consequently, showing higher frequencies. Even if today's computer codes have very much eased the computational effort, the dynamic analysis of railway bridges is a task that should always be undertaken by expert engineers, especially in the assessment of singular or large structures.

5.2 Numerical modeling of harvesters

A typical vibrating harvester is composed of a substructure, which is typically a metal, an active material (piezoelectric/piezomagnetic) and the proof mass (Figure 17). Due to the mechanical vibration exerted by the passage of a train, the harvester is subjected to mechanical oscillations. As commented, active materials under mechanical causes produce electromagnetic effects that are used to generate power from residual sources. This generation can be used in SHM, a relatively new discipline that uses arrays of sensors/actuators to control the mechanical displacements and to ensure a proper operation of the structure.

Figure 17 shows a sketch of a classic uni-morph harvester, which operates in mode 33; both poling and force directions are the same. As is common, the active material (piezoelectric or piezomagnetic) is bounded by two electrodes, see [17].

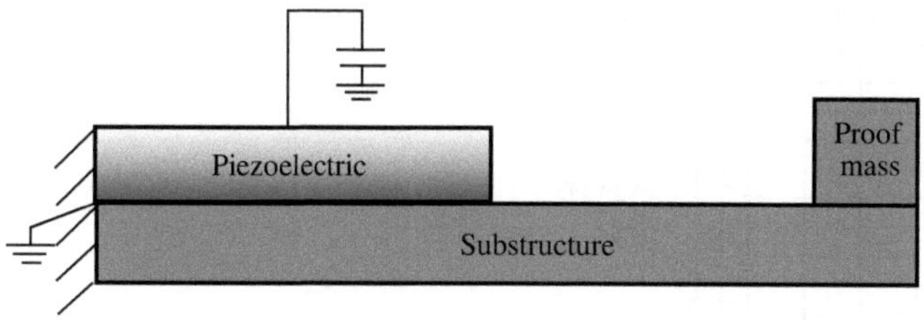

Fig. 17: Sketch of a vibrating harvester composed of piezoelectric/piezomagnetic, substructure and proof mass. Mechanically, it is a cantilever fixed-free beam. Electrically, the electric/magnetic potential is set to zero on the bottom electrode and the top one is connected to a battery.

In the present chapter, this harvester is modeled by the FE formulation described in Section 4. For this purpose, a piezoelectric/piezomagnetic beam of dimensions $30 \times 1 \times 0.5$ [mm] is simulated. Furthermore, the polarization/magnetization is applied along the thickness, and the material properties are taken from [9]:

$$[\mathbf{C}] = \begin{bmatrix} 116 & 77 & 78 & 0 & 0 & 0 \\ & 116 & 78 & 0 & 0 & 0 \\ & & 162 & 0 & 0 & 0 \\ & \text{– sym –} & & 89 & 0 & 0 \\ & & & & 86 & 0 \\ & & & & & 86 \end{bmatrix} \times 10^9 \text{ [Pa]} \tag{35}$$

$$[\boldsymbol{e}^{\mathrm{v}}] = \begin{bmatrix} 0 & 0 & 0 & 0 & 0 & 11.6 \\ & 0 & 0 & 0 & 11.6 & 0 \\ -4.4 & -4.4 & 18.6 & 0 & 0 & 0 \end{bmatrix} \text{ [mPa/V]} \tag{36}$$

$$[\boldsymbol{e}^{\varphi}] = \begin{bmatrix} 0 & 0 & 0 & 0 & 0 & 5.5 \\ 0 & 0 & 0 & 0 & 5.5 & 0 \\ 5.8 & 5.8 & 7 & 0 & 0 & 0 \end{bmatrix} \times 10^2 \text{ [mPa/A]} \tag{37}$$

$$[\boldsymbol{\epsilon}] = \begin{bmatrix} 11.2 & 0 & 0 \\ 0 & 11.2 & 0 \\ 0 & 0 & 12.6 \end{bmatrix} \times 10^{-9} \text{ [F/m]} \tag{38}$$

$$[\boldsymbol{\mu}] = \begin{bmatrix} 5 & 0 & 0 \\ 0 & 5 & 0 \\ 0 & 0 & 10 \end{bmatrix} \times 10^{-6} \text{ [H/m]} \tag{39}$$

Numerically, the harvester is meshed by using a structured mesh of 1000 eight-node elements. The mechanical boundary conditions are fixed free to represent the cantilever beam, while both electric and magnetic potentials are set to zero on the bottom electrode, as is shown in Figure 17. Figure 18 shows a three-dimensional view of the FE mesh used for modeling the active material. Notice that the substructure is not represented.

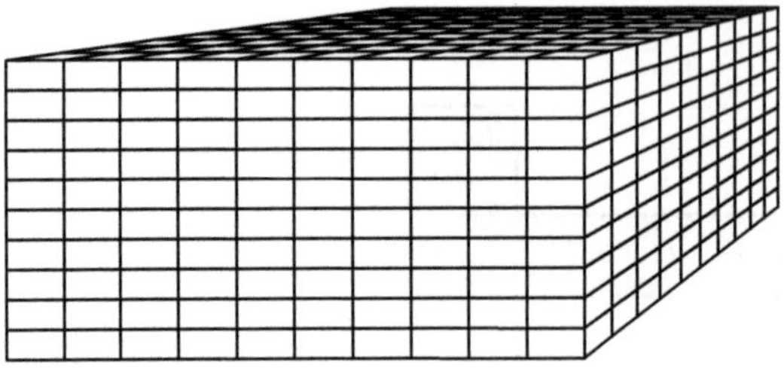

Fig. 18: Perspective view of the FE mesh used for modeling the active material.

The time-history displacement shown in Figure 14 is prescribed in FEAP at the right tip of the beam to obtain the generated electric/magnetic potentials:

- On the one hand, Figure 19 plots the time-history of produced voltage obtained for the harvester made out of piezoelectric material. As observed, the voltage increases with time due to the resonance of the cantilever harvesters and, consequently, the energy produced increases.

Due to this strong increase, the design of resonator harvesters is a challenge for the experimental community. In this sense, the present numerical formulation can be used as a "virtual laboratory". As observed, a non-negligible potential drop of approximately 12 [mV] is obtained with this simple application. Therefore, the use of several harvesters in a railway bridge produces enough energy to monitor the structure.

- On the other hand, Figure 20 shows the scalar magnetic potentials produced by the piezomagnetic harvester. Again, the cantilever harvester presents resonances for the same mechanical vibrations.

To sum up, it is observed that both piezoelectric/piezomagnetic harvesters show a similar behavior. Nevertheless, the voltage drop is one order of magnitude greater than the magnetic one, monitoring the electric field is easier, and magnetic materials are difficult to miniaturize. In conclusion, the production of energy from mechanical vibration with the former material in high-speed railway bridges looks like the best option.

With regard to the use of the energy production and despite the fact that the potential drop is 12 [mV], this signal could be applied to supply sensors and actuators made out piezoelectric materials in order to analyze the mechanical displacements of the bridge. As commented, these devices, which produce clean energy, could replace the traditional electro-chemical batteries that pollute the environment. Finally, vibrating harvesters produce alternating current (AC) and, consequently, the use of rectifiers is required. In this sense, the simplest way to rectify the AC is to connect the harvesters with diode junctions, which results in a high efficiency (approximately 84%).

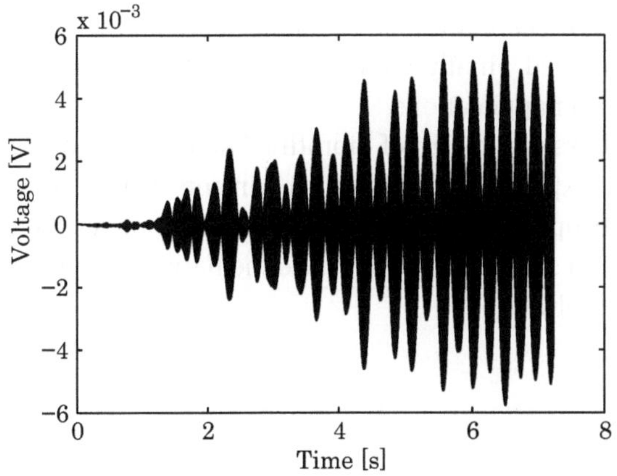

Fig. 19: Produced electric voltage versus time for a piezoelectric harvester.

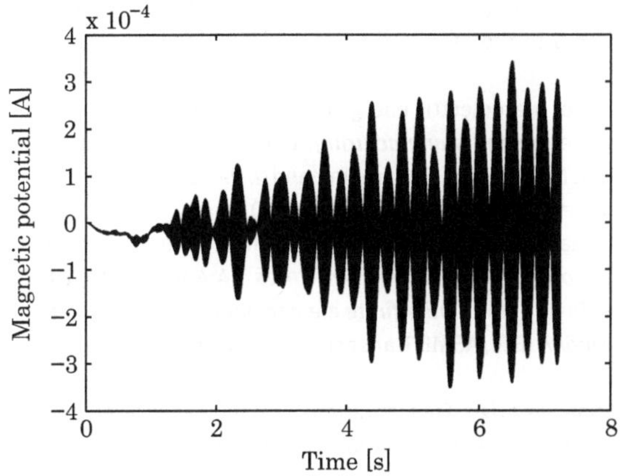

Fig. 20: Produced scalar magnetic potential versus time for a piezomagnetic harvester.

6 Concluding remarks

This chapter has reported a finite element formulation to simulate active materials that couple up to four fields of physics. For the sake of clarity, the chapter outlines the governing equations of several physics branches: continuum mechanics, electrodynamics and thermodynamics. Furthermore, a brief description of the finite element method is also reported.

From a practical point of view, the chapter contains a revision of the main variables and orders of magnitudes of vibrating high-speed railway bridges. For this vibrating application, time-history values are used to obtain the orders of magnitude of vibrating harvesters. In particular, two cantilever beams made out of piezoelectric and piezomagnetic materials are simulated. Finally, it is concluded that the former are more appropriate for vibrating harvester applications.

In short, this numerical tool allows the design of vibrating harvesters by using a numerical laboratory in contrast to expensive laboratory experiments. For this purpose, this tool could be combined with optimization and probabilistic techniques such as genetic algorithms and Monte Carlo simulation to optimize the harvesters and to perform sensitivity analyses, respectively.

Bibliography

[1] A. McWilliams. *Smart Materials and Their Applications: Technologies and Global Markets*. BCC Research, 2015.

[2] J. Hruska. To the stars: After a 25 year hiatus, Nasa restarts plutonium production, 2013. http://www.extremetech.com/extreme/150756-to-the-stars-after-a-25-year-hiatus-nasa-restarts-plutonium-production.

[3] H. S. Kim, J.-H. Kim, and J. Kim. A review of piezoelectric energy harvesting based on vibration. *International Journal of Precision Engineering and Manufacturing*, 12(6):1129–1141, 2011.

[4] P. Cahill, N. A. N. Nuallain, N. Jackson, A. Mathewson, R. Karoumi, and V. Pakrashi. Energy harvesting from train-induced response in bridges. *J. Bridge Eng.*, 19(9), 2014.

[5] J. L. Perez-Aparicio, H. Sosa, and R. Palma. Numerical investigations of field-defect interactions in piezoelectric ceramics. *International Journal of Solids and Structures*, 44:4892–4908, 2007.

[6] R. Palma, J. L. Pérez-Aparicio, and R. L. Taylor. *Non-linear finite element formulation applied to piezoelectric materials with Debye memory*. IEEE/ASME Transactions on Mechatronics, submitted, 2017.

[7] P. Museros, E. Moliner, and M. D. Martínez-Rodrigo. Free vibrations of simply-supported beam bridges under moving loads: Maximum resonance, cancellation and resonant vertical acceleration. *Journal of Sound and Vibration*, 332:326–345, 2013.

[8] P. Museros and E. Moliner. *Vibration of simply supported beams under a single moving load: A detailed study of cancellation phenomenon*. International Journal of Mechanical Sciences, in press, 2015.

[9] J. L. Perez-Aparicio, R. Palma, and R. L. Taylor. Multiphysics and thermodynamic formulations for equilibrium and non-equilibrium interactions: non-linear finite elements applied to multi-coupled active materials. *Archives of Computational Methods in Engineering*, 23(3):535–583, 2016.

[10] J. L. Jiménez, I. Campos, and M. A. López-Mariño. Maxwell's equations in material media, momentum balance equations and force densities associated with them. *The European Physical Journal Plus*, 128(46):1–6, 2013.

[11] O. C. Zienkiewicz, R. L. Taylor, and J. Z. Zhu. *The Finite Element Method: The Basis*. Elsevier Butterworth-Heinemann, 2005.

[12] R. L. Taylor. *FEAP A Finite Element Analysis Program: User Manual*. University of California, Berkeley, 2010.

[13] L. Frýba. *Dynamics of railway bridges*. Thomas Telford, 1996.

[14] Y. B. Yang, J. D. Yau, and Y. S. Wu. *Vehicle-bridge interaction dynamics with applications to high-speed railways*. World Scientific Publishing, 2004.

[15] European Rail Research Institute (ERRI). Ponts-Rails pour vitesses > 200 km/h, 2000. ERRI D-214/RP9.
[16] A. Doménech, P. Museros, and M. D. Martínez-Rodrigo. Influence of the vehicle model on the prediction of the maximum bending response of simply-supported bridges under high-speed railway traffic. *Engineering Structures*, 72:123–139, 2014.
[17] F. U. Khan and I. Ahmad. Review of Energy Harvesters Utilizing Bridge Vibrations. *Shock and Vibration*, 2016:21, 2016.

Christian Viehweger

Solar energy harvesting for wireless sensor systems

Abstract: Wireless sensor systems are often supplied by solar cells. Solar power is one of the most predictable sources in energy harvesting and, therefore, it is often preferred. The optimal use of the incoming energy is substantial, therefore, the usage of the cells has to be improved by impedance matching and maximum power point tracking. This contribution gives an overview on the determination of the incoming power and typical methods to improve the utilization of the cells. As today a huge number of maximum power point tracking methods exist, which differ just in very small aspects, the main and most important methods are described here.

Keywords: Wireless sensor network, energy harvesting, solar cells, maximum power point tracking, energy management

1 Introduction

The power supply for wireless sensor systems (WSN) is a challenge today. Energy-efficient microcontrollers with low-power sleep modes are available. Low-power sensors and wireless communication technologies are well developed, and there is a high number of possible applications. Still, a cable-based supply is often not possible or wanted, either because there is no access available, like for monitoring systems in agriculture, forests or waters, or the cables would be disturbing, like in industrial applications. The use of batteries is often not an appropriate replacement. Especially costs are a problem, as the batteries have limited energy and, therefore, have to be changed, which results in a high maintenance effort. For these reasons, the use of ambient energy to power the systems is a good method, if applicable.

Solar energy harvesting is one of the most important methods to supply a WSN. Compared to other sources, it represents, in general, a more predictable source [1]. With sufficient knowledge about the parameters, like the place of application, the type of solar cell or the positioning, it becomes possible to estimate the incoming energy with good accuracy. This information is essential for the design of a WSN.

Typically, a WSN is designed to require as little energy as possible. This means highly efficient parts and programming, the use of sleep modes and reduced functionality. Additionally, also the costs should be low to enable a high number of deployed nodes. Compared to traditional solutions, the power supply from energy harvesting shows a volatile behavior. Battery or cable based, the supply can be designed to ex-

---

Christian Viehweger, Chemnitz University of Technology

https://doi.org/10.1515/9783110445053-002

actly match the load demand. With an ambient energy transducer, the system needs to have a high safety factor and the capability to store energy to bridge periods with less energy income. It would be easy to use just a larger energy harvester, but this would signify several drawbacks. For example, the consumption of more space could prohibit the application, the size may affect the object to be monitored and, especially, the higher costs are problematic. For these reasons, it is very useful to have a source that is more predictable, so that the energy harvester can be designed efficiently.

Besides the determination of the expected energy income, also an efficient use of the power is necessary. In the case of solar cells, this entails maximum power point tracking (MPPT) to reach the best yield. The last step of the design is the development of a storage concept, using the information of the source and the load. Considering the state of research and technology for solar cells and their use, they are a good opportunity to avoid batteries for sensor systems.

The following sections describe the incoming energy and the typical MPPT algorithms.

2 Assessment of the incoming energy

The most common method for the assessment of the incoming energy to characterize a solar cell is the air mass model. It describes the amount of air that the sunlight has to pass trough (Figure 1). The position outside the atmosphere is referred to as AM 0. The zenith, where the light passes the whole atmosphere once, is called AM 1.0. With an increasing angle, more air mass has to be passed through. For standard test conditions, an angle of $48.2°$ is considered. This represents AM 1.5. The air mass is calculated with [2]:

$$\mathrm{AM} = \sec \theta_z \,. \tag{1}$$

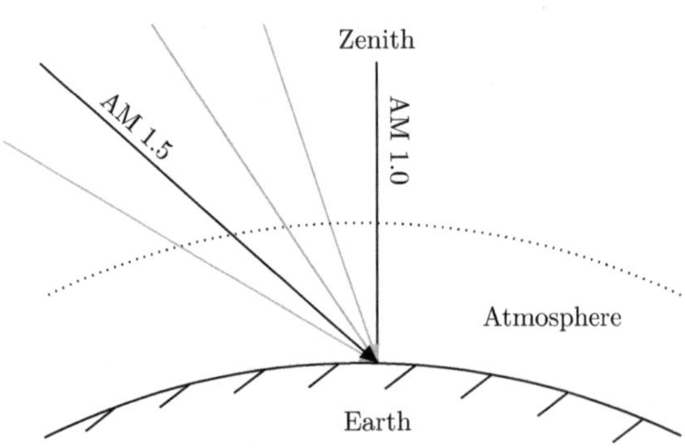

Fig. 1: Incoming power on the solar cell. Angle dependent air mass model.

The reason for the consideration of the path length in the atmosphere is that different substances influence the power. For example, water in different aggregate states, ozone or particles cause absorption and scattering, which reduces the energy that reaches the Earth's surface and, therefore, the solar cell. The more distance that has to be covered, the higher the effect of those substances. Figure 2 shows the incoming power per square meter depending on the wavelength for a standard atmosphere. A standardization is important, as the conditions are not the same everywhere. For example, air pollution strongly depends on the place of application. Also, the spectral distribution matters, as, for example, the transmission coefficients of different parts of the air are not constant.

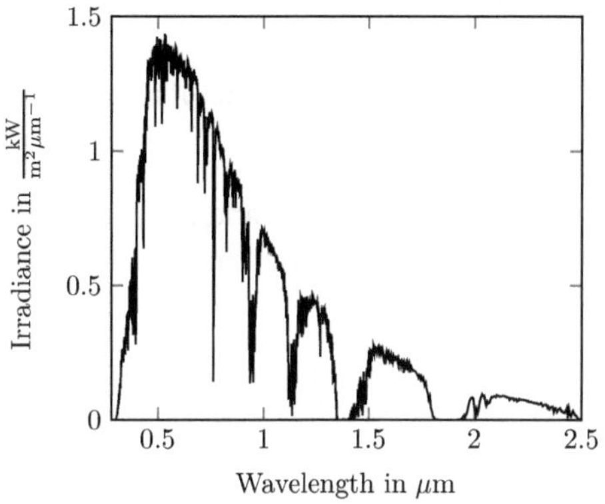

Fig. 2: Spectral irradiance for AM 1.5 [3].

In addition to the direct illumination, also scattered light has to be considered. Indirect illumination due to scattering in the atmosphere or on the ground contributes to the total incoming power, but the exact value depends on the environment and, therefore, cannot be predefined. For the measurement of the amounts of direct and indirect illumination, typically pyranometers are used. They measure the hemisphere typically using a thermopile. Direct and indirect light can also be distinguished. For this, the sun is mechanically covered during the measurement, which gives indirect illumination. Afterwards, a second measurement records the total power. By subtracting the indirect from the total, the direct power can be obtained.

Because of the strong dependency on the position, many models to simulate the incoming power exist. In general, they require a high computing power and special input data, like air composition or satellite data. For this reason, it is not useful to apply them on a wireless sensor node. Measurement data and recorded values are often available for many positions, especially from weather services.

3 Maximum power point tracking

MPPT describes technologies to maximize the energy yield from a solar cell. As it is a non-linear source, special aspects have to be considered.

The equivalent circuit diagram using two diodes is the most common one for a solar cell (Figure 3). It combines the parallel resistance that represents the leakage, the series resistance for the different conductivities within the cell and two diodes, where one represents the more ideal and the other the non-linear behavior.

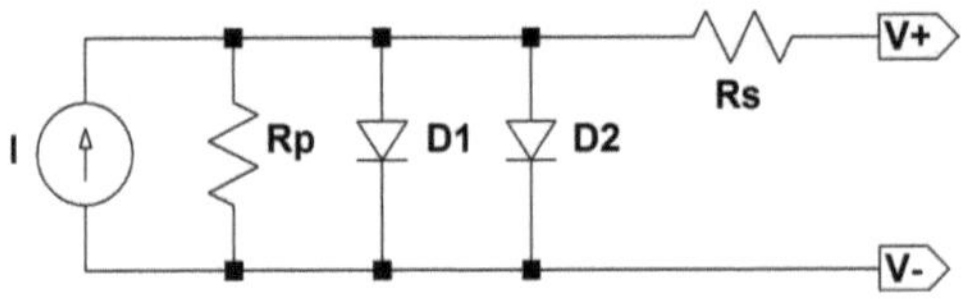

Fig. 3: Double-diode model of a solar cell.

Using the Shockley equation for the diodes, the thermal voltage U_T and the cutoff currents I_{01} and I_{02}, the current of the solar cell according to the model can be calculated by [4]:

$$I = I_{ph} - I_{01} \cdot \left(e^{\frac{V_{pv} + I_{pv} \cdot R_S}{m_{d1} \cdot V_T}} - 1 \right) - I_{02} \cdot \left(e^{\frac{V_{pv} + I_{pv} \cdot R_S}{m_{d2} \cdot V_T}} - 1 \right) - \frac{V_{pv} + I_{pv} \cdot R_S}{R_P} . \quad (2)$$

To solve this implicit equation, a number of different numerical procedures, for example the Newton method, can be used. The resulting $I-V$ curve can be seen in Figure 4. The non-linearity caused by the diodes results in a point of maximum output power, which varies depending on the incoming light. This point has to be maintained to guarantee the efficient use of the solar cell. The detailed improvement of the output power depends on the type of the cell, the illumination and, therefore, the slope of the curve.

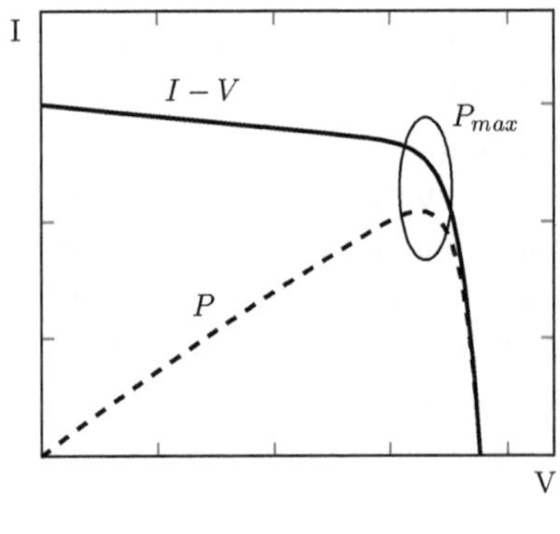

Fig. 4: Characteristic $I-V$ curve of a solar cell.

3.1 The most common methods

In most applications, the calculation of the ideal working point of the solar cell on the sensor node is not possible. It would require detailed information, for example, about the incoming light and the temperature, which are normally not available. Also, the typical sensor nodes do not have the calculation power to solve the implicit equation.

For these reasons, MPPT for wireless sensor nodes is often based on trial and error, with a measurement of the output power before and after a variation of the working point. Many different MPPT methods are available. They differ in their complexity, time and energy demand and the information used.

Best fixed voltage

To use this method means to determine the optimal operation point before the real application. Therefore, the system has to be characterized under conditions that are as similar to the later use as possible. Afterwards, the corresponding voltage will be fixed at this point. The system is kept in this state and every little deviation from the predetermined condition causes a deviation from the current MPP. Therefore, it is not an accurate method, but it is the easiest to implement and at least better than avoiding MPPT completely. It can be the default method for a system.

Lookup tables

Using a lookup table is an empirical method, where optimal values for I and V for different environmental conditions are measured and stored in a list. The node measures the conditions itself and uses the values from the table for the solar cell.

This method is more accurate, the more environmental data is available for comparison. At the same time, the sensor node also needs to have the capability to measure the values itself, which results in a huge demand for sensors and measurements. To apply extra sensors to the node just for the MPPT is not useful. Therefore, this method is only suitable for nodes that have multiple sensors anyway. In this case, the data measured can be directly used also for the power management. Additional measurements only for the tracking should be avoided due to the energy consumption. The advantage of this method is that the power supply of the electronics can continue at all time, as the solar cell does not need to be disconnected for power measurement.

Constant voltage/current

In a good approximation, the ratio between the voltage at MPP and the open-circuit voltage among different environmental conditions remains similar. The same applies to the current. Within this method, the idea is to keep the ratios

$$k_V \approx \frac{V_{MPP}}{V_{OC}} \tag{3}$$

for the voltage or

$$k_{\mathrm{I}} \approx \frac{I_{\mathrm{MPP}}}{I_{\mathrm{SC}}} \tag{4}$$

for the current, always constant. This means that the ratios have to be determined for a specific solar cell, afterwards the open-circuit voltage or the short-circuit current have to be measured and then an approximation for the MPP can be derived. Typically, the ratios are between 0.7 and 0.85 [5, p. 88]. The main advantage of this method is the ease of use; on the other hand it is not very accurate and, moreover, for the measurement, the solar cell has to be disconnected from the circuit.

Perturb and observe

This method might be the most common today. In general, it entails changing the status of the system and evaluating its reaction. The voltage of the solar cell is changed in one direction by a defined step size. Each time the output power is measured. As long as the power increases, the adaption will continue in the same direction. If the output decreases from one step to the next, the direction will change. This way, the working point alternates around the MPP; the real MPP will be reached by coincidence, but will never be kept.

A huge problem is the potential misalignment of the system if the light conditions change. If the solar power on the cell increases while the MPPT is stepping, it will continue in this direction, even if the real MPP is on the opposite side of the curve. In this situation, the output power increases not because the *perturb and observe* approaches the MPP, but it increases because of the light conditions, while the tracker diverges from the optimal working point. The tracker detects the misalignment only if the source is stable, which can result in a time delay. For the same reason, varying sunlight conditions might be a problem for this method.

The advantage of this method is the ease of use, that it is independent from the source and needs no additional information, just the output power of the cell. Also, it can react to all variations of the environmental conditions, as it just tracks the power. The most important negative aspects are that the real MPP will not be reached and that with each step it requires a power measurement, which for some systems means the disconnection of the source. In general, this method is quite accurate and does not have high hardware demands, which is why it is often used.

Hill climbing

Hill climbing is similar to perturb and observe, with the difference that with this method not the voltage but the frequency is the controlled variable. Typically, switching converters are used for their improved efficiency. For them, the frequency of the switching controls the workload of the solar cell. The advantages and disadvantages of the method are the same as with perturb and observe.

Incremental conductance

In this method, the progress of the power over the voltage is calculated. The first step is to calculate the conductance for a defined time step [6]:

$$\frac{\mathrm{d}I}{\mathrm{d}V} = \frac{I}{V}\,. \tag{5}$$

Using this information, the current position on the curve in relation to the MPP becomes clear, as with:

$$\frac{\mathrm{d}I}{\mathrm{d}V} = -\frac{I}{V} \quad \left(\frac{\mathrm{d}P}{\mathrm{d}V} = 0\right) \tag{6}$$

$$\frac{\mathrm{d}I}{\mathrm{d}V} > -\frac{I}{V} \quad \left(\frac{\mathrm{d}P}{\mathrm{d}V} > 0\right) \tag{7}$$

$$\frac{\mathrm{d}I}{\mathrm{d}V} < -\frac{I}{V} \quad \left(\frac{\mathrm{d}P}{\mathrm{d}V} < 0\right) \tag{8}$$

it can be detected whether the current working point is before, after or on the MPP. For this reason, the incremental conductance method is fast, as the direction is directly known, and accurate, as the real MPP can be reached. A disadvantage is that both voltage and current have to be measured and analyzed. Therefore, the method consumes more resources than other MPPT algorithms.

Parasitic capacitance

Parasitic capacitance is an enhancement of incremental conductance, where the barrier layer capacitance C_p is also considered. This improves the accuracy with just minimally higher demands for calculation power. A drawback is that C_p has to be known.

AI methods

With the increasing resources of modern microcontrollers, methods of artificial intelligence are growing in importance. Especially neural networks and fuzzy logic are often discussed in the literature. With fuzzy solutions, the convergency to the MPP becomes more flexible, as depending on the status, the individual steps onto the maximum power can be varied. Huge steps far from the MPP and small steps close to it guarantee both fast convergency and a small error.

The use of a neural network is more complex. For most applications, the training of the network cannot be done on the sensor system itself. On the one hand, it would imply a big computational effort, and on the other hand, the training takes a lot of time and, therefore, energy, which is very critical. Training the network outside the real application requires detailed knowledge about the environment. Also, the adaption of the system if the conditions change is critical, if the training is not done on the device itself.

AI methods, therefore, highly depend on the system and its capabilities. They need a high computation power and as much information as possible to find the correct MPP.

4 Conclusion

The consideration of the incoming energy and its efficient use is important for the design of solar powered WSN. The incoming energy determines whether a solar-based solution can be successful. Therefore, the correct positioning and knowledge about the environmental conditions at the place of installation are important. Afterwards, the energy has to be converted efficiently by using an appropriate MPPT algorithm. The choice of method depends on the available sensor data and the total power of the system. For a low overall power income, also the MPPT has to consume as little power as possible. Therefore, the allowed complexity of the control system strongly depends on the conditions.

Bibliography

[1] C. Viehweger and O. Kanoun. *Modellbasierte Bestimmung der zeitabhängigen Eingangsleistung von Solarzellen für das Energiemanagement drahtloser Sensorsysteme*. Technisches Messen, De Gruyter, 2016. doi:10.1515/teme-2015-0111.

[2] M. J. Reno, C. W. Hansen, and J. S. Stein. *Global Horizontal Irradiance Clear Sky Models: Implementation and Analysis*. Sandia Report, Sandia National Laboratories, SAND2012-2389, 2012.

[3] Renewable Resource Data Center (RReDC). Reference Solar Spectral Irradiance: Air Mass 1.5, ASTM G173-03 Tables.

[4] C. Viehweger. *Modellbasiertes Energiemanagement für die intelligente Steuerung solarversorgter drahtloser Sensorsysteme*. PhD thesis, Universitätsverlag Chemnitz, 2017. ISBN 978-3-96100-022-7.

[5] A. Goetzberger and V. Hoffmann. *Photovoltaic Solar Energy Generation*. Springer Verlag, 2005. ISBN 3-540-23676-7.

[6] M. A. Eltawil and Z. Zhao. MPPT techniques for photovoltaic applications. *Renewable and Sustainable Energy Reviews*, 25:793–813, 2013. ISSN 1364-0321, doi:10.1016/j.rser.2013.05.022.

Zdenek Hadas and Jan Smilek

Efficiency of vibration energy harvesting systems

Abstract: This chapter is focused on an efficiency analysis of vibration energy harvesting systems. Vibration energy harvesting systems convert the ambient kinetic energy of mechanical vibrations into useful electricity. Nowadays, vibration energy harvesting technologies are used in various engineering applications (automotive, transportation, aeronautics, smart buildings and cities, heavy and petrochemical industries, etc.), and the power gain of energy harvesting systems is still improving. Although the levels of ambient mechanical energy are very low, an efficient kinetic energy harvesting technology can be used as an autonomous power supply for modern ultra-low-power electronics, wireless sensing and embedded self-monitoring systems.

Keywords: Efficiency, Vibration, Energy Harvesting, Linear Model, Equivalent Circuit, Electromagnetic Energy Harvester, Output Power, Figure of Merit

1 Introduction

Most kinetic vibration energy harvesting systems have limitations in the resonance operation, otherwise these harvesting systems provide significantly lower outputs. The magnitude and frequency stability of input mechanical vibrations are fundamental for successful energy harvesting operation. However, the improvement of these harvesting systems could be useful for a wider usage of vibration energy harvesting systems in ultra-low-power applications, which are embedded in vibratory environment with a low magnitude and a variable frequency of mechanical vibrations. For this reason, this chapter analyzes the influence of individual parameters on effective vibration energy harvesting.

The efficiency of vibration energy harvesters is a frequently discussed issue [1]. Efficiencies of power electronics and power management circuits for energy harvesting applications are well known, and these results have been published several times [2]. The efficiency of energy harvesting electronics is usually in the range of 60–95%. On the other hand, the efficiency of vibration energy harvesting systems alone has not been widely discussed [3]. The main aim of this chapter is to provide information about efficiency analyses of vibration energy harvesting systems.

Zdenek Hadas, Jan Smilek, Brno University of Technology, Czech Republic

https://doi.org/10.1515/9783110445053-003

2 Vibration energy harvesting systems

The vibration energy harvesting system consists of a mechanical resonator, an electromechanical energy transducer and electronics (power management, load and energy storage element) [4]. The diagram of a kinetic vibration energy harvesting system is shown in Figure 1. The mechanical resonator is the key element of the whole energy harvesting system that is embedded in a vibrating structure. The electromechanical transducer converts the input kinetic energy of the mechanical resonator into electricity during its operation. The parameters of the mechanical resonator significantly affect the output electrical power. Based on the character of the vibrations and the maximal dimensions of the resonator, a physical principle and a design of the electromechanical energy transducer is chosen.

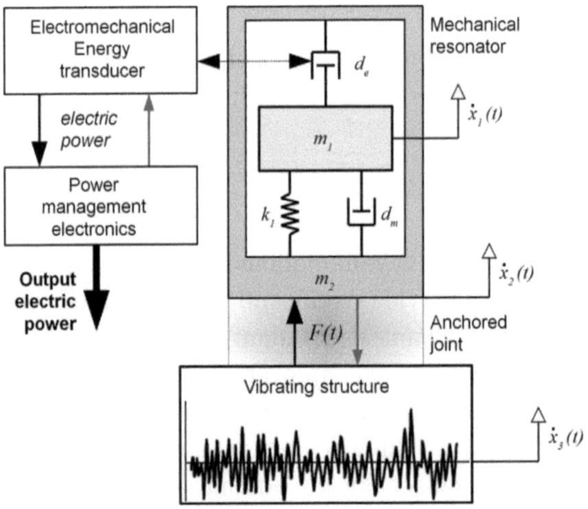

Fig. 1: Principle of the vibration energy harvesting system.

The relative movement of the seismic mass inside the resonator is exploited for conversion of energy into electricity by one or a combination of several available physical principles of the electromechanical conversion. The vibration energy harvesters usually use either an electromagnetic, piezoelectric, electrostatic or magnetostrictive principle of electromechanical energy conversion [5]. The harvested electric power is rectified and stabilized using power management electronics, and the output electrical power is consumed or stored in energy storage elements.

2.1 Mechanical resonators of vibration energy harvesting systems

A precise mechanism for resonance operation with a very high quality factor is a fundamental part of a vibration energy harvesting system (Figure 1). This mechanical resonator is based on a spring suspension, with stiffness k_1, of a moving seismic mass m_1. A frame of the resonator m_2 is anchored to the vibrating structure and, therefore, the mechanical resonator is excited by ambient mechanical vibrations of the vibrating structure and the harvester frame x_2, and it provides the relative movement x_1 against frame m_2. The frame is anchored to the vibrating structure, and interaction forces can be observed due to operation under vibrations. The relative oscillation x_1 of the seismic mass m1 against the frame is found by solving the second-order differential equation of motion (1) and it is inversely proportional with mechanical damping d_m.

$$m_1\ddot{x}_1 + d_\mathrm{m}\dot{x}_1 + d_\mathrm{e}\dot{x}_1 + k_1 x_1 = -m_1\ddot{x}_2 \tag{1}$$

The overall harvested electrical power is extracted from the energy harvesting system by an electromechanical transducer. The extraction provides an electromechanical damping effect, which is depicted as damper d_e. The maximal harvested electric power (2) is generated when energy is dissipated by mechanical damping forces inside the harvester, depicted as damper d_m, and electromechanical conversion forces are in equilibrium. The maximal electric power depends on the seismic mass, squared magnitude of vibration acceleration A_v and on the mechanical quality factor Q_m. The harvested electric power is inversely proportional to operating frequency. Nevertheless, the vibration energy harvesting system has to operate in resonance operation Ω for generating the theoretical maximum of electric power [6],

$$P_\mathrm{e} = \frac{m_1 \cdot A_\mathrm{v}^2}{8 \cdot \Omega} \cdot Q_\mathrm{m} \tag{2}$$

The amplitude of the relative oscillation movement in the resonance operation depends on the magnitude of input vibrations and on the quality factor. The mechanical quality factor Q_m provides information about a ratio of the output amplitude displacement X_1 to the input amplitude displacement X_2. This value is a function of the mechanical damping, mass and the resonance frequency, expression (3). A very high quality factor of the mechanical resonator is fundamental for generating useful energy from mechanical vibrations [7],

$$Q_\mathrm{m} = \frac{X_1}{X_2} = \frac{m_1 \cdot \Omega}{d_\mathrm{m}} \tag{3}$$

2.2 Complex model of a vibrating structure with an energy harvesting system

The efficiency of the energy harvester is the ability to avoid wasting input energy when obtaining the desired electrical energy. This means that an efficient vibration energy harvesting system uses the input energy of mechanical vibrations and provides maximal obtainable output power. The harvested electric power can be measured on an electrical load, but a definition of the input power might not be clear for the vibration energy harvesting system, the function of which is based on the resonance operation. For this reason, analysis of the complex model of a vibrating structure with an energy harvesting system is very important. The energy flow with the force interactions is shown in Figure 2.

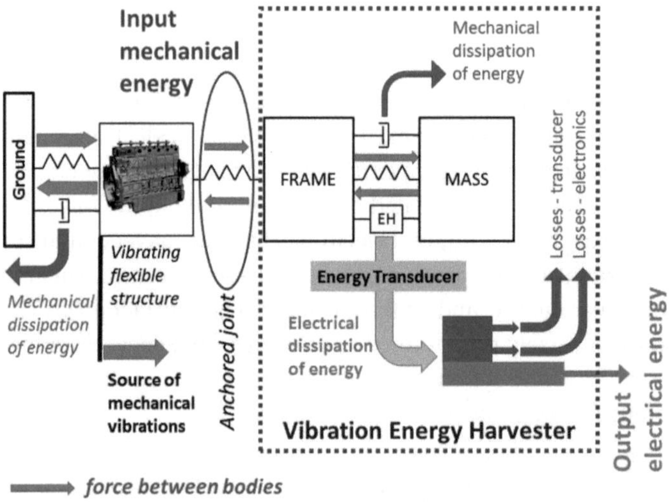

Fig. 2: Energy flow in the complex model of a vibrating structure with an energy harvesting system.

The vibration energy harvesting system is usually used in a force driven vibratory environment where a flexible vibration structure is excited by a source of mechanical vibrations, which are transmitted through the flexible structure to a place with an anchored vibration energy harvester. The mechanical energy that is accumulated in the vibrating flexible structure could be very high, e.g., the vibrations of a bridge structure, a building, a train, an aircraft, etc., and only a small part of this kinetic energy is consumed during the harvesting of electrical energy. The input mechanical energy into the vibration energy harvesting system has to be defined for efficiency calculation. The definition of an adequate input of mechanical energy has been discussed for a long time. An approach, which was defined under the project Metrology for Energy

Harvesting[1] respects all the aspects of the energy flow through systems, Figure 2, and it will be used in this chapter.

2.3 Efficiency of vibration energy harvesting systems

The input mechanical energy into the vibration energy harvesting system can be defined as the work of forces between the frame of the harvester and the vibrating structure. The mechanical energy, which is a time integral of the power caused by these forces, is defined as the input mechanical energy. The mechanical energy is converted into useful electricity, and the output electric power is observed on the electrical load, Figure 3. The time integral of the output power gives the output electrical energy.

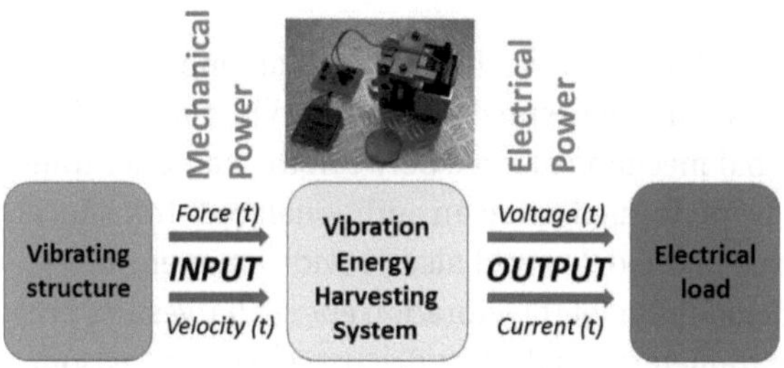

Fig. 3: Input and output power of the vibration energy harvesting system.

The output electric power P_{electric} is consumed on the electrical load. The output electric power is a product of the output voltage and the current flowing through the electrical load (4). This useful electric power P_{electric} and electric losses inside the electromechanical transducer dissipate the energy from the oscillating system of the mechanical resonator, Figure 1,

$$P_{\text{electric}} = u(t) \cdot i(t) \tag{4}$$

The input mechanical power P_{mech} is transferred to the vibration energy harvesting system through the harvester frame, Figure 2. P_{mech} is defined as a product of the internal mechanical force $F(t)$ in the anchored joint and the velocity of mechanical vibrations $\dot{x}_2(t)$, see expression (5),

$$P_{\text{mech}} = F(t) \cdot \dot{x}_2(t) \tag{5}$$

The efficiency of the vibration energy harvesting system can be calculated for harmonic electrical signals and the harmonic course of mechanical vibrations in the form

1 http://projects.npl.co.uk/energy_harvesting/

of a ratio between the input mechanical power and the output electric power, as shown by expression (6),

$$\eta = \frac{P_{\text{electric}}}{P_{\text{mech}}} \tag{6}$$

However, common mechanical vibrations and electrical outputs have the character of non-harmonic signals. Therefore, the energy harvesting efficiency has to be calculated in the form of the electrical and mechanical energy ratio, see expression (7). This definition of the efficiency can be used for the analyses of vibration energy harvesting systems [8] that operate in an environment with non-harmonic vibrations, random moving oscillations or mechanical shocks [9],

$$\eta = \frac{E_{\text{electric}}}{E_{\text{mech}}} = \frac{\int_{t_1}^{t_2} u(t) \cdot i(t)\,\mathrm{d}t}{\int_{t_1}^{t_2} F(t) \cdot \dot{x}_2(t)\,\mathrm{d}t} \tag{7}$$

The output voltage and current through the electrical load are commonly measured. The velocity of mechanical vibrations can be measured by common vibrometers. However, the measurement of the internal mechanical force between the harvester frame and the vibrating structure has to be performed using an additional force transducer. Efficiency analyses, which are based on model, could also provide an interesting information for gaining a better understanding of the vibration energy harvesting principle. The influence of individual parameters can then be discussed. For this reason, a complex model of the vibration energy harvesting system is created. This chapter will present only analyses and models of the electromagnetic vibration energy harvester, but the same approach can be used for piezoelectric, magnetostriction or electrostatic harvesters.

3 Model of an electromagnetic energy transducer

The electromagnetic principle of electromechanical conversion is used in the presented vibration energy harvesting system, and it will be used for efficiency analyses. The electromagnetic transducer employed consists of an oscillating magnetic circuit (seismic mass) and a self-bonded air coil that is fixed to the frame. As the mechanical resonator of the harvester is excited by mechanical vibrations, it produces the oscillation of the magnetic circuit with permanent magnets against the fixed coil. This movement is described by the motion equation (1). Due to Faraday's law, the oscillation movement provides a change of the magnetic field through the coil, and it induces voltage on the coil. The voltage u_i (8) depends on the number of the coil turns N, the active length of the coil l, the magnetic flux density B through the active coil turns and on velocity of the relative movement from equation (1),

$$u_i = N \cdot l \cdot B \cdot \dot{x}_1(t) \tag{8}$$

In the case of the resistive load R_L being connected, a current i flows through the coil and causes a negative feedback in the form of an electromagnetic damping force, which depends on the parameters of the electromagnetic damper d_e (9), where Z is the impedance of the whole electrical circuit. The inductance of the self-bonded air coil is assumed to be very low, so that a simplified model, where the impedance includes only the resistance of the coil and the load, can be used,

$$d_e = \frac{(N \cdot l \cdot B)^2}{Z} \tag{9}$$

The electromagnetic damping force results from the feedback of the current flowing through the coil and affects the relative oscillations of the magnetic circuit against the fixed coil. This feedback force (10) can be expressed as an electromagnetic damping force, which is the same as an electrical damping force that depends on the parameters of the magnetic circuit and the coil. Both expressions of this feedback force are shown in the equation (10), where parameter c_{FL} provides information about electromagnetic coupling (11),

$$d_e \cdot \dot{x}_1(t) = c_{FL} \cdot i(t) \tag{10}$$

$$c_{FL} = N \cdot l \cdot B \tag{11}$$

The electromagnetic coupling is used in the equation for Faraday's law, and the equation (8) can be rewritten in the form (12),

$$u_i = c_{FL} \cdot \dot{x}_1(t) \tag{12}$$

The expression of the electromagnetic damping force (10) can be used in the second-order differential equation of the mechanical resonator. The final motion equation is in the form of (13),

$$m_1\ddot{x}_1 + d_m\dot{x}_1 + c_{FL}i(t) + k_1x_1 = -m_1\ddot{x}_2 \tag{13}$$

4 Efficiency analysis based on vibration energy harvesting simulation

4.1 Mechanical model of the vibration energy harvesting system

Both coupled electromechanical equations (12) and (13) with the electromagnetic coupling c_{FL} provide a linear energy harvesting model that can be used for efficiency analyses of the vibration energy harvesting system. A linear model of the whole complex mechanical system is used for the calculation of the internal mechanical force in the vibration energy harvesting system, Figure 2. A mechanical model with three degrees of freedom (DOF), which include the vibrating flexible structure with the source

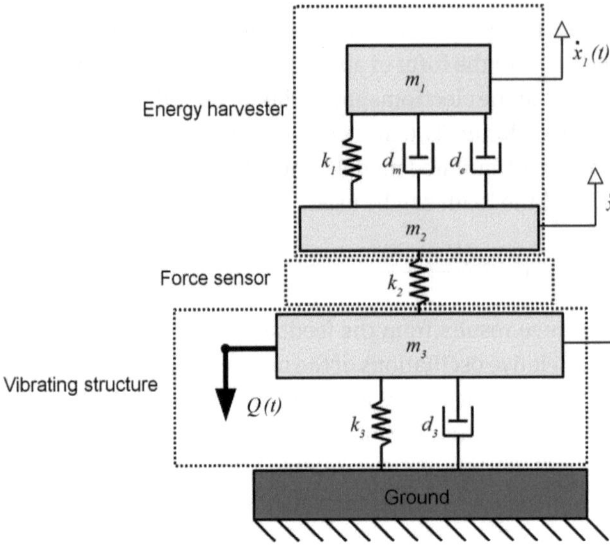

Fig. 4: 3-DOF linear mechanical model of a mechanical system with a vibration energy harvesting system.

of the vibration, the force sensor and the vibration energy harvesting system, is shown in Figure 4. The vibrating flexible structure has only one DOF with respect to the value of the first natural frequency of the flexible structure. In this case, the vibrating flexible structure is depicted as 1-DOF mass-spring-damper system (m_3, k_3, d_3), and it represents the mechanical part where the vibration energy harvesting system is anchored.

The presented linear model is excited by the source of vibrations Q, which provides an energy flow into the vibration energy harvesting system, represented by parameters (m_1, k_1, d_m, d_e). The resonance frequency of the energy harvesting system is tuned up with respect to the value of the first natural frequency of the vibrating flexible structure. The linear model of the electromagnetic transducer (10) is integrated into this mechanical model for calculation of the dissipation feedback provided by the electromagnetic damper.

The required internal force for the efficiency calculation is modeled by a force sensor with stiffness k_2. The model of the force sensor is in the form of a mass-spring system (m_2, k_2), and it is integrated between the vibrating structure and the frame of the energy harvester m_2. The input force into the vibration energy harvesting system is calculated as shown by expression (14). The stiffness of the force sensor k_2 is very high and corresponds to the stiffness of the screw joint between the frame and the vibrating structure. Therefore, it can be assumed that the vibrations $x_2(t)$ and $x_3(t)$ are practically identical for the energy harvesting application. The stiffness k_2 then provides information about the input mechanical force.

$$F(t) = k_2(x_2(t) - x_3(t)) \tag{14}$$

The calculated internal force between the vibrating structure and the harvester frame (14), and the velocity of the excitation vibrations provide the information about the input mechanical energy into the energy harvesting system.

This complex mechanical system is excited by the model of external force $Q(t)$, and it provides mechanical vibrations of the vibrating structure, which correspond to real vibrations of engineering systems.

4.2 Equivalent RLC circuits of the vibration energy harvesting system

The three-mass mechanical system with the vibration energy harvester, Figure 4, can be expressed in a circuit form using an equivalent RLC circuit. The Kirchhoff equations are calculated for each loop [10]. This model can easily be connected to the circuit of the electromagnetic converter (12) with the electrical load, where the output electrical energy is calculated. The electromechanical model of equivalent RLC circuit is shown in Figure 5, and it consists of three equivalent electrical subcircuits. The physical analogy of electro-mechanical elements is depicted. Currents through individual branches of the RLC circuit are mapped to the velocities. The voltages are then mapped to mechanical forces. The inductors are equivalent to the masses, the resistors to the dampers and the capacitors to the compliances of the mechanical springs. The compliance of a mechanical spring is inversely proportional to its stiffness k, and it is used as the capacitance in the equivalent electrical RLC circuit, as is shown in Figure 5.

The equivalent RLC circuit that corresponds to the vibrating structure is excited by the source of an input voltage, which is equivalent to the external mechanical force $Q(t)$. The middle equivalent RLC subcircuit provides information about the force sensor and about the velocity of the frame. The electrical energy, which flows through this equivalent RLC circuit, presents the input mechanical energy to the vibration energy

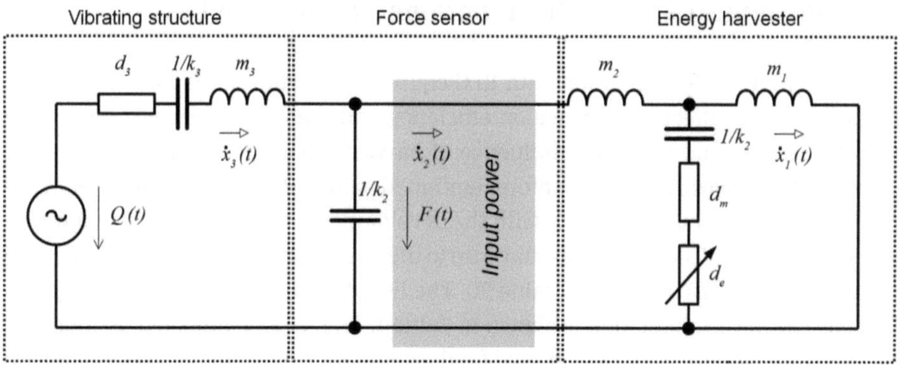

Fig. 5: Equivalent RLC circuit of the vibrating flexible structure, force sensor and vibration energy harvesting system.

harvester, which is presented as the last equivalent RLC circuit, corresponding to the model (13). The current in the parallel branch of this RLC subcircuit represents the relative velocity of the seismic mass of the harvester. This velocity is used in the linear model of Faraday's law, and the induced voltage (12) is calculated [11]. This induced voltage (12) represents the input to the electrical circuit of the electromagnetic vibration energy harvester, which is shown in Figure 6, and the harvested output power on a resistive electrical load is calculated.

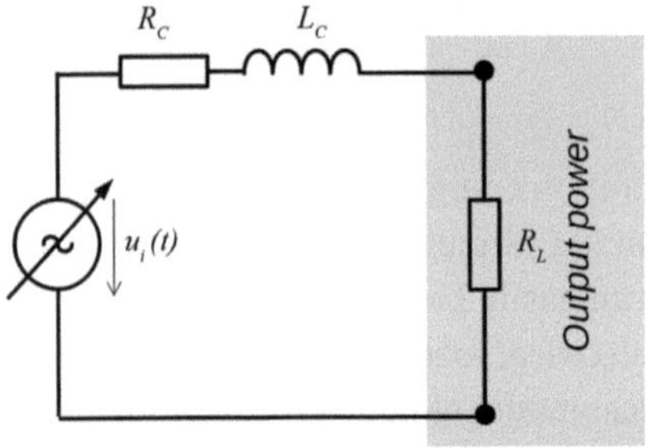

Fig. 6: Equivalent RCL circuit of the electromagnetic energy transducer.

Thereafter, the harvested electrical energy from the model in Figure 6 and the input mechanical energy to the vibration energy harvester, which flows through the equivalent RLC circuit of the force sensor, Figure 5, represent the inputs to the efficiency analysis. The value of the efficiency of the linear vibration energy harvesting system is then calculated according to the equation (7).

5 Analyses and simulation results of equivalent RLC circuits

5.1 Mechanical and electrical parameters for analyses

The main aim of the simulation with RLC equivalent circuits is the identification of parameters that affect the conversion efficiency and the output power of the linear vibration energy harvester. The influence of harvester parameters on the resonance operation was investigated using aforementioned model. Parameters employed in the model follow from our experience with the development of electromagnetic vibration energy harvesters. The model was tuned up to the operation frequency of 17 Hz and the quality factor of this system has value 90. The list of parameters used for the equivalent electrical RLC simulation is shown in Table 1. The list of parameters influencing the electromagnetic coupling and used for the simulation of Faraday's law is shown in Table 2. The mass m_3 is significantly higher compared to the seismic mass m_1; therefore the harvester operation does not affect the input mechanical vibrations.

Tab. 1: Parameters of the equivalent RCL model for efficiency analyses.

Mechanical system		Equivalent electrical system		Value
Parameter	Unit	Parameter	Unit	
Seismic mass m_1	g	Inductor 1	mH	50
Quality factor	–	Quality factor	–	90
Mechanical damping d_m	Ns/m	Resistor 1	Ω	0.0593
Harvester stiffness k_1	N/m			570
Mechanical compliance $1/k_1$	m/N	Capacitor 1	mF	0.00175
Mass m_2	g	Inductor 2	mH	75
Force sensor stiffness k_2	N/µm			1050
Mechanical compliance $1/k_2$	µm/N	Capacitor 2	µF	0.00095
Mass m_3	kg	Inductor 2	H	100
Stiffness k_3	N/µm			7.994
Mechanical compliance $1/k_3$	µm/N	Capacitor 3	µF	0.125
Damping d_3	Ns/m	Resistor 3	Ω	565.5
Frequency of vibration f	Hz	Resonance operation f	Hz	17
Velocity of vibration RMS	m/s	Current loop 2 RMS	A	107

Tab. 2: Parameters of electromagnetic coupling and electrical circuit.

Electrical circuit		
Parameter	Unit	Value
Magnetic flux density B	T	0.4
Coil turns N	–	1000
Active length of coil l	mm	40
Diameter of wire d	mm	0.1
Inner coil resistance R_C	Ω	115

5.2 Analyses of efficiency and output power for mechanical parameters

The first analysis focused on the dependence between the output harvested power and the conversion efficiency. The model was excited with the sinusoidal external mechanical force $Q(t)$ and the vibrating structure m_3 moving with the sinusoidal vibration with the frequency of 17 Hz and the magnitude of 100 mG RMS. The resistive load was varied in the simulations to reach the result presented. The harvested output electrical power and the calculated harvester efficiency are shown in Figure 7. The maximal efficiency of the vibration energy harvesting system is 72.6% for a resistive load of 700 Ω However, the value of maximal harvested power 9.45 mW is generated for a connected resistive load of 4.4 kΩ. The value of the efficiency for the maximal harvested power is 47.2%, meaning that during the operation with the optimal load, a larger fraction of the total input energy is dissipated elsewhere and then on the load, compared to

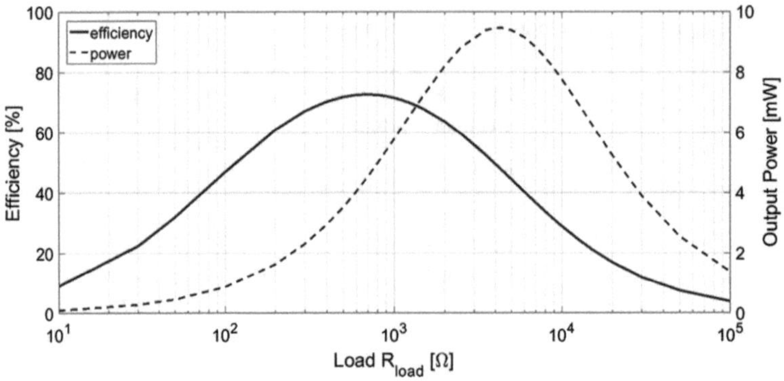

Fig. 7: Results of efficiency and power analysis for various resistive loads.

the operation with load of 550 Ω where the mechanical and electrical energy ratio is maximal. However, the total harvested power output is higher with the optimal load.

The simulation results are in accordance with the published study about the optimal resistive load [12] of vibration energy harvesting systems. The maximal harvested power on the optimal resistive load R_L does not exactly correspond to the equilibrium of both mechanical and electrical damping. Due to electrical losses on the coil resistance, the value of optimal resistive load R_L shifted slightly to achieve the maximal output power. The optimal resistive load can be calculated using expression (15), and its value is used for other efficiency analyses. The optimal load in a general form is frequency dependent, and the expression given here corresponds only to the resonance operation, which is shown in Figure 8 for output power and voltage;

$$R_{\text{Optim}} = \frac{Q_m \cdot c_{FL}^2}{m_1 \cdot \Omega} + R_C = \frac{c_{FL}^2}{d_m} + R_C \tag{15}$$

During the first analysis of the other parameters that could affect the efficiency the influence of the harvester frame mass m_2 was calculated. The result is that it does not affect the value of the conversion efficiency. On the other hand, increasing the seismic mass m_1 causes a drop in the efficiency. However, the harvested output power is proportional to the mass, expression (2), and the harvested power grows with the mass, Figure 9.

Another parameter present in the expression (2) is the mechanical quality factor. A large number of analyses was done with varied values of the mechanical quality factor. The calculated efficiency dependency on the mechanical quality is shown in Figure 10. The maximal value of the efficiency is reached for the quality factor 80. However, the output power rises with the increasing quality factor, see Figure 10. Both power results for varied mass and mechanical quality factor correspond to the theoretical model (2).

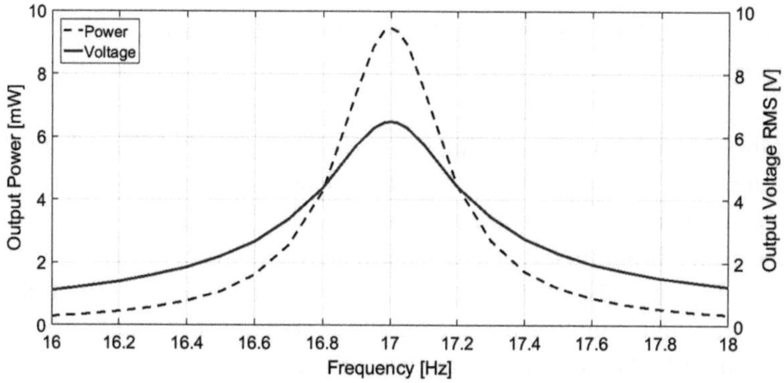

Fig. 8: Resonance operation of the vibration energy harvester.

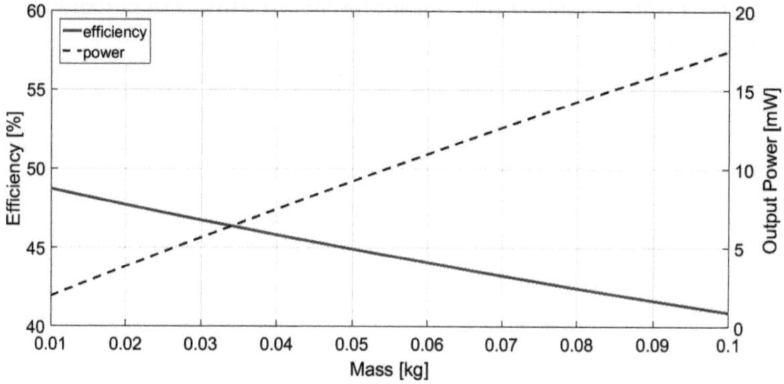

Fig. 9: Influence of the seismic mass.

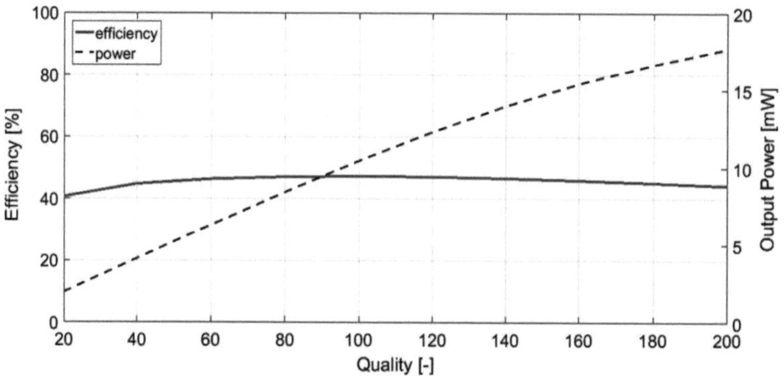

Fig. 10: Influence of the mechanical quality factor.

5.3 Analyses of efficiency and output power for coupling parameters

The influence of the electromagnetic coupling parameters, such as the magnetic flux density B, the number of coil turns N and the active length of the coil l, were investigated. Most electric machines are designed with the aim of maximal magnetic flux, and the same design is required for electromagnetic vibration energy harvesters, which is obvious in Figure 11. The number of coil turns and the active length of the coil correspond to the same coupling parameter, and it is the overall length of coil wire. For this reason, only the dependency on coil turns is shown in Figure 12. The electromagnetic vibration harvester has to be designed with an adequate value of the coupling factor for optimal harvesting of electricity, which is predicted by Figures 11 and 12. Thereafter, all parameters have only a minor impact in a range of percentages on the calculated efficiency.

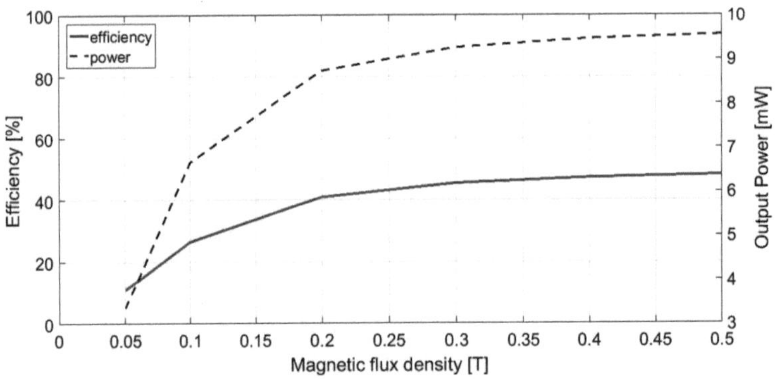

Fig. 11: Influence of the magnetic flux density through a coil.

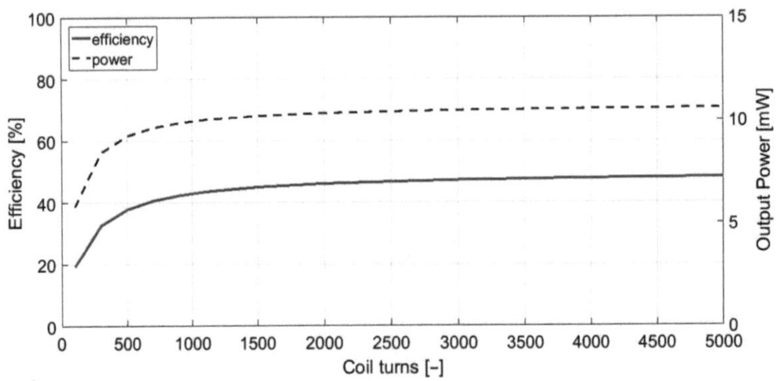

Fig. 12: Influence of the number of coil turns.

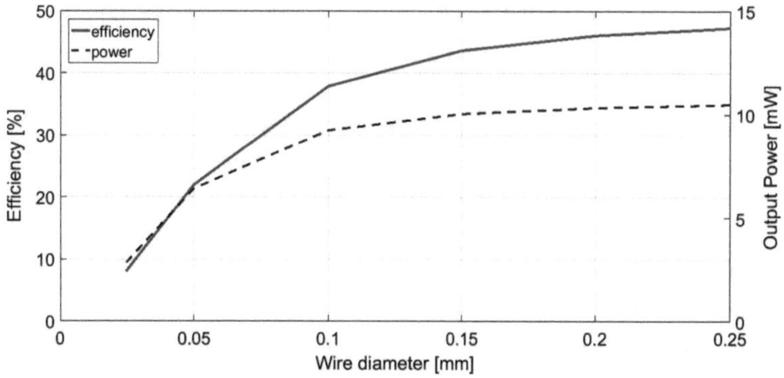

Fig. 13: Influence of the coil wire diameter.

The design of the coil is very important for optimal harvesting of electricity. A higher number of coil turns provides only higher voltage but does not have the effect of efficient energy harvesting. Also, the coil area and volume are usually limited by an air gap dimension of the magnetic system, and thinner wire may be used for an adequate number of coil turns. However, a smaller diameter of the coil wire provides higher electric losses on the coil due to inner coil resistance, and energy harvesting systems with thinner wires may provide higher voltage, but the output power and efficiency will be lower; this is obvious in Figure 13. An advanced optimization methods could be very effectively used for the optimal design of coil parameters, e.g., artificial intelligence methods or self-organizing algorithms [13].

6 Comparing the performance of vibration energy harvesters

As the above text shows, the efficiency of the energy harvester can be used as an important indicator to assess the design of the electromechanical transducer. It does not, however, provide useful information about the actual performance of the energy harvester with respect to its parameters and, thus, cannot be used to compare different designs, sizes and choices of energy conversion principles of tested vibration energy harvesters.

Teams developing inertial energy harvesters for various applications publish parameters of their devices with very variable degree of details, so the comparison of different designs is not straightforward. Multiple different metrics for the purposes of benchmarking the overall performance of energy harvesters have been introduced in literature over the years:

Roundy [14] defined a metric, called an effectiveness of the harvester, taking into account the electromechanical coupling coefficient, quality factor, ratio of actual harvester density to baseline density, and the ratio of the transmission coefficient of the transduction mechanism to the maximum transduction coefficient.

A different approach was taken by Beeby et al. [15], which lies in calculating the normalized power density (NPD), where the output power of the harvester is divided by the square of acceleration magnitude and by the volume of the harvester. This approach is feasible only for devices working in resonance and does not take into the account the operation frequency range of the device.

Mitcheson et al. [16] proposed a so-called volumetric figure of merit (FoM_V) derived as a ratio of the useful power output to the theoretical maximum power obtainable by a device with the same total volume and a gold proof mass occupying half of the total available volume. The other half is left for the proof mass displacement. This approach does not take into account the bandwidth of the harvester and devices with flatter frequency response but a broader bandwidth is penalized, even though the broader operation frequency range might be advantageous in some applications. Mitcheson, therefore, proposed a modification of his original volume FoM_V, multiplying the FoM_V by a fractional 1 dB bandwidth, in contrast with the commonly used 3 dB.

Another way of defining the figure of merit was presented by Sebald et al. [17], who proposed using the maximum obtainable power divided by the square of the input acceleration magnitude and multiplied by the bandwidth. This version of FoM, however, does not take the size of the device into consideration, so comparing devices of different volumes would be difficult.

Najafi et al. [18] used a FoM defined as a modification of normalized power density, where the NPD is multiplied by the 1 dB bandwidth of the harvester. This makes their metric useful for comparing harvesters operating at similar frequencies, as it does not use the fractional bandwidth.

A more recent paper, by Ruan et al. [19], proposed a similar metric to that of Najafi, utilizing the normalized power density, multiplied by the inverse of the harvester's Q factor. By definition, the Q factor corresponds to a fraction of 3 dB bandwidth and resonance frequency. This makes the way Ruan's approach compensates for flatter but wider frequency responses of some harvesters similar to Mitcheson's modification of FoM, just with the 3 dB instead of the 1 dB fractional bandwidth being considered.

Interesting metrics mainly for nonlinear harvesters were proposed by Mallick et al. [20]. Instead of relying on the normalized power density calculated as a function of the maximum harvested power, they proposed calculating a power integral in the whole frequency range, and calculated normalized power integral density (NPID) instead.

Different from other definitions of the figure of merit, Arroyo et al. [21] proposed using a normalized power defined as a ratio of the real output power and the theoretical limit power to evaluate the harvester's performance. The power normalization uses three key parameters of a generic harvester: the coupling factor, the loss coefficient and the mechanical quality factor.

Very similarly, Balato et al. [22] presented the harvester ideal utilization factor (HIUF). This metric quantifies how close to the optimum the harvester is electrically loaded, also taking into account the diode bridge rectifiers used for power conditioning.

While other FoM definitions focus on the harvesters themselves and their comparison, the normalized power approach and the HIUF serve as tools for assessing how close to its optimal point a particular harvester is working given its parameters.

A summary of the metrics used is presented in Table 3. Variables used by different authors have been unified to give a better possibility of a direct comparison of their metrics. It should also be noted that some authors take into consideration the volume of the harvester as reported, regardless of whether or not the volume reported includes the necessary free space for the displacement of the proof mass. Other authors explicitly state that the volume needs to be defined as the total space occupied by the harvester during its operation. In many cases though, depending on the data published for each device, the values reported also include the space for the displacement of the proof mass.

7 Conclusions

The efficiency of the vibration energy harvesting system was presented, defined and analyzed. The efficiency of the vibration energy harvesting system provides the information about the input mechanical and output electrical energy ratio. The linear equivalent electrical RLC circuit with N electromagnetic transducer was used for the analyses of the individual parameters of vibration energy harvesting systems. The mass of the harvester frame does not affect the efficiency due to self-oscillation around the equilibrium point with no energy transfer. The linear equivalent electrical RLC circuits provide the information that electrical energy is mainly wasted in mechanical damping (the parameter of the quality factor) and in the coil resistance (the parameter of the electro-mechanical transducer). The coupling factor has to be designed with optimal values. Other parameters have a minor influence on the energy harvesting system's efficiency. Nevertheless, these parameters have a significant impact on the harvested output power. The output power also depends on the seismic mass and on the operation frequency. Furthermore, the vibration energy harvester should operate at maximal output power point, which is the task for power management electronics.

The peak of maximal output power is not aligned with the maximal efficiency peak and, furthermore, the output power peak is determined by the parameters of the energy harvesting system and the excitation mechanical vibrations. Modern power management electronics often include a maximal power point tracking function,

Tab. 3: Summary of metrics used for vibration energy harvesters.

Reference	Metric	Expression
S. Roundy [14]	Effectiveness (e)	$e = (k^2)Q^2\left(\frac{\rho}{\rho_0}\right)\left(\frac{\lambda}{\lambda_{max}}\right)$
S. P. Beeby et al. [15]	NPD	$NPD = \frac{P}{A_0^2 V}$
P. D. Mitcheson et al. [16]	FoM_V	$FoM_V = \frac{P}{1/16 Y_0 \rho_{Au} V_S^{4/3} \omega^3}$
P. D. Mitcheson et al. [16]	FoM_{BW}	$FoM_{BW} = FoM_V \cdot \frac{\delta\omega_{1\,dB}}{\omega}$
G. Sebald et al. [17]	FoM	$FoM = \frac{P_{lim}}{A_0^2} \cdot \frac{f_2 - f_1}{f_0}$
K. Najafi et al. [18]	FoM	$FoM = \frac{P}{A_0^2 V} \cdot \Delta f$
J. J. Ruan et al. [19]	FoM	$FoM = \frac{P}{A_0^2 V_S} \cdot \frac{1}{Q}$
D. Mallick et al. [20]	NPID	$NPID = \frac{P_f}{A_0^2 V}$
E. Arroyo et al. [21]	Normalized power ($\underline{P}$)	$\underline{P} = \frac{P}{P_{lim}}$
M. Balato et al. [22]	HIUF	$HIUF(\omega) = \frac{P_{max}(A_0, \omega) - P(A_0, \omega)}{P_{max}(A_0, \omega)}$

Where individual input parameters are represented by these variables:

A_0	harmonic excitation acceleration magnitude
P	useful power output
P_f	power integral over frequency
P_{lim}	obtainable power limit
P_{max}	maximum obtainable power with given excitation frequency
Q	quality factor
V	reported total volume of the harvester
V_S	volume of the harvester including the displacement
Y_0	harmonic excitation displacement magnitude
f_0	frequency where the maximum power is harvested
$f_{1,2}$	half-power cutoff frequencies
Δf	bandwidth
k	coupling coefficient
λ	transmission coefficient of the transduction mechanism
λ_{max}	maximum transmission coefficient
ρ	actual density
ρ_0	baseline material density
ρ_{Au}	density of gold
ω	harmonic excitation angular frequency
$\delta\omega_{1\,dB}$	1 dB bandwidth

which guarantees maximal harvesting of electricity during varied input conditions. This mode has to be implemented in power management electronics in order to reach the maximal harvested power for given input conditions. The maximal harvested power does not correspond to the most efficient transfer of the kinetic energy of vibrations into electricity. The efficiency provides information about the mechanical feedback into the vibrating flexible structure. This could be very important in the case of the mass of vibrating structure being low enough for the mechanical feedback to affect the input vibrations. This situation, where input vibrations occur on a lightweight beam, is presented in the paper [1], and an increase of the input vibration magnitude was observed. Mechanical vibrations in the industrial environment are commonly observed in lightweight structures like housing and sheathings, which makes the efficiency analyses potentially very useful. On the other hand, if the main aim of the analyses is the comparison of several different vibration energy harvesters, one of the presented FoM metrics would be more useful.

The results presented correspond to the real operation of electromagnetic vibration energy harvesters, which were tested in our lab [1]. The efficiency of the tested electromagnetic energy harvesters without electronics is in the range of 40–47%. This range depends on the nonlinear quality factor and the level of input vibrations. Basic electronics like the rectifier and the capacitor decrease the efficiency to the range of about 30–35%. Finally, the harvesters were measured with the commercial power management electronics bq25504EVM and reached an efficiency in the range of about 13–25%.

Bibliography

[1] Z. Hadas, V. Vetiska, J. Vetiska, and J. Krejsa. Analysis and efficiency measurement of electromagnetic vibration energy harvesting system. *Microsyst. Technol.*, Jan. 2016.

[2] R. Amirtharaiah, J. Wenck, J. Collier, J. Siebert, and B. Zhou. Circuits for energy harvesting sensor signal processing. In *43rd Design Automation Conference*, pages 639–644, 2006.

[3] T. Funck. Metrology for energy harvesting. In *Joint IMEKO TC11-TC19-TC20 Int. Symp*, pages 20–22. Metrological Infrastructure, Environmental and Energy Measurement and Int. Symp. of Energy Agencies of Mediterranean Countries, IMEKO-MI 2011, 2011.

[4] Z. Hadas, V. Vetiska, R. Huzlik, and V. Singule. Model-based design and test of vibration energy harvester for aircraft application. *Microsyst. Technol.*, 20(4–5):831–843, 2014.

[5] J. A. Paradiso and T. Starner. Energy Scavenging for Mobile and Wireless Electronics. *IEEE Pervasive Comput.*, 4(1):18–27, 2005.

[6] C. B. Williams and R. B. Yates. Analysis of a micro-electric generator for microsystems. *Sensors Actuators A Phys.*, 52(1–3):8–11, 1996.

[7] Z. Hadas, C. Ondrusek, and V. Singule. Power sensitivity of vibration energy harvester. *Microsyst. Technol.*, 16(5):691–702, 2010.

[8] A. E. Kubba and K. Jiang. Efficiency enhancement of a cantilever-based vibration energy harvester. *Sensors (Basel).*, 14(1):188–211, 2013.

[9] P. Li, S. Gao, and H. Cai. Modeling and analysis of hybrid piezoelectric and electromagnetic energy harvesting from random vibrations. *Microsyst. Technol.*, 21(2):401–414, 2015.

[10] M. A. Erismis. Design and modeling of a new robust multi-mass coupled-resonator family with dynamic motion amplification. *Microsyst. Technol.*, 19(8):1105–1110, 2013.

[11] Z. Hadas, V. Singule, C. Ondrůšek, and M. Kluge. Simulation of vibration power generator. In *Recent Advances in Mechatronics*, pages 350–354. Springer, 2007.

[12] F. Khan, B. Stoeber, and F. Sassani. Modeling and simulation of linear and nonlinear MEMS scale electromagnetic energy harvesters for random vibration environments. *Sci. World J.*, 2014, 2014.

[13] Z. Hadas, J. Kurfurst, C. Ondrusek, and V. Singule. Artificial intelligence based optimization for vibration energy harvesting applications. *Microsyst. Technol.*, 18(7–8):1003–1014, 2012.

[14] S. Roundy. On the Effectiveness of Vibration-based Energy Harvesting. *J. Intell. Mater. Syst. Struct.*, 16(10):809–823, 2005.

[15] S. P. Beeby, R. N. Torah, M. J. Tudor, P. Glynne-Jones, C. R. S. T. O'Donnell, and S. Roy. A micro electromagnetic generator for vibration energy harvesting. *J. Micromechanics Microengineering*, 17(7), 2007.

[16] P. D. Mitcheson, E. M. Yeatman, G. K. Rao, A. S. Holmes, and T. C. Green. Energy Harvesting From Human and Machine Motion for Wireless Electronic Devices. *Proc. IEEE*, 96(9):1457–1486, 2008.

[17] G. Sebald, H. Kuwano, D. Guyomar, and B. Ducharne. Experimental Duffing oscillator for broadband piezoelectric energy harvesting. *Smart Mater. Struct.*, 20(10):102001, 2011.

[18] K. Najafi, T. Galchev, E. E. Aktakka, R. L. Peterson, and J. McCullagh. Microsystems for energy harvesting. In *2011 16th International Solid-State Sensors, Actuators and Microsystems Conference, TRANSDUCERS'11*, pages 1845–1850, 2011.

[19] J. J. Ruan, R. A. Lockhart, P. Janphuang, A. V. Quintero, D. Briand, and N. D. Rooij. An automatic test bench for complete characterization of vibration-energy harvesters. *IEEE Trans. Instrum. Meas.*, 62(11), 2013.

[20] D. Mallick, A. Amann, and S. Roy. Interplay between electrical and mechanical domains in a high performance nonlinear energy harvester. *Smart Mater. Struct.*, 24(12):122001, 2015.

[21] E. Arroyo, A. Badel, F. Formosa, Y. Wu, and J. Qiu. Comparison of electromagnetic and piezoelectric vibration energy harvesters: Model and experiments. *Sensors Actuators A Phys.*, 183:148–156, 2012.

[22] M. Balato, L. Costanzo, and M. Vitelli. Resonant electromagnetic vibration energy harvesters: The harvester ideal utilization factor. In *2016 IEEE International Power Electronics and Motion Control Conference, PEMC 2016*, 2016.

Martin Götz

Energy management concepts for wireless sensor nodes

Abstract: Energy management for wireless sensor nodes and wireless sensor networks differs from classical approaches focusing on the management of the storage only. It depends on the application, the energy income, harvesting technologies and sources. Techniques like maximum power point tracking, voltage and current conversion, energy income forecast and modeling, consumption profile tuning or power flow control are used to manage the energy efficiently. The implementation requires the modeling of the consumption on the one hand and the energy income on the other. The consumption profile can be based on current, voltage, power and energy measurements or model generated forecasts. The energy income is modeled by the temperature, vibration characteristics, irradiation or wind speed. Because the profiles do not match, an energy buffer and an energy management logic are used to match them together.

Analog, discrete and switched control architectures are used to perform the energy management task. This book chapter provides an overview over the techniques used in different areas of a wireless sensor node and possible complexity levels. It shows the interaction of these domains and the integration into a node-wide or network-wide holistic energy management. Using these techniques makes energy harvesting from ambient energy sources feasible. It enables and supports the latest developments in industrial areas, the environmental monitoring domain, the "Internet of Things" or smart dust networks.

Keywords: energy management, energy conservation, scheduling, energy optimization, maximum energy extraction, maximum efficiency tracking, energy harvesting, wireless sensor networks, sensor nodes, smart sensors, embedded systems, smart dust, Internet of Things, smart city

1 Introduction

Energy harvesting enables devices to use energy from their environment. Transducers like solar cells, Peltier elements or cantilevers harvest energy from light, temperature gradient or vibration energy sources. The active development in the domain lead to a significant improvement of the technology. The efficiency of energy harvesters has increased, and their size as well as their cost have decreased. Energy harvesting is suitable for a variety of applications and more domains than ever before. It is used in the "Internet of Things" (IoT) [1–6], environmental monitoring [7–9], the industrial [10] and many more domains.

Martin Götz, University of Technology Chemnitz

https://doi.org/10.1515/9783110445053-004

Energy harvesting is applied from simple active remote inductive sensors [11–13] over wireless sensor nodes up to complex wireless sensor networks [14] and smart dust [15–17].

Systems powered by energy harvesting can operate without human intervention for a certain time. If this time is long enough their operation is "maintenance free". Such systems can be deployed in inaccessible, dangerous, mobile or hostile areas. Wireless sensor-actor networks, typically named "wireless sensor networks" (WSN), scale easily when they are maintenance free. Maintenance free systems are used in areas where no power supply or cable-based communication is available such as IoT [1–5], environmental monitoring, the industrial domain and many more application fields. Forbes gives an overview over estimations by McKinsey, General Electric, IHS, Gartner and others in the field of IoT [18]. The predictions range from several billion to a trillion devices in the year 2050.

Classical approaches in these domains use batteries or coin cells, which need to be replaced or recharged on a regular basis. These batteries limit the lifetime, and their development speed does not correspond with that of microelectronics. Future approaches like wireless sensors in inaccessible areas or IoT and smart dust networks requiring a high number of devices create a demand for alternative solutions. Energy harvesting can be one of the solutions extending system lifetime on batteries or completely replacing the battery as an energy source.

Both approaches require an energy management to balance the energy income from the different sources, manage unmatched energy generation and consumption profiles, transform voltage and currents from the harvesters and so on.

Developing a good energy management overcompensates for the cost for energy storages, batteries or maintenance caused without such a strategy. Figure 1 shows a

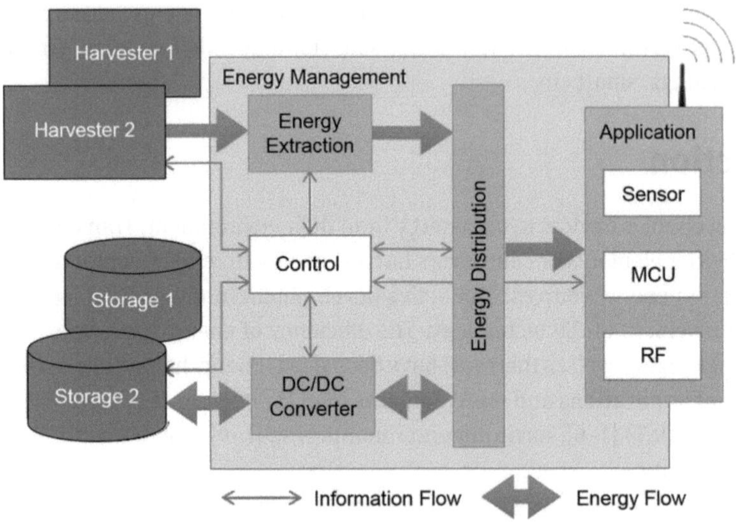

Fig. 1: Block diagram of a harvesting node.

block diagram of a harvesting node. The energy management provides rectification, impedance matching and maximum power point tracking (MPPT) to extract the energy in an optimized manner from the harvesting source. If multiple harvesters are connected the management coordinates the energy combination as well. An internal energy distribution system routes the energy into energy storage buffers, long-term energy storages or to the application. It converts between different voltage levels typically by using switched DC/DC converters. The control of the energy management system monitors energy availabilities, consumptions, charge levels and voltages to control the energy distribution. These power management strategies range from simple discrete setups, especially to recover from power outages, to complex microcontroller-based approaches and vary strongly with the needs of the application, the node design, available energies and the consumption profiles.

2 Energy harvesting from ambient power sources

Energy converters are used to convert ambient energies into electrical energy. Figure 2 proposes different energy source classifications. The energy can be classified after the origin it is generated from. This can be a natural one like solar energy or a derived one like wind, heat or water flow. The by-process is the second category, comprising machine vibrations, dissipated heat, fluid flows, moving magnets or pressure changes, e.g., by human movement. The third energy origin is active transmission, where energy is pumped to drive an energy harvester. Examples are light/laser energy pumping [19], e.g., to drive high-precision electric field detectors without cables, acoustic energy, radio waves [20] or magnetic energy pumping, e.g., by oscillating coils [21, 22].

One origin can produce different forms of energy, and the same form of energy can be created by different origins. The categories of this classification are shown in the second row of Figure 2.

Based on the available energy, the frequency, the form of energy and the application, one or more conversion techniques are applicable. The categories of this classification are shown in row three of Figure 2.

Each conversion technique uses one physical conversion principle classified in row 4 of Figure 2. All classifications are important for the design of an energy management:

- the energy origin needed to derive the energy availability;
- the energy form to select appropriate conversion techniques;
- the conversion technique to identify the used conversion principle;
- the conversion principle to choose the right energy extraction methods (Figure 1).

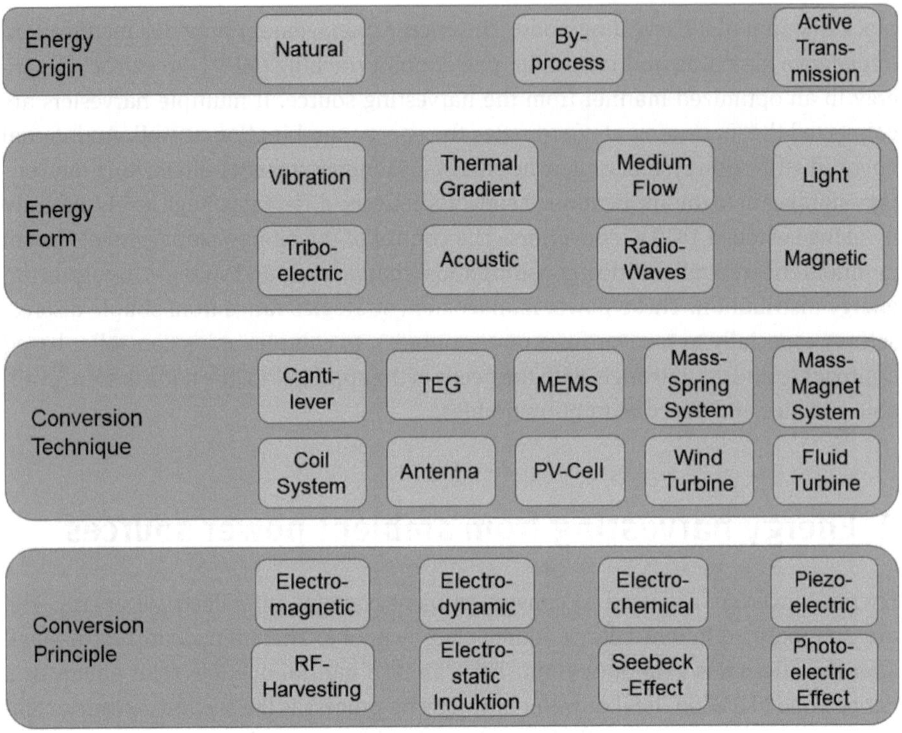

Fig. 2: Energy source classification. Microelectromechanical systems are abbreviated as MEMS and thermoelectric generators as TEG.

The requirements regarding the application, environment and size of a planned sensor node limit the possible elements in each class. One task in the design of energy management is to select appropriate combinations and dimension of multiple energy harvesters. Each harvesting input can be classified by one category of each row in the end.

3 Wireless power transfer

Wireless power transfer is used in order to:
- transfer greater powers than available from the ambient;
- transfer power over a special form of energy to avoid measurement influences;
- transfer power to hazardous, isolated or inaccessible areas;
- realize sensor nodes which communicate over the modulation of the sender's signal.

Transferring high powers is the most developed area of wireless power transfer. Inductive charging is already well established for mobile devices like electric toothbrushes, mobile phones, induction heaters and so on. They usually use the inductive energy transmission principle, which only works in the vicinity of the sending coil. This is a disadvantage concerning the powering of distant nodes but an advantage regarding influences to other systems. The energies transferred by this technique can be very high. Intelligent energy management systems allow a pulsed sensing of a device and only switch the full power on, when the correct device is recognized. This saves a lot of energy on the sending side and reduces negative effects on human health, incompatible electronic devices and the environment.

Wireless power transfer is also used to electrically decouple sensor systems from the environment. This technique has been used in communication systems for a long time to electrically decouple signals, e.g., to avoid ground loops in high fidelity or professional audio systems. New research uses the same technique to power remote systems. Warneke [19] creates smart dust powered by a laser beam. The system harvests the energy with a miniaturized on-board solar cell and communicates over laser light as well. Patents like [23] show that such techniques can be used to eliminate electromagnetic effects in the observed frequency range. Field influencing effects by electrically conducting wires are eliminated as well.

Wireless nodes operate in special areas like vacuum vessels, liquids, explosion-hazard areas, in space, under high pressures, thermo- or diffusion-isolated vessels and so on. Wireless power transfer is used to power systems when the ambient energy is insufficient for energy harvesting. Nodes are typically powered by inductive power transfer, light, and fluid or air movement.

Nodes can use their powering signal and modulate information into it. This has been used for several years in radio frequency identification (RFID) technology [24]. It is also possible to use this technique for RF sensing [13, 25] or inductive sensing using coils [26]. Simple circuits contain a coil and a capacitive or resistive sensor and a tuning capacity. When this setup is powered by a pumping coil, the receiving oscillating circuit provides a frequency-dependent feedback to the sending coil. By analyzing the detuning of the oscillation, the system can calculate the sensor's value of the wireless node. Yang et al. show such a system in [11]. Similar setups can be realized with the modulation of light by micro-mirrors or optical fiber cables. Pressures, strains, sounds, structure-borne sound or vibration can be measured as discussed in various publications and patents: [12, 27–30]. The measurement principle is robust, not prone to power failures, relatively easy to implement and has been used for a long time. The effort for such systems comes from the design tuned to the application and the related signal processing.

New approaches like passive WiFi described by Kellogg [31] focus on the powering from and modulation of radio-frequency (RF) signals. This technique also does not require any sending power from the measuring device. The communication is based on weakening or reflecting of electromagnetic waves.

4 Energy storage

Even if some node types powered by wireless power transfer or by the use of passive sensing capabilities that do not need energy storage elements [12, 27–30], most sensor nodes today contain one or more energy storages. The energy storages discussed in this section do not cover storage elements contained in subsystems like inductors storing the energy in the magnetic field around the inductor in DC/DC converters. The ones discussed are located before or after a certain block or form an own block. Figure 10 shows a typical block diagram of a sensor node including a storage element. The storage element is used to buffer energy for non-continuous energy input or temporally changing the energy requirements of the load. Such buffer storages range from small capacitors to rechargeable batteries. Bigger storage capacities are required if a harvesting system needs to operate while the harvested source is not or is only limitedly available for a longer time. This can be during the night or a half-year period for natural energy origins. Some process-originated energy sources, like the water faucet-based sensor node discussed in [32], need to face long down times as well. The purpose of the system is to scan for a user and open a water valve if one is detected. The sensor node needs to scan regularly for users without being powered by the thermoelectric generator (TEG) which uses the temperature difference of a water stream. In such cases, appropriately designed energy storages are essential. The mainly used storage principles are electrical or chemical. Thermal or mechanical storage principles like flywheels are typically not used due to low conversion efficiencies, losses like friction in flywheels and the size of the nodes.

5 Loads

Loads are all systems that need to be supplied with power. There are loads needed for the energy management and loads needed for the application. Loads for the energy management are unwanted and decrease the conversion efficiency from the harvester to the application loads. Nevertheless, they are needed for a high quality energy management that, in contrast, can increase the overall system efficiency. Some loads, including energy management and application loads, require controlled and stable voltages or currents. Stabilization is usually achieved by introducing capacitors or inductivities. Regulated voltages and currents are created with the help of DC/DC converters.

Typical loads required by the application are:
- sensors
- user interfaces
- microcontrollers
- RF communication components

Sensors might be directly driven by the energy source as described in the section on wireless power transfer (3). Discrete logic-based load implementations are rarely used, because they provide low flexibility and imply a high development effort. Most of today's sensor nodes use a microcontroller to control the measurement and perform the communication. This approach has the advantage of being very flexible, consuming low power and handling complex processing or control tasks in a very short time. The design process for such node is relatively low, as no complex discrete or analog circuitry needs to be designed and adjusted. Through RF communication, they can communicate with their surroundings, transmit measured data, control subsystems and do measurement accumulation or signal processing themselves. Additionally, they can make their own control or transmission decisions based on the measured values. Sleep modes in low-power microcontrollers enable them to use much less energy when they are not needed. Microcontrollers can even take part in the energy management and implement complex management strategies. Microcontrollers and RF transceivers have special requirements on the supply voltage. The supply voltage has to be in a certain range and must not exceed a maximum ripple voltage, especially when using integrated analog to digital converters (ADCs) or digital to analog converters (DACs). Voltage information obtained by ADCs is essential for microcontroller-based energy management. Whenever the voltage cannot be stabilized, or the system cannot provide the minimal operating voltage, the system can "die" due to non-functioning energy management. Without special precautions it can no longer recover from this "dead" state. The next section on power management strategies deals with these and other problems and shows possibilities to avoid such dead states.

6 Energy management approaches

6.1 Active and passive energy management

In the energy management domain, it is distinguished between active and passive approaches. Passive refers to the ability of the energy management to operate only with the power from the harvester and no additional energy source. An example of this is the Graetz bridge rectifier that uses only passive components to transform AC to DC current. Passive systems often have a lower efficiency compared to active ones, because they have static thresholds defined by physical properties like the forward threshold voltages of the diodes of the Graetz bride. Non-linear properties are often adverse and limit the efficiency. On the other hand, they are actively exploited like by the use of Z-diodes being utilized to create quasi regulated voltages and, thereby, switching points for the energy management. The border to the active energy management is fuzzy, so that also solutions with comparators or operational amplifiers might be categorized as passive energy management, when they generate all needed voltages themselves. Such systems usually contain multiple stages with different efficiencies. Further information about this is provided in Section 6.3 "Multi-level energy management."

Active energy management needs additional power to drive the control stages, comparators, operational amplifiers or even the microcontrollers involved in the management. This power comes either from a precharged buffer capacitor or a backup battery. The control abilities, adaptiveness to a changing environment, energy availability or demands and their flexibility is much higher compared to passive solutions.

6.2 Microcontroller-based energy management

Microcontroller-based energy management is a special form of active management. It includes one or more microcontrollers in the management. Such a setup offers a great amount of flexibility, might be updated over the air with real-time data, perform forecasts itself and calculate adaptively new management strategies. It provides the ability to integrate various types of sensor information and aggregate them. Even if such a complex energy management can be realized with active electronic components, the use of a microcontroller is usually preferred because it is flexible, cheap in comparison, easily integrable and usually already provides ADCs, comparators, internal voltage references, communication solutions and a high calculation power to track and control energy flows.

Being contained in the active management domain, microcontroller-based solutions also need initial power to be operational. Microcontrollers require a special supply power in a certain voltage range, often 1.8 to 3.6 V. Special microcontrollers like the MSP430L092 [33] can operate on voltages down to 0.9 V. A voltage supply without ripple is required when internal ADCs and comparators are used.

6.3 Multi-level energy management

As discussed in the preceding sections, Figure 3 shows a categorization of passive, active and microcontroller-based management approaches. Passive solutions consume the lowest power but provide the lowest flexibility. The efficiencies of active and microcontroller solutions are much better, but their power demand is higher and can have special requirements on the supplied voltage.

For all active solutions, the depletion of energy buffers leads to the death of the sensor node. This can occur during an unforeseen energy outage of the harvested energy source, a depleted backup battery or an inflexible energy management.

Backup battery change cycles or the recharging or replacement of death sensor nodes lead to maintenance effort. Using passive energy management only provides low efficiency. Combining both approaches and switching between them leads to a multi-level energy management. This combines the advantages of active and passive

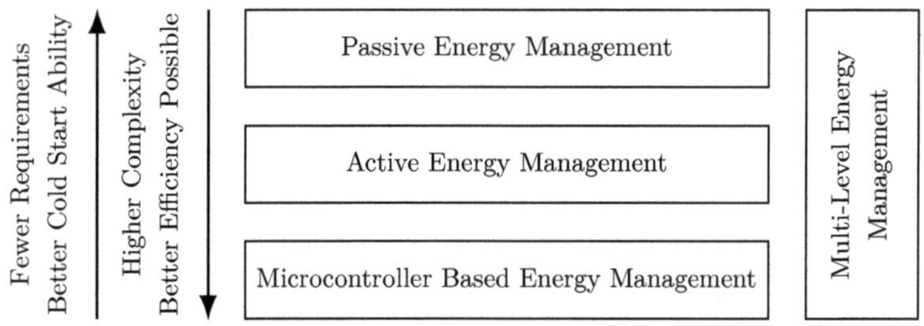

Fig. 3: Overview over energy management implementations.

approaches without the drawbacks. On the other hand, the design is quite complex, and the advance and fall-back between the stages must be additionally designed. A smart hybrid approach for passive rectification with active elements to power from an electromagnetic energy harvester is discussed in [34].

Another approach is to use hybrid harvesting with one harvester dedicated to powering the energy management only and another one for efficient extraction of the larger portion of energy, as is described by Rohan in [35].

7 Extraction power management strategies

While a classical power management approach focuses on energy storages only, the approach for harvested energy driven sensor nodes can use the additional components discussed in the following sections. Figure 1 shows the parts of the sensor node where energy management is employed.

7.1 Maximum power point

Each source has a characteristic current to voltage output behavior, which can be visualized in a current over voltage diagram, as shown for a solar cell in Figure 4(a). The intersection on the y-axis marks the short-circuit current observed when the two output terminals of the solar cell are short-circuited. The intersection with the x-axis is the open-circuit voltage, which is observed, when the terminals are not connected. Multiplying voltage and current from the diagram leads to the power output, which can be drawn for each working point and, therefore, each voltage. Figure 4(b) shows this power behavior for the same solar cell. It can be seen that the power during the short-circuit current and the open-circuit voltage is zero. This is because either the voltage or the current component is zero, leading to a zero in a multiplication with every value. The graph shows that there is a maximum power point at 0.925 V sourcing a maximum of 0.377 mW. Keeping the solar cell operating on this voltage guarantees the maximum power transfer.

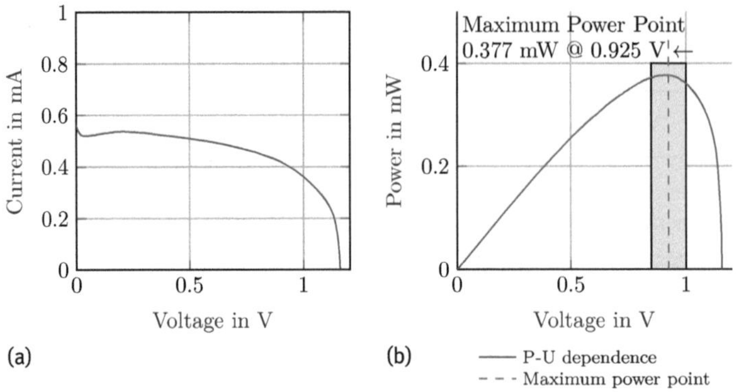

Fig. 4: Output characteristics of a solar cell. (a) Current over voltage output characteristics. (b) Power over voltage characteristics.

7.2 Maximum power point tracking

MPPT is a technique to extract and transfer the highest possible power from a source to the load. Therefore, the load impedance needs to be "matched" to the source impedance. This is achieved by using exactly the complex conjugate of the source impedance as the load. The actual load usually does not fit this requirement. Depicted in Figure 4 is the current versus voltage and the power versus voltage measurement of a photo-voltaic cell used as harvester. The maximum power point at a voltage of 0.925 V is highlighted in the diagram. The duty of the MPPT is to maintain this voltage in order to extract the maximum power. Typically, a combination of a (parasitic) capacitance and a DC/DC converter is used as a voltage control element of the MPPT and as a matched load of the energy harvester. Switching DC/DC converters are used, because they have a better transfer efficiency compared to classic voltage regulators.

Because of the switching voltage control design, a voltage/current ripple is introduced at the output of the harvester. This ripple leads to a deviation of the working point around the maximum power point of the energy harvester. This behavior is shown in Figure 4(b) as a gray box around the ideal voltage for maximal power extraction. The bigger the buffering capacitor, the smaller the ripple around the desired voltage for the same switching frequency. Every added capacity introduces losses and needs to be initially charged, on the other hand. The energy stored in this capacitor cannot be used in the later stages.

Another way to reduce the deviation from the ideal working point is to increase the switching frequency, with the drawback of a higher power consumption of the DC/DC circuits, as the gate capacities of the switching elements must be charged and discharged more often.

How strongly the input voltage changes is influenced by a third factor: the current provided by the harvesting source. Higher currents lead to greater voltage changes on the same capacitor sizes.

To control the duty cycle of the DC/DC converter, the MPPT needs to regularly measure the actual voltage and, depending on the MPPT technique, the open-circuit voltage. The more often this is measured, the smaller the voltage ripple around the optimal working point can be. The MPPT control consumes energy in each measurement and control cycle, so that too-high update rates are not feasible.

Finally, the MPPT algorithm used can have a big influence. The tracking is needed to adapt to changing environment conditions, leading to a shift of the maximum power point. Sophisticated algorithms find the maximum power point more accurately and adapt better to changing conditions and energy availabilities. This increases the extraction efficiency on the one hand, but costs additional energy needed to perform the measurements, provide voltage references, do the calculations and set the controls, on the other hand.

Famous MPPT algorithms are "perturb and observe", "hill climbing", particle swarm optimization or fuzzy logic and artificial intelligence-based algorithms, described in [36] and [37].

If the environment conditions are relatively constant, no tracking is needed. This is no longer a "tracking", it is more maximum power point controlling after an initial definition. A large portion of available MPPT chips use this technique, but the products are still named "maximum power point trackers". For these products, the static maximum power point is set via resistors during the deployment.

Table 1 concludes the findings and gives a reference on designing the non-algorithmic part of the MPPT.

7.3 Maximum energy extraction

Extracting the maximum power is very common in the discussion of efficient energy usage on wireless sensor systems. Nevertheless, it is not the most efficient way of transferring energy from limited resources. Maximum energy extraction is a neglected way of energy extraction, although it is very important for sources with limited amounts of energy over time.

Figure 5 shows an electrical circuit comprising a voltage source with an inner resistance of 50 Ω and a resistive load. To extract the maximum power as described in the last section, the load must be matched to the inner impedance. This is achieved by using the complex conjugate of the source impedance as the load impedance. In the example discussed, the matched load will be 50 Ω.

Figure 6 shows the extracted power over the load resistance. It can be seen that the power transferred to the load, indeed, has a maximum at 50 Ω. The dashed line shows the power turn over in the inner resistance R_I. At 50 Ω, the power spreads half

Tab. 1: Decision matrix for choosing the right MPPT.

Advantage(s)	Disadvantage(s)
MPPT implementation	
– Better power extraction – Better extraction efficiency possible	– Additional energy requirement for MPPT – Backup energy needed for start from depleted energy buffers
High DC/DC converter duty cycle	
– Closer working range around optimal working point – Better extraction efficiency	– Higher power demand for DC/DC converter
Big input capacitors	
– Smaller voltage ripple – Better extraction efficiency	– Leakage current losses – Initial energy for charging the capacitor is lost
Fast MPPT measurement and control reaction	
– Smaller voltage ripple – Better extraction efficiency	– Higher power demand for MPPT
High current from energy harvester	
– More energy available at the same voltage	– Faster charging of input capacitors – Faster voltage changes of input capacitors – Lower energy extraction efficiency
Sophisticated MPPT algorithm	
– Better finding of maximum power point – Better extraction efficiency	– Additional energy requirement for MPPT – Sometimes stable voltages from regulators and comparators introduce additional power consumption – Higher cost and design complexity

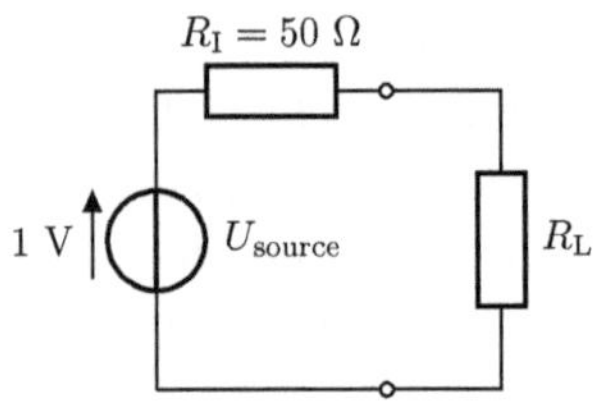

Fig. 5: Impedance matching circuit based on resistive components.

and half to the load and inner resistance. The dotted line shows the overall power delivered by the source. It can be seen that the maximum power is sourced in the case of a short circuit. The higher the load resistance, the less energy is provided by the source. For the energy efficiency it is interesting, how much of the sourced power is actually delivered to the load, as shown in the lower graph of Figure 11. The energy extraction efficiency at the matched resistance is not the highest. The efficiency increases with

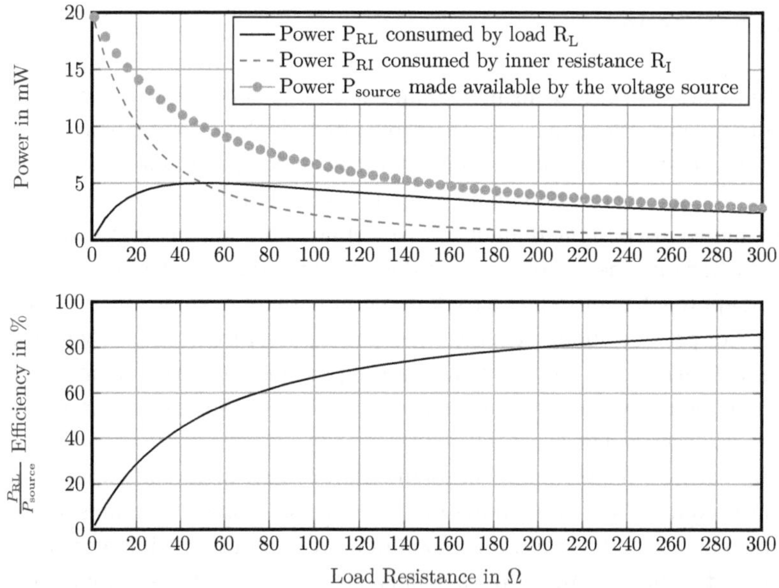

Fig. 6: Diagrams from the simulation of matching circuits from Figure 5. Upper diagram: emitted power from source and power distribution to internal and external impedances depending on the load impedance. Lower diagram: efficiency of energy transferred to the load divided by the energy emitted by the source versus the load impedance.

the load resistance. At an infinite load resistance, the efficiency will be 100%, because no current flows and no power is consumed at all. It is obvious that this may also not be the desirable case, but the example shows why there are often much higher load resistances than inner resistances used.

As [38] describes, an example is typical battery use. The load resistance is usually much higher than the inner resistance of the battery. With that configuration, not the maximum power is extracted, but the battery will live much longer.

Energy efficiency is important for a lot of harvesting setups, but is still not in focus. It provides opportunities to improve the energy usage on limited resources. One cause is that the harvesting system is often cut between the source energy part, e.g., in a TEG the thermal energy and the electrical part. Simulations, tests and setups are performed without regarding the mutual influence on heat flows, thermal inertia, available energies and internal losses. There are multiple effects that contradict the general energy preserving approach (in the example of an thermoelectric generator):

- High currents in the TEG cause joule heating in all elements.
- If the energy origin (Figure 2) does not constantly provide new energy, the harvestable amount for one cycle is limited.
- A high extraction current causes a decrease of the temperature difference at the element leading to a lower TEG voltage and, thereby, often lower extraction efficiencies of following energy processing stages.

7.4 Hybrid harvesting

Hybrid harvesting is a principle using more than one energy harvester to power a system. This additional harvester can be used to recover from depletion states or to power the energy management system during startup. It can provide a supplementary energy provision, e.g., during day and night, when the other harvester provides no energy. A simultaneous use of more than one harvester as a combined energy source is needed as well when one energy source alone does not provide enough power. Rarely seen is a hybrid approach to introduce a backup power path for the case that the main harvester does not provide enough energy. Most backup harvesters employ wireless power transfer (see Section 3). The decision regarding an appropriate hybrid management approach depends on the use case as well as the energy source and harvester characteristics, as described in 2.

An example can be given regarding one of the characteristics: the matching of two maximum power points. Figure 7(a) shows the power versus voltage diagrams of two different harvesters. For both curves, the maximum power point is highlighted. If an MPPT is to be used, either one MPPT for each harvester can be used or a common one for both harvesters. Using one MPPT for each means that each working point can be controlled separately. This is useful if the maximum power points are located at different voltages, as is often the case for inductive (low maximum power point voltage) and piezoelectric harvesters (high maximum power point voltage). It is also useful if changing environment conditions lead to an independent change of the maximum power points. The operating region of distinct MPPTs is visualized as gray box around the maximum power point of each harvester in Figure 7(b). From the figure, it can be seen that separated MPPT only makes sense if the operating areas of the two MPPTs are small.

Otherwise, the outputs of both harvesters can be connected and one MPPT can regulate the voltage of both to a tradeoff-working-point-region, as is shown in Figure 7(b) as a dashed box around the maximum power points of both harvesters. This

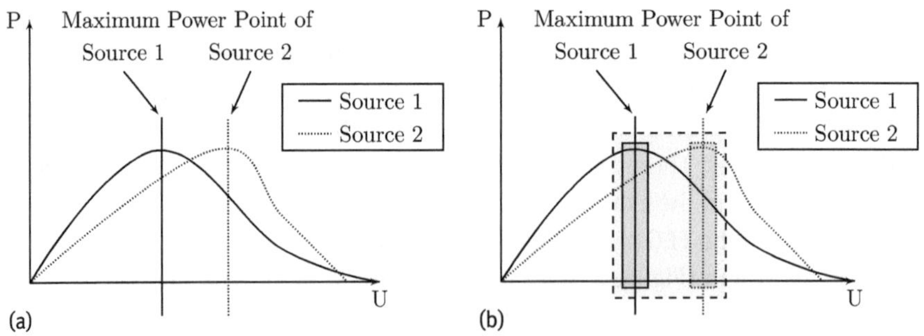

Fig. 7: Power versus voltage of two different energy harvesters. (a) Maximum power points of two different sources. (b) Operation regions of distinct or shared MPPT.

approach has the advantage of a lower complex circuitry and an introduced quiescent current but extracts power from the harvesters with a lower efficiency.

As discussed for the matching of the power points, typical parameters of the harvesters and the energy sources influencing the design of the hybrid energy management are:

- Current alternation:
  - only alternating current (AC)
  - only direct current (DC)
  - some AC and some DC
- Waveform shape
  - sinusoidal
  - multi-sinusoidal
  - noisy
- Power output amount
- Energy availability

Depending on these, groups with shared components may be formed. For each combination, several parts of the energy extraction can be shared or exclusively used.

Figure 8 shows a generalized diagram for various hybrid energy management approaches on DC sources. Opening or closing the switches lead to the use or bypass of different blocks. For AC sources, a rectification or an energy extraction stage like P-SSHI for piezoelectric harvesters has to be added after the harvester, if a DC output is required. Like the MPPT on DC sources, these stages can be shared or separately realized for each harvester. Diodes can be employed to prevent energy flow back to the harvesters, but they introduce voltage drops or quiescent currents, if they are realized as active diodes [39]. New approaches use mechanical switches to avoid quiescent currents completely [40, 41]. Publications already present hybrid energy management solutions of different complexity. They span from very simple approaches for light and thermal energies [42, 43] over general approaches [44] to review articles [45]. Uluşan et al. [46] claim to have created the first hybrid harvester circuit to simultane-

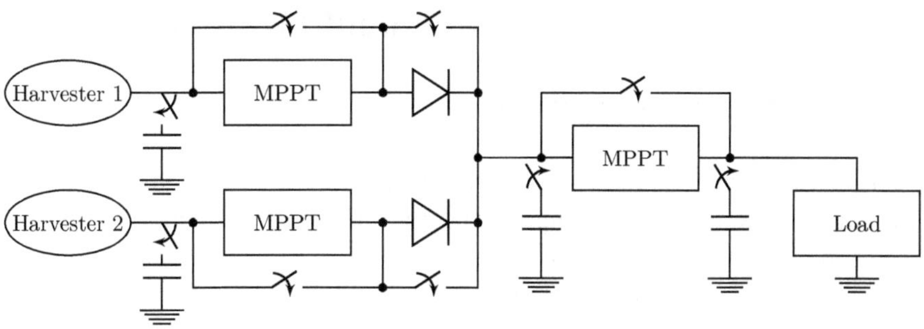

Fig. 8: General hybrid energy management system structure for DC sources.

ously extract energy from three independent sources and delivers a single DC output. A promising approach for the harvesting from multiple DC energy harvesters with different impedances is presented by Kang et al. [47].

7.5 Non-linear energy management

Non-linear energy management is used to mechanically or electrically match energy converters/energy harvesters to the energy source. A typical example is vibration harvesters designed for an optimal frequency by fitting the eigenfrequency to the frequency of the energy source, so that the harvester operates in a resonant working point. When the energy source working point fluctuates or changes over time, the extraction is no longer in an optimal working point. An often used approach to face this problem is non-linear energy harvesters using magnets instead of mechanical springs to create oscillation. This widens the area of resonance but lowers the maximum general output of the harvester. Mechanical tuning approaches are discussed in [48–50]. Figure 9 shows typical setups. In 9(a), magnets are used to generate a bi-stable harvester. Such an approach is discussed in [51]. Repellent setups like the one shown in Figure 9(c) are possible as well. They might be integrated in a microelectromechanical system (MEMS) as described by [52–54]. It is also possible to use two or more magnetically coupled cantilevers with anti-phase bistable behavior [55]. Figure 9(b) shows an approach using clamps to change the oscillation frequency of the cantilever when the amplitude is high. This approach is investigated in [56].

New electrical tuning approaches [57] utilize capacitive or magnetic tuning to change the resonance frequency of piezoelectric generators. [58] proposes electrical bias voltages to tune the stiffness or other parameters of the harvester dynamically by the electrical side. This was proved to be possible at least with piezoelectric converters. It is expected that further research in this field has the potential to apply this technique also to other energy transducers.

8 Core system energy management strategies

8.1 Energy storage management

As discussed in the Introduction 4 and shown in Figure 10, energy storage elements are used as buffers between the power input and output. They are also required as buffers between two stages to filter a voltage or current ripple. If a previous block outputs a current, a storage element can collect the charges and transform the voltage into a useful range for the successor block.

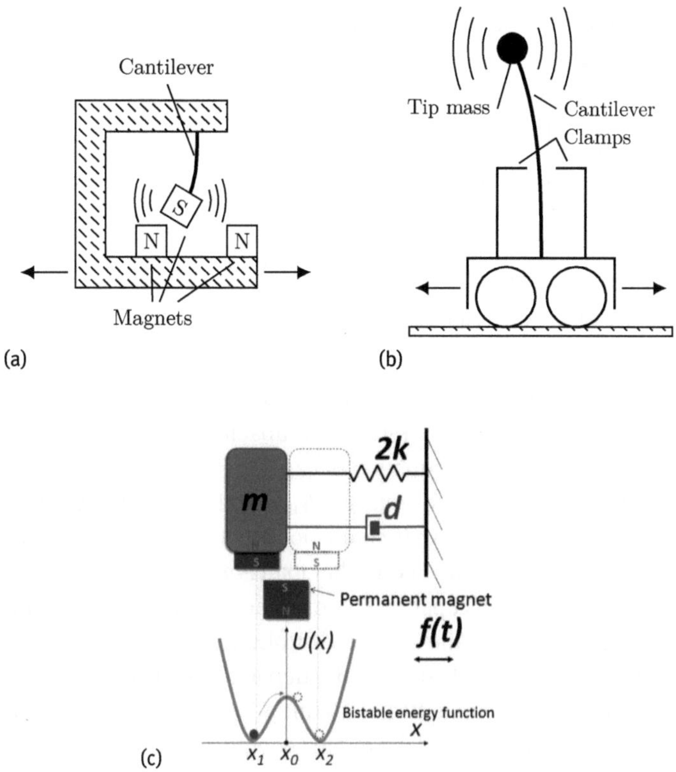

Fig. 9: Typical non-linear setups. (a) Cantilever with attractive magnets arrangement form a bistable system. (b) Cantilever with clamps reduces the beam length for high amplitudes. (c) Cantilever with an arrangement of repellent magnets forms a bistable system.

This technique is needed for MPPTs regulating the voltage on the input side and outputting currents transformed by a DC/DC converter. A storage element like a capacitor collects the incoming currents. The voltage over the storage element rises, and the energy becomes accessible to the following stages, as soon as a usable voltage range is maintained.

Figure 10 shows that different blocks, e.g., harvester, storage and load, can operate on different voltage levels. One task of energy management design is to match the voltage ranges of neighboring blocks. The voltage ranges of two neighboring blocks can overlap. If this is the case, they can operate in the shared region. For example, microcontroller loads are often used. As depicted in Figure 10, many of them need an input voltage between 1.8 and 3.6 V, but the storage can operate on different voltages.

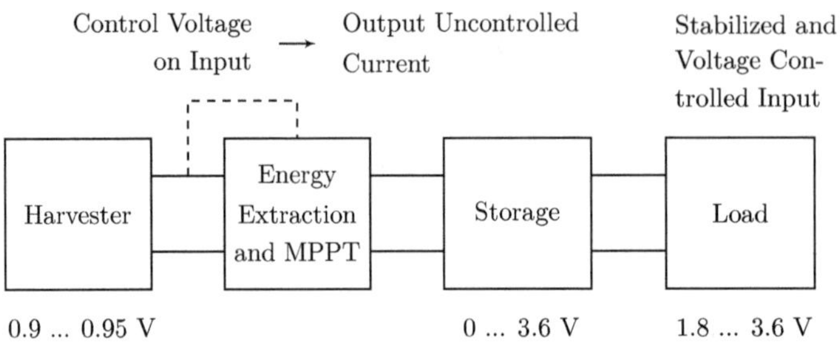

Fig. 10: Various voltage levels in a wireless sensor node.

As shown in Figure 10, a simple way to prevent destruction of the load is to limit the maximum voltage of the storage to 3.6 V. If the storage is a capacitor, equation (1) gives the energy E_{cap} stored in the capacitor as a function of capacity C and the voltage U. The energy that can be extracted between two voltages U_{begin} and U_{begin} can be calculated by equation (2). The shared voltage range for both blocks is 1.8 to 3.6 V. The maximum energy that can be accessed by the microcontroller load is given by equation (3). If the stored energy is not sufficient, the capacity C can be increased. The maximum possible capacity C_{max} can be limited by the size, cost or physical constraints of the sensor node and the components. Figure 11 depicts the dependence of the energy of a capacitor versus the voltage for two capacities of 10 and 47 μF. The green shaded area shows the voltage range of 1.8 to 3.6 V and the related energy range for a capacitor with 47 μF.

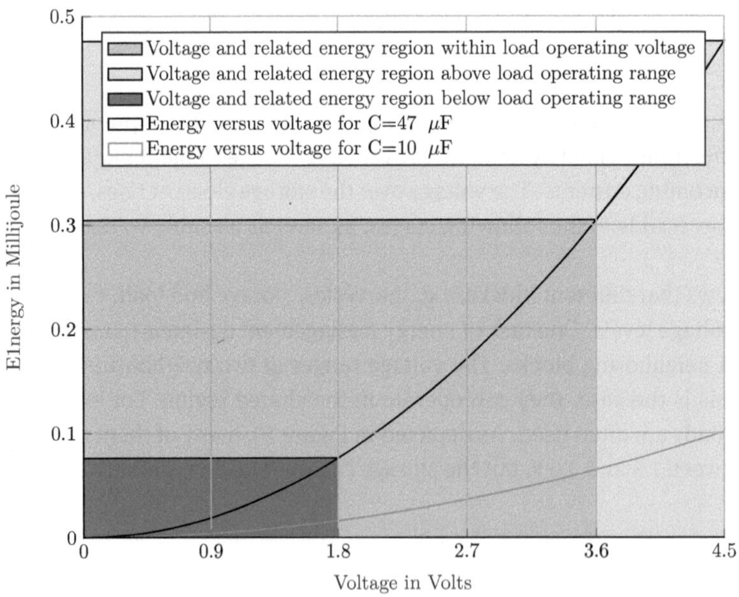

Fig. 11: Stored energy vs. voltage stored in a capacitor.

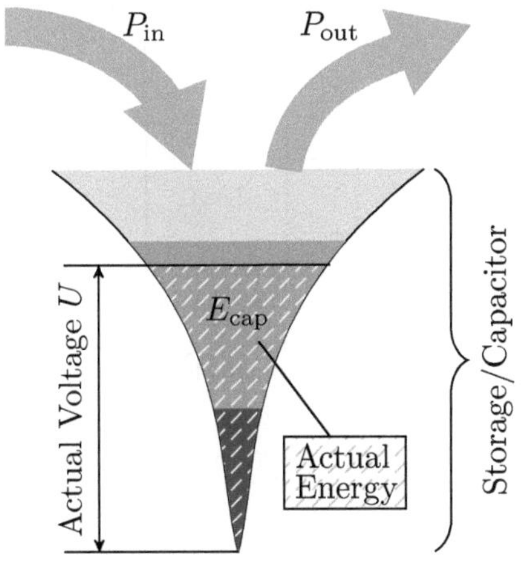

Fig. 12: Various voltage levels in a wireless sensor node.

Tab. 2: Classification of energy budget states of the storage of an energy harvesting driven node with typical scenarios when they occur.

Energy budget class	Typical scenario
1: $P_{in} \gg P_{out}$	Depleted node charging up
2: $P_{in} > P_{out}$	Sleep with active harvester
3: $P_{in} = P_{out}$	Simple sensor without storage and self-stabilizing voltage
4: $P_{in} < P_{out}$	Sleep with inactive harvester
5: $P_{in} \ll P_{out}$	Active load / microcontroller / radio-frequency front end

It can be seen that more energy can be stored at higher voltages, as the voltage has a quadratic influence on the stored energy. Figure 12 shows the energy budget of the capacitor. The capacitor stores an actual energy E_{cap} which is directly related to its voltage U by equation (1). An input power from the energy harvester increases and an output energy consumed by the loads decreases, the stored energy E_{cap}. For the energy budget, there are different categories of the system state presented in Table 2. The table also gives typical scenarios in which the energy budget behaves like described in the first column.

$$E_{cap} = \frac{1}{2} \cdot C \cdot U^2 \tag{1}$$

$$E_{cap,\ voltage\ difference} = \frac{1}{2} \cdot C \cdot \left(U_{end}^2 - U_{begin}^2\right) \tag{2}$$

$$E_{cap,\ usable} = \frac{1}{2} \cdot C \cdot \left[(3.6\,V)^2 - (1.8\,V)^2\right] \tag{3}$$

$$E_{cap,\ bulk} = \frac{1}{2} \cdot C \cdot (1.8\,V)^2 \tag{4}$$

All power P_{out} consumed by the load must be either provided by the harvester P_{in} or has to be retrieved by the stored energy E_{cap}, which is limited to $E_{cap,usable}$ (green area) by the overlapping voltage ranges of the neighboring blocks as described in equation (3).

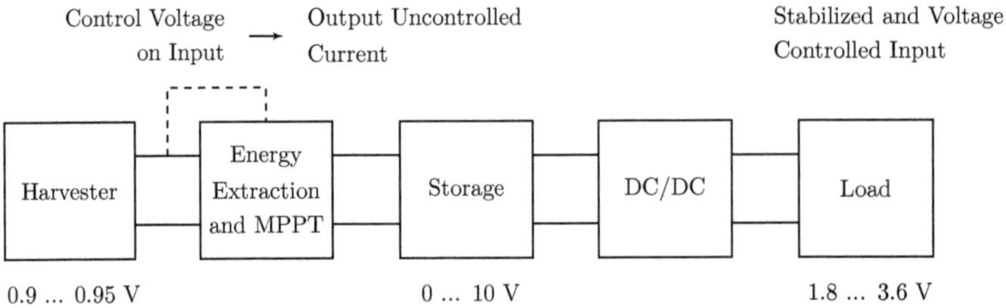

Fig. 13: Block diagram of a wireless sensor node with a DC/DC converter before the load.

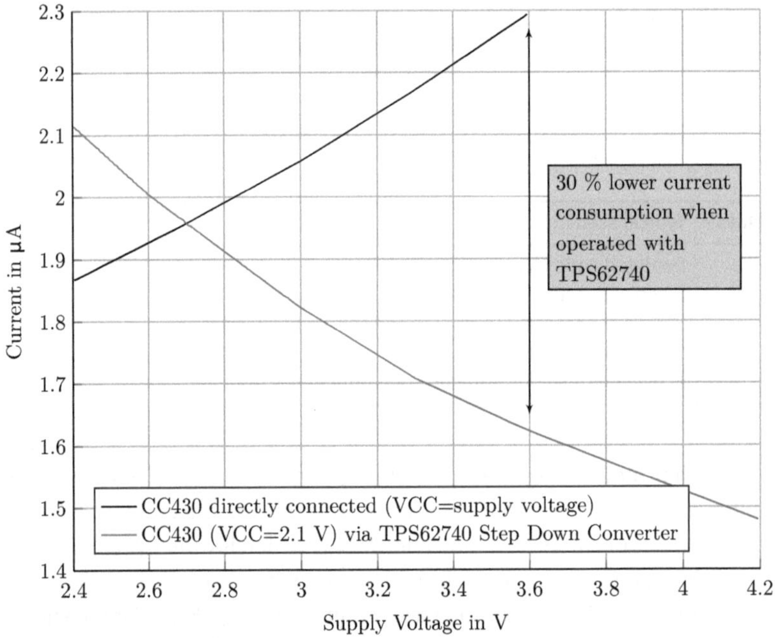

Fig. 14: Power consumption of a microcontroller with and without voltage regulation. Up to 2.7 V it is better to drive the microcontroller directly. From 2.7 V voltage controller and microcontroller in combination consume less energy [59].

By using a buck DC/DC converter, as proposed in Figure 13 it is possible to transform higher voltages to lower voltages. This extends the usable storage region of the capacitor to higher voltages at which greater energies can be stored. This region is marked in yellow in the Figures 11 and 12. As the energy E_{cap} stored in a capacitor increases quadratically with the voltage U as shown in equation (1). Much more energy can be stored and accessed during long times of negative energy budget.

Furthermore, such a setup offers the possibility to regulate the voltage for the load to the lower limit of the loads range. Most loads consume less power at lower voltages. As depicted in Figure 14, using a buck DC/DC converter in conjunction with the load saves energy when the input voltage crosses the threshold, at which the introduced additional power consumption of the DC/DC converter is compensated by the saving of the combination of the buck DC/DC converter and load. Such a setup is shown as a block diagram in Figure 13. The multi-level energy management approach described in 6.3 can deactivate the buck DC/DC converter to conserve energy when its operation is not necessary.

A boost DC/DC converter can be used to transform lower voltages to higher voltages. Using it between the storage and the loads lets the load access the energy stored in the region marked in red in Figures 11 and 12. This approach enables a fast startup of the load, when the storage has been depleted.

Both approaches can be integrated into a buck-boost DC/DC converter enabling the load to access a much wider voltage range and much more stored energy from the same capacitor. If the system operates near the threshold voltages often, both setups can be integrated into one with an additional switch using or bypassing the DC/DC converter. Advanced solutions use dual input multiple output (DIMO) or multiple input multiple output (MIMO) architectures to reduce losses added by adding DC/DC converters. Figure 15 shows an overview of setups from direct supply up to DC/DC converter solutions with multiple inputs and outputs.

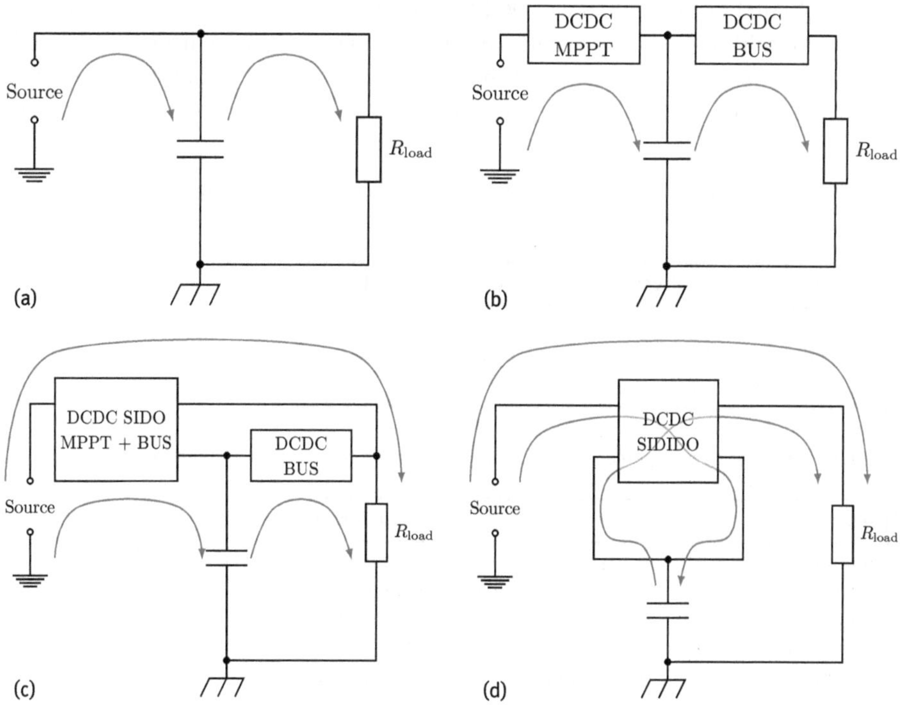

Fig. 15: Energy management using no, multiple or combined DC/DC converters. Adapted from [60]. (a) using no DC/DC converter. (b) using distinct DC/DC converter for storage and retrieval of energy. (c) using distinct DC/DC converter for storage and bypass. (d) using one DC/DC converter for all operations.

The designer should always keep in mind, that more complex approaches consume energy for the management itself and may not start from depleted batteries. That is why the energy management in is designed as complex as needed, not as complex as possible. To maintain advantages of different complexity levels multi-level energy management described in 6.3 is used.

8.2 Multi-storage management

Using a capacitor as buffer only is the simplest way of storing energy. For long-term storage, capacitors do not provide enough capacity. Instead, double-layer-capacitors, super-capacitors or rechargeable batteries are used. They have an increased response time but also a higher capacity compared to capacitors. Rechargeable batteries provide the highest capacity but also the highest response time compared to capacitors and super-capacitors. Table 3 compares different types of capacitors. Table 4 compares supercaps and batteries. Table 5 provides an overview over different kinds of rechargeable batteries.

All kinds of capacitors, as well as nickel-cadmium (NiCd) and nickel-metal-hydride (NiMH) accumulators can be fully discharged. Other chemistries like lead batteries or lithium-ion (Li-ion) batteries are destroyed if they are deep–discharged. All presented storage elements are sensitive against too high voltages. Particularly rechargeable batteries are prone to overcharging and can create gases, burn or ex-

Tab. 3: Comparison of capacitors, electrolytic capacitors electrochemical storages for 1 MJ of energy storage [61].

Specifications	Electrostatic	Electrolytic	Electrochemical
Mass (kg)	200,000	10,000	30–100
Volume (ml)	$140 \cdot 10^6$	$2.2 \cdot 10^6$	$0.02 - 0.1 \cdot 10^6$
Response time (s)	10^{-9}	10^{-4}	≈ 1

Tab. 4: Characteristics of supercapacitors compared to Li-Ion cells; Li-Ion: Lithium-ion [62].

Specifications	Supercapacitor	Lithium-ion (general)
Charge Time	1–10 s	10–60 min
Cycle life	1 million or 30,000 h	500 and higher
Cell voltage	2.3 to 2.75 V	3.6 to 3.7 V
Specific energy (Wh/kg)	5 (typical)	100–200
Specific power (W/kg)	Up to 10,000	1,000 to 3,000
Cost per Wh	$20 (typical)	$2 (typical)
Service life (in vehicle)	10 to 15 years	5 to 10 years
Charge temperature	−40 to 65 °C	0 to 45 °C
Discharge temperature	−40 to 65 °C	−20 to 60 °C

Tab. 5: Characteristics of commonly used rechargeable batteries; NiMH: Nickel-metal-hydride; NiCd: Nickel-cadmium; DoD: depth of discharge – specifies how much of the actual capacity is used.; C: e.g., 2 C for an 40 mAh battery equals a discharge current of 80 mA [62, 63]

Specifications	Lead acid	NiMH	NiCd	Li-Ion		
				Cobalt	manganese	Phosphate
Specific energy (Wh/kg)	30–50	45–80	60–120	150–250	100–150	90–120
Internal resistance	Very low	Very Low	Low	Moderate	Low	Very Low
Cycle life (80% DoD)	200–300	1,000	300–500	500–1,000	500–1,000	1,000–2,000
Charge time	8–16 h	1–2 h	2–4 h	2–4 h	1–2 h	1–2 h
Charge tolerance	High	Moderate	Low	Low. No trickle charge		
Self-discharge/month (room temp)	5%	20%	30%	< 5% Protection circuit consumes 3% / month		
Cell voltage nominal	2 V	1.2 V	1.2 V	3.6 V	3.7 V	3.2–3.3 V
Charge cutoff voltage (V/cell)	2.40 V Float 2.25 V	Full charge detection by voltage signature		4.20 V typical; some go to higher V		3.60 V
Discharge voltage	1.75 V	1.0 V		2.5–3.0 V		2.5 V
Peak load current	5 C	20 C	5 C	2 C	> 30 C	> 30 C
Best current	0.2 C	1 C	0.5 C	< 1 C	< 10 C	< 10 C
Charge temperature (°C)	−20 to 50	0 to 45 C		0 to 45		
Discharge temperature (°C)	−20 to 50	−20 to 65		−20 to 60		
Maintenance requirement	3–6 months (toping chg.)	Full discharge every 90 days when in full use		Maintenance free		
Safety requirements	Thermally stable	Thermally stable, fuse protection		Protection circuit mandatory		
Coulombic efficiency	~ 90%	~ 70% slow chg. ~ 90% fast chg.		99%		
Toxicity	Very high	Very high	Low	Low		
Cost	Low	Moderate		High		

plode. Depending on the used battery chemistry charging and safety circuits are required. Integrated battery management circuits are available as one chip solution for common battery chemistries. They prevent dangerous battery states like overcharging, deep-discharge and provide a current and charge control as well as temperature monitoring. Still, they cannot prevent battery deep discharges by quiescent currents and self discharge. This remains a task for a superordinate energy management and planning system.

Non-rechargeable primary batteries provide the highest capacity, as shown in Figure 16. Primary batteries tend to have a high internal resistance, which is usually no problem for low power requirements. "Primary alkaline and lithium batteries can be stored for up to 10 years with minimal capacity loss" [62, 174]. Primary batteries do not need a complicated circuitry and provide start-up power for nodes if all other energy buffers are depleted. Disadvantages are the slow discharging over time, even if no power is needed, and the related maintenance effort to replace depleted batteries.

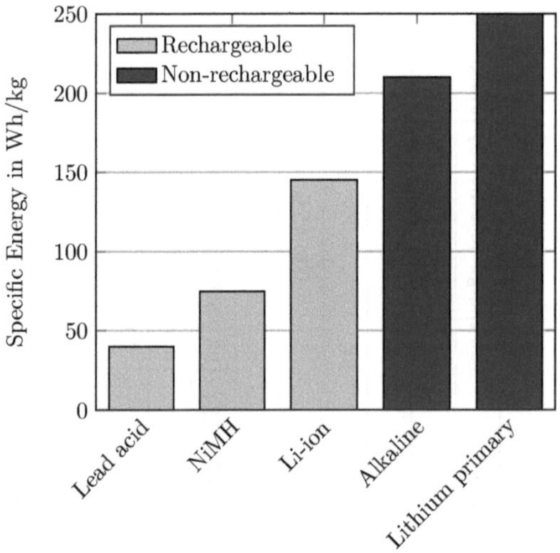

Fig. 16: Specific energy comparison of secondary and primary batteries [62].

An energy management system can manage several rechargeable and non-rechargeable energy storages. For this, it can simply switch/multiplex or use energy converters to transfer energy between them. Different architectures like the one shown for one storage in Figure 15 are possible. The energy management implementation can vary from passive to real-time data and prediction-based approaches, implemented with a microcontroller. Multi-level implementations can lower the energy consumption overhead of not needed conversion stages and can provide start-up abilities for a start from depleted energy storages.

Common approaches combine the advantages of long-term storages with short-time storages to incorporate multiple storage stages and control the use and energy routing between them. Because of this, the advantages of a short time to operation, cold-start-abilities, as well as long term energy storage can be combined.

8.3 Voltage supervision

Figure 12 shows the energy budget at an energy storage. If the input power is greater than the output power, more energy can be stored and the storage voltage increases. In the opposite case, the energy and storage voltage decrease. A deeper description is contained in Section 8.1. The connected load can only operate in the green area. In the red area, the load is not operational, but it can be still connected to the energy path and consume power. This consumed power does not serve any purpose. For some loads, especially microcontrollers, the power consumption in the region around their activation voltage is much higher compared to the power consumption when they are not activated or when they are active. This effect is caused by state changes of internal voltage references, switches, metal oxide semiconductor field effect transistors (MOSFETS), as well as the start of crystals upon start-up. Most microcontrollers also introduce a noticeable consumption when they are not used at all. This can be demonstrated by calculating a virtual-direct-current-resistance as shown by equation (5). If the voltage is regulated to U, the mean virtual resistance can be calculated by either calculation of the mean over time of the current, as shown in equation (6) or by calculating the virtual resistance for each sample and calculating the mean over time for all the resistances, as shown in equation (7). Figure 17 visualizes the mean virtual resistance for a common low-power microcontroller at different voltages. The microcontroller examined is able to operate from a minimal voltage of 1.8 V. The diagram is split into a left part, where the microcontroller is not running and a right part, where the microcontroller resides in a typical deep sleep mode. The measurements of the right part are cleaned for the transient start-up current. The measurement of the direct current was started 1 second after the powering of the microcontroller. The values represent the static mean virtual resistance. The parts are separated by a dashed red line. The lower the virtual resistance, the higher the direct current power consumption at the same voltage, as can be calculated by equation (8) and is shown in Figure 18.

$$R_{\text{virt}} = \frac{U}{i} \tag{5}$$

$$R_{\text{virt,mean1}} = \frac{U}{I_{\text{mean}}} \tag{6}$$

$$R_{\text{virt,mean2}} = \text{mean}(R_{\text{virt}}) \tag{7}$$

$$P_{\text{mean}} = R_{\text{virt,mean}} \cdot U^2 \tag{8}$$

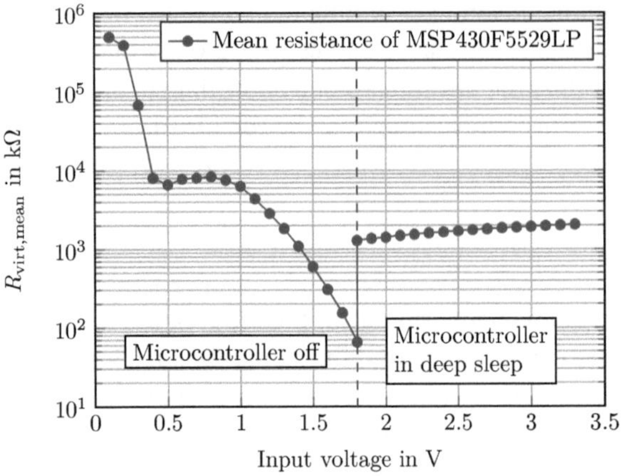

Fig. 17: Virtual resistance of Texas Instruments MSP430F5529LP. Virtual resistance shows the DC resistance seen at the terminal calculated from the actual current consumption by $R = U/I$. Displayed is the mean value of the virtual resistance over 4 seconds for the off-voltages and > 0.8 seconds for the on-voltages. The lower resistance at 1.8 V is measured when the microcontroller is off. The higher virtual resistance is measured when the microcontroller is in deep sleep.

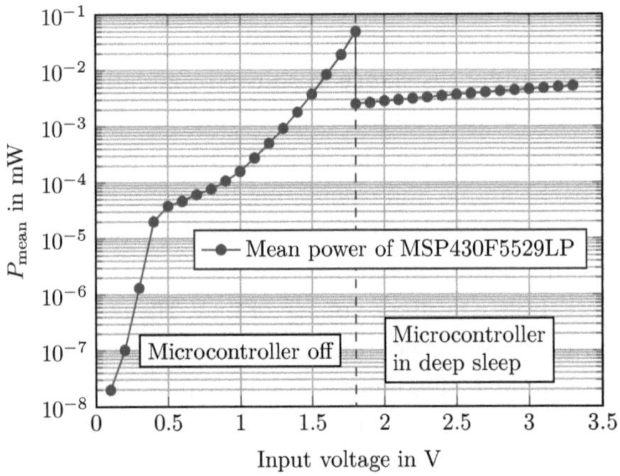

Fig. 18: Power consumption of Texas Instruments MSP430F5529LP. Displayed is the mean value of the direct current power over 4 seconds for the off-voltages and > 0.8 seconds for the on-voltages. The higher power at 1.8 V is measured when the microcontroller is off. The lower power is measured when the microcontroller is in deep sleep.

In conclusion, it can be seen that an unready load, especially an unready microcontroller still consumes energy. Even worse, the consumption before the start peaks to higher levels than in the sleep mode. If the load needs to cross the threshold to switch on, an even higher power is demanded for the start-up. If a sensor node operates on the buffer storage without sufficient input power from the harvester, it might get stuck before the activation of the load. Even if the power is so high as to let the load run all

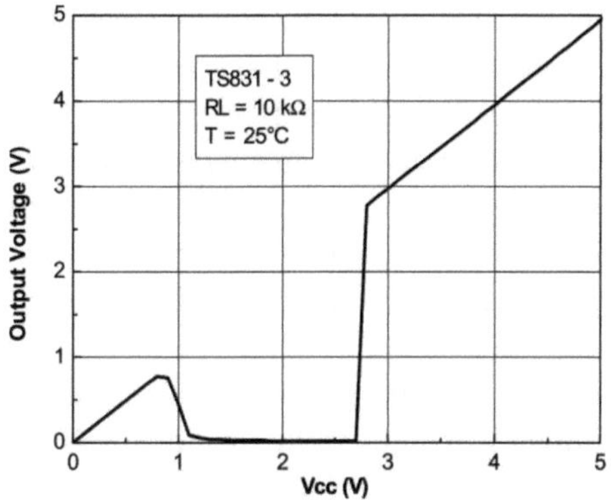

Fig. 19: Output voltage versus supply voltage V_{CC} of the TPS831 voltage supervisor [75].

the time in deep sleep and wake it up from time to time (the planned operation mode), it can get stuck before the activation, as each attempt to cross the activation threshold results in a high power consumption and as result in a dropping voltage on the input. By this control loop, the load is prevented from activation, and the node is prevented from operating in the desired manner. The problem occurs whenever the input voltage drops below the activation voltage of the loads. Even if enough input power can be provided, the loads constantly consume power during the charging process and delay the time to operation.

To avoid this problem, special circuits are used to disconnect the loads when they are not able to operate and only increase the power consumption. As this technique is in strong development, it is known by different names, such as the start-up circuit [64–66], the switching circuit [67], the trigger circuit [68, 69], the control unit [70], the under-voltage lockout circuit [71], the buffer circuit [72], and the wake-up circuit [73].

Industrial solutions like the TPS831 often lack a defined switching and a related low quiescent current at low voltages, as shown in Figure 19. For a low-power monitoring from depleted storages, sometimes also called the “cold-start” ability, more advanced fully passive circuits need to work from zero volts onwards with a low quiescent current over the whole voltage range. For a defined switching behavior and low power consumption during the transition, a switching hysteresis is needed. A promising fully passive, bistable “trigger” circuit combining these advantages is proposed by Alghisi et al. [74].

8.4 System optimization

Figure 1 from the introduction into power management strategies gives an overview of the different parts of the energy management of a harvested energy driven wireless sensor node. Not only the subsystem, but also their integration, can be managed.

Subsystems that are not required can be switched off or bridged to increase the overall system efficiency. Depending on the energy availability, voltage levels, energy income and consumption, as well as their forecasts, different setups can be used.

This approach gives the opportunity to extract the maximum energy from a defined harvesting source. Nevertheless, it is rarely used as the design is complex, incorporates additional components and needs a great insight into all domains of the energy management. The multi-level energy management described earlier can also be categorized as system optimization.

A system optimization approach can activate or bypass a variety of extraction stages, from simple passive, single diode solutions over multi-diode solutions, MPPTs, extraction circuits like parallel synchronized switching harvesters on inductors (P-SSHI), as discussed in [76]. It can decide what energy harvesters are used in hybrid setups. It can route the energy directly from the input stage to the load, to load and storage, storage only and so on, as described earlier in the energy storage section. It can activate or deactivate DC/DC converters and voltage supervisors before the load, influence the consumption profile of the load by operation modes, the route energy between multiple loads and so on.

The system level energy management optimizes the system with respect to the overall efficiency. Each part of the node, each converter and even harvesting inputs can use more energy than provided or than improves the efficiency. The management keeps track of that and activates subsystems only if this improves the efficiency. The system management needs to be designed in a staged architecture, if the node should still be able to start from depleted energy storages. Because of the complexity, the system optimization is usually realized with the help of a microcontroller program.

9 Wireless energy management strategies

9.1 Wireless communication/energy conservation in link and network layers

Wireless communication has the greatest impact on the energy consumption of a typical wireless sensor node and is, therefore, considered as a part with high optimization potential. Sending data consumes more energy than receiving data, and using the RF front end at all consumes more energy than only using the microcontroller. The goal of energy management in this domain is to minimize the power consumption introduced by the wireless communication system. Several approaches for a reduction of power consumption exist in this domain.

One is to minimize the number of transmissions as they consume the most power. This can be achieved by preprocessing data on the sensor node in order to compress it and eliminate unnecessary information. Control decisions upon sensor data can be made directly within the sensor node without communicating them wirelessly. Data can be sent in a low duty cycle up to defined measurement data thresholds, leading to an increased send rate. It is also possible to send no data at all until an emergency threshold is exceeded, and the sensor node reports critical measurement values only. Intelligent approaches implement the same forecast algorithm on the sensor node and the receiver. Measured values are only transmitted in the case the forecast deviates more than a set threshold.

If the wireless sensor node should not only send but also receive control commands, there are different approaches when to listen and, thereby, keep the receiver active. There are synchronized and unsynchronized approaches. The synchronized ones need time synchronization, so that the sender needs to send only if the respected receiver is listening. For this, they need to stay active or utilize real time clocks, which wake the microcontroller and receiver regularly. Unsynchronized methods send the data multiple times to the receiving node until it must have been recognized. This consumes a lot of energy on the sender node. It introduces a lot of additional communication on the channel, which leads to a lowering of the available bandwidth and an increase of the round-trip times or a total communication lock of the network in the worst case as well. This approach is a simple and robust solution in star networks where the central node is supplied by the electricity grid.

Advanced architectures can reduce the power consumption on the sending and receiving sides without synchronization by using "wake-up receivers". These circuits allow wakening a sensor node only when data is available. There is no need to wake up regularly or listen for long periods of time. The wake-up message is received by the "wake-up circuit" and wakes the microcontroller for the reception of the message from the sleep mode. This is done using pin interrupts, enabling the use of the deeper, more energy-saving, sleep modes. Such a system has no need of a real time clock, which additionally saves power. Wake-up receivers may even parse simple addresses being transmitted with the wake-up message (see also [77–80]). This technique enables waking of parts or single nodes within the network but increase the complexity, on the other hand. Higher complexities lead to an increased power consumption of the wake-up circuit itself. Recent research has been concentrating on creating fully passive wake-up receivers, which harvest the energy they need directly from the RF energy. Further information can be found in [81, 82]. Wake-up receivers may also use non-RF communications methods as a trigger. [83] discusses acoustic wake-up receivers. Light flashes and pressure changes can be also used for wake up and provide initial energy. Despite their possible potential, research covering such systems has not been published yet.

Optimizations in the different network layers are performed to fit the communication scheme and reduce the amount of bytes sent that are not actual data, which is called "overhead". Most modifications are performed in the lowest network layer, the "media access control" (MAC) layer.

Sensor deployment is another important part of the energy consideration. Placing the sensors closer together, using a directional radio, adaptive sending powers or avoiding walls, reduces the power consumption. The power consumption forecast for a deployed scenario is an important factor for the design of the wireless sensor node, its supplies and energy buffers.

To face the problems introduced by constraints in the deployment, like long transmission distances or distributed sensing setups, WSNs employing routing abilities are formed.

9.2 Energy aware routing, multi-hop deployments and clustering

Routing WSN contain multiple sensor nodes in a non-star network architecture, like a tree or mesh architecture. Wireless sensor nodes in such deployments get roles assigned. Central nodes extract the information from the whole network or parts of it, translate them into other networks or store them. They are called, "operator", "gateway" or "coordinator" nodes. Nodes not forwarding data from other nodes are called endpoints, and nodes forwarding data from other nodes are called routers. Figure 20 shows an overview of a small network with these WSN types. In a homogeneous network, all nodes consist of the same hardware. The opposite applies for heterogeneous network architectures.

The energy management in the WSN domain covers tasks like intelligent deployment during the design phase, adaptive load, power and topology control. Comparable to the point to point communications discussed in the last section, data reduction is in such setups a network wide task. Energy management approaches are data compression, reduction, prediction and network processing. So-called "hot-spot handling" is used to avoid sensor nodes collecting data from multiple surrounding nodes to deplete their energy storages too fast and risk the loss of certain sensor nodes. Synchronization and wake-up receivers are used to enable power saving sleep modes also

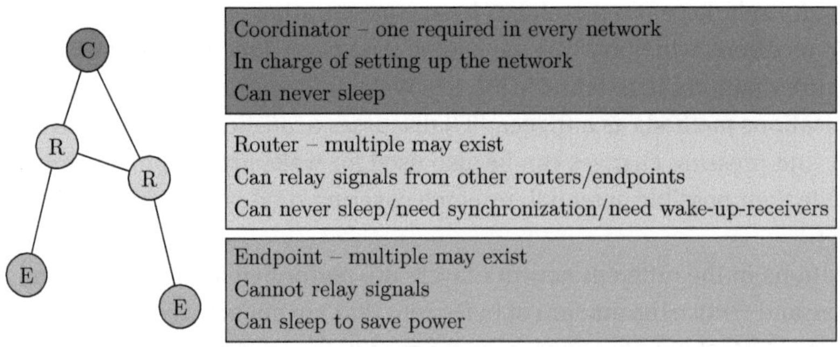

Fig. 20: Typical roles of nodes in a WSN. Image adapted from [84].

for routers or endpoints with receiving tasks. Typical optimized parameters are the energy and power consumption, as well as the node and network lifetime or the percentage of active nodes over a set time span.

Spatially changing deployments with moving endpoints, routers or gateways is discussed in a under the term mobility. Dynamic network forming, energy optimization, routing, caching and load balancing offer possibilities to improve the networks and wireless sensor nodes energy consumption and lifetime in mobile networks.

10 Conclusion

Energy management is an important part of the design and operation of wireless sensor nodes. In comparison to classical energy management, which concentrates only on the storage, the energy management in systems powered by harvested energies contains several energy management subsystems interacting mutually. A holistic approach is needed to consolidate energy availability profiles, load usage profiles, application constraints and unpredictable environmental influences. The final energy management design comprises the system structure defining the subsystems used, like MPPT active, passive or multi-level energy management, energy transducers and their size, energy storage, the parametrization of the components and the control logic being implemented in hardware and software. Scheduling based on measurements and forecasts enable the efficient use of the available energy and energy buffers.

A good energy management enables long-lasting sensor nodes and WSNs driven by energy harvesting. It facilitates the creation of totally maintenance free nodes, which can be deployed in harsh environments, contaminated, dangerous or hazardous areas. One example is the sensor network for power line monitoring [85] operating on a high electric potential, communicating wirelessly and being inaccessible/maintenance free. Special applications like highly sensitive measurement systems for electrical fields or fully floating or galvanic separated sensors become possible, because such systems can be powered and communicate by light, vibration, (ultra)sound and other non-electrical domains.

In some domains, like in space, energy harvesting and wireless communication is the only possibility for long-lasting and cheap operation. For extraterrestrial systems, the power consumption, usually realized by solar energy with photo-voltaic harvesters, is not the main focus. The problem here is the waste heat, which can only be radiated as there is no convection in space. This shows that there exist other limitations demanding an intelligent energy management.

Sophisticated energy management concepts do not only deal with static applications. They are employed in environments with changing conditions and adapt automatically. This ability is of special interest in mobile deployments. Further research is being carried out on passive solutions working in the start-up phase, the fusion of measurement, prediction and external data, as well as the use of multiple energy sources.

Acknowledgment: Parts of this work were done within the "Landesinnovations-stipendium" funded by the Sächsische Aufbaubank (SAB) and the European Social Fund (ESF).

Europäische Union

Europa fördert Sachsen.
ESF
Europäischer Sozialfonds

Bibliography

[1] L. D. Xu, W. He, and S. Li. Internet of Things in Industries: A Survey. *IEEE Transactions on Industrial Informatics*, 10(4):2233–2243, 2014.

[2] C. Perera, C. H. Liu, and S. Jayawardena. The Emerging Internet of Things Marketplace From an Industrial Perspective: A Survey. *IEEE Transactions on Emerging Topics in Computing*, 3(4):585–598, 2015.

[3] O. Hersent, D. Boswarthick, and O. Elloumi. *The Internet of Things - Applications to the Smart Grid and Building Automation*. John Wiley & Sons, 2012.

[4] L. Atzori, A. Iera, and G. Morabito. The Internet of Things: A survey. *Computer Networks*, 54(15):2787–2805, 2010.

[5] H. Sundmaeker, P. Guillemin, P. Friess, and S. Woelfflé, editors. *Vision and Challenges for Realising the Internet of Things*. Publications Office of the European Union, Luxembourg, 2010. doi:10.2759/26127.

[6] J. Gubbi, R. Buyya, S. Marusic, and M. Palaniswami. Internet of Things (IoT): A vision, architectural elements, and future directions. *Future Generation Computer Systems*, 29(7):1645–1660, 2013.

[7] C. Zet, C. Fosalau, D. Petrişor, and I. Hogaş. Study of the power consumtion of a landslide sensor node. In *2016 IEEE 7th Annual Ubiquitous Computing, Electronics Mobile Communication Conference (UEMCON)*, pages 1–5, Oct. 2016.

[8] D. Anurag, S. Roy, and S. Bandyopadhyay. Agro-sense: Precision agriculture using sensor-based wireless mesh networks. In *2008 First ITU-T Kaleidoscope Academic Conference - Innovations in NGN: Future Network and Services*, pages 383–388. IEEE, May 2008.

[9] A. Mainwaring, D. Culler, J. Polastre, R. Szewczyk, and J. Anderson. Wireless sensor networks for habitat monitoring. In *Proceedings of the 1st ACM International Workshop on Wireless Sensor Networks and Applications*, pages 88–97. Acm, 2002.

[10] M. Tijero, X. Eguiluz, J. Elizalde, V. Diez, A. Arriola, I. Aranburu, A. Zarketa, M. Martinez-Agirre, A. Martin-Mayor, and M. M. Bou-ali. Vacuum packaging and semipassive chips for wireless temperature monitoring in industrial applications. *IEEE Sensors*, Oct. 2017.

[11] B. Yang, F. Meng, and Y. Dong. A coil-coupled sensor for electrolyte solution conductivity measurement. In *Proceedings of 2013 2nd International Conference on Measurement, Information and Control*, volume 01, pages 57–60, 2013.

[12] *Radio frequency telemetry system for sensors and actuators*, 2003. US-Klassifikation 343/895, 340/572.1, 340/572.7, 343/866, 343/741, 343/742; Internationale Klassifikation H01Q7/00, H01Q1/22; Unternehmensklassifikation H01Q7/005, H01Q1/22, H01Q1/2225; Europäische Klassifikation H01Q1/22C4, H01Q1/22, H01Q7/00B.

[13] C. Strangfeld, S. Johann, M. Müller, and M. Bartholmai. Embedded passive RFID-based sensors for moisture monitoring in concrete. *IEEE SENSORS*, 2017. doi:10.1109/ICSENS.2017.8234166.

[14] F. Wu, C. Rüdiger, and M. Yuce. Real-Time Performance of a Self-Powered Environmental IoT Sensor Network System. *Sensors*, 17(2):282, 2017.

[15] T. Fischer, A. Agarwal, and H. Hess. A smart dust biosensor powered by kinesin motors. *Nature Nanotechnology*, 4(3):162–166, 2009.

[16] B. A. Warneke and K. S. J. Pister. An ultra-low energy microcontroller for Smart Dust wireless sensor networks. In *2004 IEEE International Solid-State Circuits Conference (IEEE Cat. No.04CH37519)*, pages 316–317 Vol.1, Feb. 2004.

[17] J. M. Kahn, R. H. Katz, and K. S. J. Pister. Emerging challenges: Mobile networking for "smart dust". *Journal of Communications and Networks*, 2(3):188–196, 2000.

[18] L. Columbus. Roundup Of Internet Of Things Forecasts And Market Estimates, 2016, 2016.

[19] B. Warneke, M. Last, B. Liebowitz, and K. S. J. Pister. Smart dust: communicating with a cubic-millimeter computer. *Computer*, 34(1):44–51, 2001.

[20] *Powerharvester Receivers - Powercast Co.*, 2017.

[21] S. Stoecklin, T. Volk, A. Yousaf, and L. Reindl. A maximum efficiency point tracking system for wireless powering of biomedical implants. *Procedia Engineering*, 120:451–454, 2015.

[22] S. Stöcklin, A. Yousaf, and L. M. Reindl. adaptive Elektronik zur effizienten drahtlosen Energieversorgung biomedizinischer Implantate. *Tagungsband*, pages 92–98, 2016.

[23] D. P. Hilliard. Optically powered amplifier used by an electromagnetic field sensor to amplify an electrical signal from an antenna, 1995. US Patent 5,389,782.

[24] E. Brard. *Process for radiotelegraphic or radiotelephonic communication*, 1930. US-Klassifikation 455/39, 455/121; Internationale Klassifikation H04B1/54; Unternehmensklassifikation H04B1/54; Europäische Klassifikation H04B1/54.

[25] R. Khalifeh, M. S. Yasri, B. Lescop, F. Gallée, E. Diler, D. Thierry, and S. Rioual. Development of Wireless and Passive Corrosion Sensors for Material Degradation Monitoring in Coastal Zones and Immersed Environment. *IEEE Journal of Oceanic Engineering*, 41(4):776–782, 2016.

[26] M. Demori, M. Masud, M. Baù, M. Ferrari, and V. Ferrari. Passive LC Sensor Label with Distance-Independent Contactless Interrogation, 2017.

[27] J. N. Fields, C. K. Asawa, O. G. Ramer, and M. K. Barnoski. Fiber optic pressure sensor. *The Journal of the Acoustical Society of America*, 67(3):816–818, 1980.

[28] J. A. Bucaro. Optical fiber acoustic sensors. In B. Bendow and S. S. Mitra, editors, *Fiber Optics*, pages 641–655. Springer US, 1979. doi:10.1007/978-1-4684-3492-7_33.

[29] T. G. Giallorenzi, J. A. Bucaro, A. Dandridge, G. H. Sigel, J. H. Cole, S. C. Rashleigh, and R. G. Priest. Optical fiber sensor technology. *IEEE Transactions on Microwave Theory and Techniques*, 30(4):472–511, 1982.

[30] E. Li, X. Wang, and C. Zhang. Fiber-optic temperature sensor based on interference of selective higher-order modes. *Applied Physics Letters*, 89(9):091119, 2006.

[31] B. Kellogg, V. Talla, S. Gollakota, and J. R. Smith. Passive wi-fi: Bringing low power to wi-fi transmissions. In *13th USENIX Symposium on Networked Systems Design and Implementation (NSDI 16)*, pages 151–164. USENIX Association, 2016.

[32] C. Beisteiner and B. G. Zagar. Thermo-electric energy harvester for low-power sanitary applications. *Proceedings SENSOR 2013*, pages 471–476, 2013.

[33] *Low voltage microcontroller MSP430L092*, 2017.

[34] H. Uluşan, K. Gharehbaghi, Ö. Zorlu, A. Muhtaroğlu, and H. Külah. A fully integrated and battery-free interface for low-voltage electromagnetic energy harvesters. *IEEE Transactions on Power Electronics*, 30(7):3712–3719, 2015.

[35] R. Dayal and L. Parsa. Modified electromagnetic microgenerator design for improved performance of low-voltage energy-harvesting systems. *IET Power Electronics*, 6(9):1751–1758, 2013.

[36] J. Ram, N. Rajasekar, and M. Miyatake. Design and overview of maximum power point tracking techniques in wind and solar photovoltaic systems: A review. *Renewable and Sustainable Energy Reviews*, 73:1138–1159, 2017.

[37] M. Kermadi and E. M. Berkouk. Artificial intelligence-based maximum power point tracking controllers for photovoltaic systems: Comparative study. *Renewable and Sustainable Energy Reviews*, 69:369–386, 2017.

[38] J. C. McLaughlin and K. L. Kaiser. "deglorifying"the maximum power transfer theorem and factors in impedance selection. *IEEE Transactions on Education*, 50(3):251–255, 2007.

[39] C. Peters, J. Handwerker, D. Maurath, and Y. Manoli. A Sub-500 mV Highly Efficient Active Rectifier for Energy Harvesting Applications. *IEEE Transactions on Circuits and Systems I: Regular Papers*, 58(7):1542–1550, 2011.

[40] F. Maiorca, F. Giusa, C. Trigona, B. Andò, A. R. Bulsara, and S. Baglio. Diode-less mechanical H-bridge rectifier for “zero threshold” vibration energy harvesters. *Sensors and Actuators A: Physical*, 201:246–253, 2013.

[41] F. Giusa, F. Maiorca, A. Noto, C. Trigona, B. Andò, and S. Baglio. A diode-less mechanical voltage multiplier: A novel transducer for vibration energy harvesting. *Sensors and Actuators A: Physical*, 212:34–41, 2014.

[42] Y. K. Tan and S. K. Panda. Energy Harvesting From Hybrid Indoor Ambient Light and Thermal Energy Sources for Enhanced Performance of Wireless Sensor Nodes. *IEEE Transactions on Industrial Electronics*, 58(9):4424–4435, 2011.

[43] V. Verma, A. Kane, and B. Singh. Complementary performance enhancement of PV energy system through thermoelectric generation. *Renewable and Sustainable Energy Reviews*, 58:1017–1026, 2016.

[44] L. Kumar and S. Jain. A multiple source DC/DC converter topology. *International Journal of Electrical Power & Energy Systems*, 51:278–291, 2013.

[45] A. S. Weddell, M. Magno, G. V. Merrett, D. Brunelli, B. M. Al-Hashimi, and L. Benini. A Survey of Multi-source Energy Harvesting Systems. In *Proceedings of the Conference on Design, Automation and Test in Europe*, DATE ’13, pages 905–908, San Jose, CA, USA, 2013. EDA Consortium.

[46] H. Uluşan, S. Chamanian, W. M. P. R. Pathirana, Ö. Zorlu, A. Muhtaroğlu, and H. Külah. Triple hybrid energy harvesting interface electronics. *Journal of Physics: Conference Series*, 773:012027, 2016.

[47] T. Kang, S. Kim, C. Hyoung, S. Kang, and K. Park. An Energy Combiner for a Multi-Input Energy-Harvesting System. *IEEE Transactions on Circuits and Systems II: Express Briefs*, 62(9):911–915, 2015.

[48] V. R. Challa, M. G. Prasad, Y. Shi, and F. T. Fisher. A vibration energy harvesting device with bidirectional resonance frequency tunability. *Smart Materials and Structures*, 17(1):015035, 2008.

[49] E. S. Leland and P. K. Wright. Resonance tuning of piezoelectric vibration energy scavenging generators using compressive axial preload. *Smart Materials and Structures*, 15(5):1413–1420, 2006.

[50] C. Peters, D. Maurath, W. Schock, F. Mezger, and Y. Manoli. A closed-loop wide-range tunable mechanical resonator for energy harvesting systems. *Journal of Micromechanics and Microengineering*, 19(9):094004, 2009.

[51] C. A. K. Kwuimy, G. Litak, and C. Nataraj. Nonlinear analysis of energy harvesting systems with fractional order physical properties. *Nonlinear Dynamics*, 80(1-2):491–501, 2015.

[52] B. Ando, S. Baglio, and C. Trigona. Autonomous sensors: From standard to advanced solutions [Instrumentation notes]. *IEEE Instrumentation Measurement Magazine*, 13(3):33–37, 2010.

[53] B. Andò, S. Baglio, C. Trigona, N. Dumas, L. Latorre, and P. Nouet. Nonlinear mechanism in MEMS devices for energy harvesting applications. *Journal of Micromechanics and Microengineering*, 20(12):125020, 2010.

[54] M. Ferrari, V. Ferrari, M. Guizzetti, B. Andò, S. Baglio, and C. Trigona. Improved energy harvesting from wideband vibrations by nonlinear piezoelectric converters. *Sensors and Actuators A: Physical*, 162(2):425–431, 2010.

[55] B. Andó, S. Baglio, L. Latorre, F. Maiorca, P. Nouet, and C. Trigona. Magnetically-Coupled Cantilevers with Antiphase Bistable Behavior for Kinetic Energy Harvesting. *Procedia Engineering*, 47:1065–1068, 2012.

[56] G. Litak and M. Borowiec. On simulation of a bistable system with fractional damping in the presence of stochastic coherence resonance. *Nonlinear Dynamics*, 77(3):681–686, 2014.

[57] L. Mateu, J. Knauer, P. Spies, and H. Zessin. D7.3 - Autonomous Tuning Methods for Piezoelectric Energy Harvesting Generators. *Proceedings Sensor 2017*, pages 540–545, May 2017.

[58] J. Dicken, P. D. Mitcheson, I. Stoianov, and E. M. Yeatman. Increased power output from piezoelectric energy harvesters by pre-biasing, 2009.

[59] T. I. Florian Feckl, Applications Engineer. Sensing the iot. *Electronic Component News*, 2015.

[60] A. Ramond, G. A. Rodriguez, H. Durou, B. Jammes, and C. Rossi. A SIDO buck converter with ultra low power MPPT scheme for optimized vibration energy harvesting and management. *Proceedings PowerMEMS 2009*, 2009.

[61] A. K. Shukla, A. Banerjee, M. K. Ravikumar, and A. Jalajakshi. Electrochemical capacitors: Technical challenges and prognosis for future markets. *Electrochimica Acta*, 84:165–173, 2012.

[62] I. Buchmann. *Batteries in a Portable World: A Handbook on Rechargeable Batteries for Non-Engineers*. Cadex Electronics Inc, Richmond, British Columbia, 3 edition, 2011.

[63] I. Buchmann. *Batteries in a Portable World: A Handbook on Rechargeable Batteries for Non-Engineers*. Cadex Electronics Inc., Richmond, British Columbia, 4 edition, 2016.

[64] N. N. H. Ching, H. Y. Wong, W. J. Li, P. H. W. Leong, and Z. Wen. A laser-micromachined multimodal resonating power transducer for wireless sensing systems. *Sensors and Actuators A: Physical*, 97-98:685–690, 2002.

[65] S. C. l. Yuen, J. M. h. Lee, W. J. Li, and P. H. w. Leong. An AA-Sized Vibration-Based Microgenerator for Wireless Sensors. *IEEE Pervasive Computing*, 6(1):64–72, 2007.

[66] B. H. Stark, G. D. Szarka, and E. D. Rooke. Start-up circuit with low minimum operating power for microwatt energy harvesters. *IET Circuits, Devices Systems*, 5(4):267–274, 2011.

[67] M. Edamoto, Y. Suzuki, N. Kasagi, K. Kashiwagi, Y. Morizawa, T. Yokoyama, T. Seki, and M. Oba. Low-Resonant-Frequency Micro Electret Generator for Energy Harvesting Application. In *2009 IEEE 22nd International Conference on Micro Electro Mechanical Systems*, pages 1059–1062, Jan. 2009.

[68] D. Marinkovic, A. Frey, I. Kuehne, and G. Scholl. A New Rectifier and Trigger Circuit for a Piezoelectric Microgenerator. *Procedia Chemistry*, 1(1):1447–1450, 2009.

[69] G. Liu, R. Fuentes, H. Koser, and T. Kaya. A self-powered power conditioning circuit for battery-free energy scavenging applications. *Analog Integrated Circuits and Signal Processing*, 83(2):203–207, 2015.

[70] J. Colomer-Farrarons, P. Miribel-Catala, A. Saiz-Vela, M. Puig-Vidal, and J. Samitier. Power-Conditioning Circuitry for a Self-Powered System Based on Micro PZT Generators in a 0.13-μm Low-Voltage Low-Power Technology. *IEEE Transactions on Industrial Electronics*, 55(9):3249–3257, 2008.

[71] G. D. Szarka, S. G. Burrow, and B. H. Stark. Ultralow Power, Fully Autonomous Boost Rectifier for Electromagnetic Energy Harvesters. *IEEE Transactions on Power Electronics*, 28(7):3353–3362, 2013.
[72] N. M. Roscoe and M. D. Judd. Harvesting Energy From Magnetic Fields to Power Condition Monitoring Sensors. *IEEE Sensors Journal*, 13(6):2263–2270, 2013.
[73] A. Romani, M. Filippi, and M. Tartagni. Micropower Design of a Fully Autonomous Energy Harvesting Circuit for Arrays of Piezoelectric Transducers. *IEEE Transactions on Power Electronics*, 29(2):729–739, 2014.
[74] D. Alghisi, V. Ferrari, M. Ferrari, F. Touati, D. Crescini, and A. B. Mnaouer. A new nano-power trigger circuit for battery-less power management electronics in energy harvesting systems. *Sensors and Actuators A: Physical*, 263(Supplement C):305–316, 2017.
[75] ST. *TS831: Micropower Voltage Supervisor Reset Active Low*, 2001.
[76] S. Lu, F. Boussaid, and M. K. Law. Efficient parallel-SSHI interface circuit for piezoelectric energy harvesting. In *2013 IEEE 11th International New Circuits and Systems Conference (NEWCAS)*, pages 1–4, 2013.
[77] S. J. Marinkovic and E. M. Popovici. Nano-power wireless wake-up receiver with serial peripheral interface. *IEEE Journal on Selected Areas in Communications*, 29(8):1641–1647, 2011.
[78] S. H. Lee, Y. S. Bae, and L. Choi. The design of a ultra-low power RF wakeup sensor for wireless sensor networks. *Journal of Communications and Networks*, 18(2):201–209, 2016.
[79] Y. Ammar, S. Bdiri, and F. Derbel. An ultra-low power wake up receiver with flip flops based address decoder. In *2015 12th International Multi-Conference on Systems, Signals & Devices (SSD)*, pages 1–5. IEEE, 2015.
[80] J. Ansari, D. Pankin, and P. Mähönen. Radio-triggered wake-ups with addressing capabilities for extremely low power sensor network applications. *International Journal of Wireless Information Networks*, 16(3):118–130, 2009.
[81] L. Huo. A comprehensive study of passive wake-up radio in wireless sensor networks. Master thesis, Delft University of Technology, 2014.
[82] J. Masuch, M. Delgado-Restituto, D. Milosevic, and P. Baltus. Co-integration of an RF energy harvester into a 2.4 GHz transceiver. *IEEE Journal of Solid-State Circuits*, 48(7):1565–1574, 2013.
[83] A. Bannoura, F. Höflinger, O. Gorgies, G. Gamm, J. Albesa, and L. Reindl. Acoustic wake-up receivers for home automation control applications. *Electronics*, 5(1):4, 2016.
[84] *XBee S2 quick reference guide/cheat sheet*, 2017.
[85] S. Voigt, J. Wolfrum, M. Pfeiffer, T. Keutel, C. Brockmann, V. Großer, S. Lissek, H. During, B. Rusek, M. Braunschweig, et al. Sensornetzwerk zum Monitoring von Hochspannungsleitungen. In *VDE-Kongress 2012*. VDE VERLAG GmbH, 2012.

Part II: **Vibration converters and hybridization**

Slim Naifar, Sonia Bradai, Slim Choura, and Olfa Kanoun

Magnetoelectric vibration energy harvesting

Abstract: Magnetoelectricity, the relation between electrical and magnetic properties in a matter, is a promising conversion mechanism for kinetic energy harvesting. Magnetoelectric (ME) converters take advantage of strong mechanical nonlinearities of magnetostrictive and piezoelectric materials for conceptually novel vibration harvesters.

This chapter presents the basics of magnetoelectricity in ME laminate composites and discusses the choice of materials. The general physical concept of ME response in conventional ME transducers and an overview of the recent developments in the state of the art of magnetoelectric energy harvesting are presented. A finite element model of magnetostriction is developed in order to investigate the ME response in the transducer in two different magnetic circuit configurations aiming to develop a model-based design method for ME converters.

Keywords: Magnetoelectric converter, magnetostriction, piezoelectric effect, vibration energy harvesting

1 Introduction

The development of low-power electronic devices has reduced power requirements for wireless sensor and communication node networks, which opens the possibility of developing self-powered autonomous systems supplied from available environmental resources. These resources are mainly solar and wind power, thermal gradients and mechanical vibration [1]. Among these energy scavenging sources, kinetic energy is interesting, since it is a potential power source that is available in indoor and outdoor applications [2]. Mechanical vibration can be converted into electrical energy through electromagnetic, magnetoelectric electrostatic and piezoelectric transducers [3]. Due to their high ME response, the possibility of using laminate composites formed by magnetostrictive and piezoelectric layers to convert magnetic to electrical energy has recently attracted attention when compared to other transduction mechanisms [4].

Magnetoelectric transducers combine piezoelectric and magnetostrictive materials to form one composite able to convert magnetic fields into electrical energy, which may improve the efficiency of the vibration energy harvesting system [5].

Slim Naifar, Sonia Bradai, Olfa Kanoun, Chair for Measurement and Sensor Technology, Technische Universität Chemnitz, 09126 Chemnitz, Germany
Slim Naifar, Sonia Bradai, Slim Choura, Laboratory of Electromechanical Systems, National Engineering School of Sfax, University of Sfax, 3038 Sfax, Tunisia

https://doi.org/10.1515/9783110445053-005

Conventional ME laminate composites are generally formed by a thin rectangular piezoelectric layer sandwiched between two magnetostrictive layers. In vibration converters, the three-layer composite is placed in a time varying magnetic field that can be induced from ambient vibration by an arrangement of neodymium magnets. The strain induced in the outer magnetostrictive layers can exceed that in the traditional d31 bending mode in the piezoelectric layer [6]. The optimum design of a vibration-based ME converter depends mainly on the choice of materials, the magnetic field configuration and the transmitted deformation to the piezoelectric layer. Therefore, a model-based design method is required instead of using a trial-and-error approach. To that end, a good understanding of the model of the ME effect in ME laminate composite used in energy harvesting systems as well as a finite element (FE) model are inevitable.

This chapter consists of four sections. An overview of the basic physical principles behind ME energy harvesting is presented in Section 2. Section 3 discusses recent development in the state of the art of ME converters. Finally, the ME effect for two types of magnetic circuit used in ME energy harvesting is analyzed using the FE method in Section 4. Results and a discussion are presented in Section 5.

2 Magnetoelectric composites

2.1 Magnetoelectric effect

The ME effect can be presented as a polarization $\vec{P}_0$ response to an applied magnetic field $\vec{H}$ or, conversely, a magnetization $\vec{M}$ response to an applied electric field $\vec{E}$ [7]. This effect was predicted in 1894 and observed for the first time in 1960 [8].

The coupling mechanism mentioned above can be written as given by equation (1):

$$\mathrm{ME} = \frac{\text{electrical energy}}{\text{magnetic energy}} \tag{1}$$

The different interacting couplings for the ME effect are presented in Figure 1.

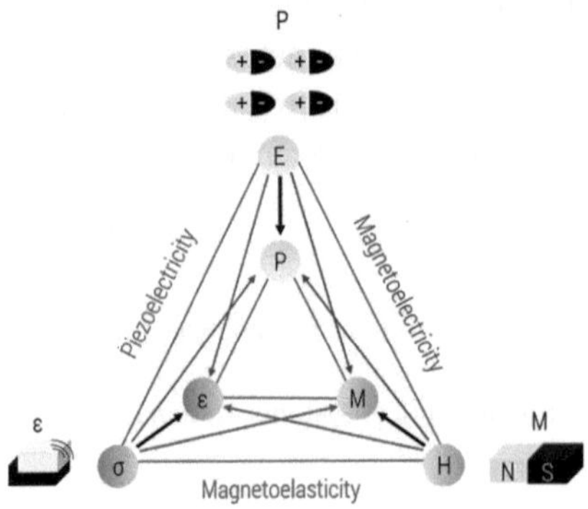

Fig. 1: Schematic illustrating different types of coupling for the ME effect. *E*, *H*, *P*, *M*, *ε* and *σ* refer to electric field, magnetic field, polarization, magnetization, strain and stress, respectively.

The ME effect is characterized by the ME voltage coefficient, α_{ME}, which relates the induced polarization P_o to the applied magnetic field H respecting equation (2) [9],

$$P_o = \frac{\alpha_{ME}}{\varepsilon_0 \varepsilon_r} H \tag{2}$$

where ε_0 and ε_r are, respectively, the vacuum permittivity and the relative permittivity.

Furthermore, the ME voltage coefficient can be defined as the change of the electric field with the change of the applied magnetic field, as shown in equation (3),

$$\alpha_{ME} = \frac{dE}{dH} \tag{3}$$

2.2 Magnetoelectric materials

In general, materials that exhibit the ME effect are classified into two categories: single-phase materials and composites (Figure 2). Magnetoelectricity has been observed as an intrinsic effect in some multiferroic materials, which possess simultaneously ferroelectric and ferromagnetic properties [10]. These single-phase ME materials have been under intensive study in the past decades [11]. In this context, materials such as Cr_2O_3 [12, 13], ME boracites [14] and phosphates [15] have been studied. However, the ME coupling in most of these materials is relatively too low to be used in practical applications [8]. Furthermore, most single-phase ME materials exhibit ME behavior only at low temperature, and they are not useful at room temperature [16].

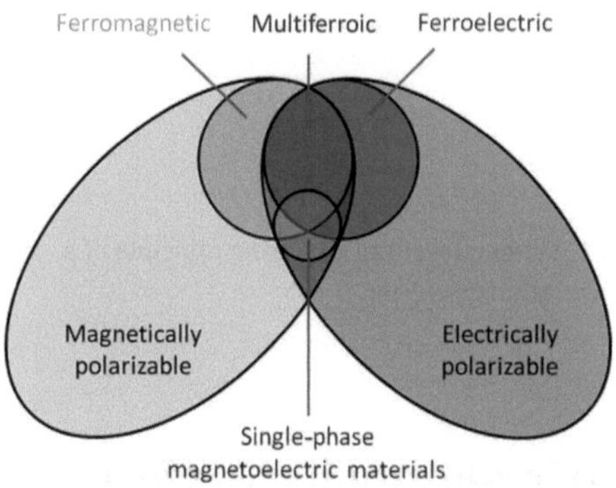

Fig. 2: Magnetoelectric materials.

To overcome the limitations of single-phase ME materials, the ME behavior has also been developed as a composite effect in multiphase systems by combining piezoelectric and magnetostrictive components[17, 18]. Various piezielectric (PE) and magnetostriction (MS) materials commonly used in ME composites are listed in Table 1.

Tab. 1: List of well-known piezoelectric and magnetostrictive materials used as constituents of ME composites [19].

PE materials	MS materials
Lead based:	**Metals:**
$Pb(Zr,Ti)O_3$ (PZT)	Fe, Co, Ni
$Pb(Mg_{1/3}Nb_{2/3})O_3$-$PbTiO_3$ (PMNT)	**Alloys:**
$Pb(Zn_{1/3}Nb_{2/3})O_3$-$PbTiO_3$ (PZN-PT)	FeNi based
(PMN-PZT)	FeCo based
(PIN-PMN-PT)	CoNi based
Lead-free:	Ni_2MnGa
$BaTiO_3$ (BTO) based	Terfenol-D ($Tb_{1-x}Dy_xFe_2$)
$(K_{1/2}Na_{1/2})NbO_3$ (KNN) based	Galfenol (FeGa), FeGaB
$Na_{1/2}Bi_{1/2}TiO_3$ (NBT) based	Samfenol ($SmFe_2$)

2.3 Magnetoelectric laminate composites

It has been reported that the structure of laminate composites has higher ME voltage coefficients than other interfacial bondings of piezoelectric (PE) and magnetostrictive (MS) constituents [19]. These composites can be prepared with different dimensions, number of layers and geometries, including discs [20], rectangles [21] and rings [22]. However, a well-known implementation of these structures used in vibration energy harvesting is to bond one PE layer to two MS layers using an epoxy resin (Figure 3).

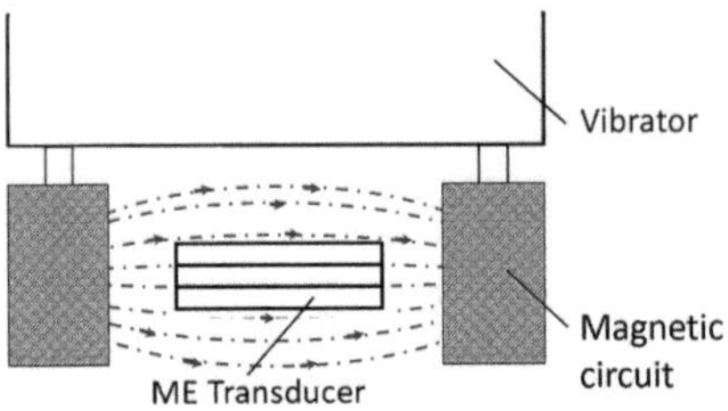

Fig. 3: Schematic of the operating principle of a ME vibration converter.

3 Overview of magnetoelectric energy harvesters

Different ME converters have been developed by numerous researchers. The influence of the arrangement of the magnets is a critical aspect for ME converters, since an optimal configuration can ensure high ME response and hence high output energy. Various arrangements of magnets in ME converters are reported in the literature, including the use of rectangular, spherical and hollow magnets. In the following, the main developed ME harvesters are detailed and classified respecting the configuration of the magnetic circuit acting on the ME transducer. Two main configurations are discussed: the magnetic circuit formed by rectangular magnets and the magnetic circuit formed by hollow magnets. ME converters based on a spherical magnet free to move on an ME transducer are not included in this study because of the very low output voltage that can be reached by using this configuration.

3.1 Rectangular magnets

In 2009, Dai et al. developed a ME converter based on cantilever beam. In this design, three layers of Terfenol-D/PZT/Terfenol-D are placed in the air gap of a magnetic circuit fixed at the end of a cantilever beam, as shown in Figure 4(a). The results prove that the converter can generate up to 1.055 mW across a 564.7 Ω load for excitation of about 51 Hz frequency with an acceleration of 1 g [23].

Another cantilever-based solution was published by Zhang et al. in 2012 [24]. With a Terfenol-D/PMN-PT/Terfenol-D transducer, there was a maximum output power of

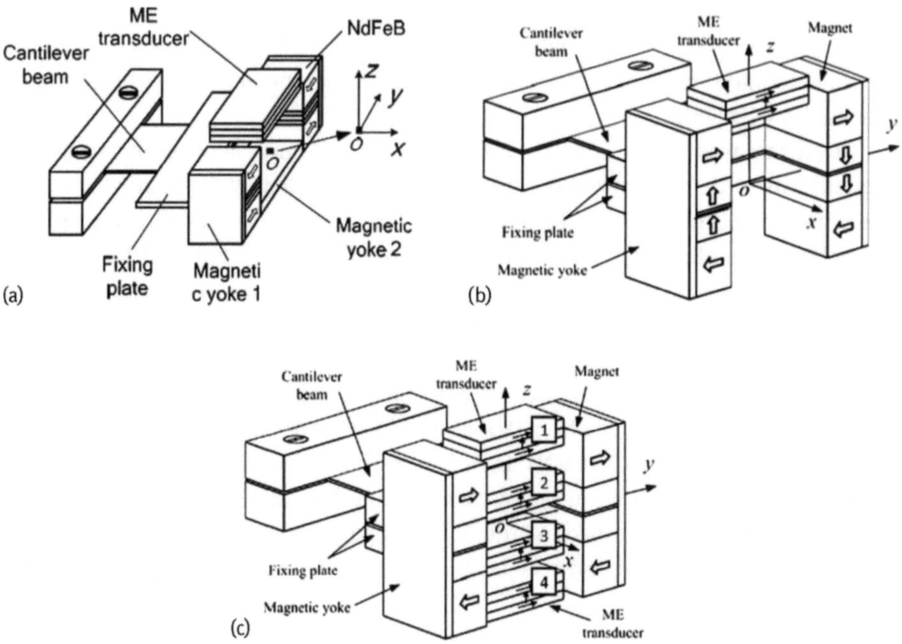

Fig. 4: Magnetoelectric energy harvesters based on a cantilever beam. (a) Single transducer reported by Dai et al., (b) single transducer reported by Zhang et al. and (c) multiple transducers reported by Dai et al. [23–25].

1.9015 mW across a 3.585 MΩ resistor at 1 g acceleration and a power of 0.7804 mW at 0.3 g acceleration at a resonant frequency of 30.6 Hz (Figure 4(b)).

For the ME converter-based cantilever beam, some research proposed multiple ME transducers by connecting several in series or in parallel. Dai et al. assembled a multiple transducer converter in 2011 [25]. As presented in Figure 4(c), up to four Terfenol-D/PMNT/Terfenol-D laminate composites are used to harvest energy from mechanical vibrations. Power outputs in the range of 1.44 mW to 7.13 mW can be generated for excitation frequencies from 26.2 to 34.8 Hz at 1 g acceleration.

3.2 Hollow magnets

Behind resonant ME converter-based rectangular magnets, researchers developed harvesters using an ME transducer placed in the cavity of a hollow magnet. In this context, Li et al. came up with a device-based cantilever beam vibration using the magnetic spring principle. A first prototype, developed in 2011, was based on a Terfenol-D/PMNT/Terfenol-D laminate composite placed in a hollow magnet, as shown in Figure 5(a). A cylindrical magnet is fixed to the end of a cantilever and placed below the hollow magnet to create a magnetic spring. The configuration of the magnets makes the resonant frequency of the harvester tunable based on the repulsive magnetic force. The prototype is capable of generating 68.8 μW under an acceleration of 0.2 g at a resonant frequency of 25.4 Hz. It is reported that the frequency tunable range can reach 4.4 Hz [26].

The device was optimized in 2013 [27]. The designed harvester uses the same concept but with twin ME transducers symmetrically separated by the cantilever beam. The schematic of the converter formed by two transducers, two cylindrical magnets, two hollow magnets, two brackets and a cantilever is given in Figure 5(b). For a rotation frequency of a 9.8 Hz range, the nonlinear converter is able to generate 517 μW, corresponding to a half-power bandwidth of 13.5 Hz when it is mounted on a wheel driven by an AC servo-actuator.

Table 2 summarizes all presented ME harvesters. The devices reviewed are capable of delivering energy outcomes from 0.0688 mW to 7.13 mW. These converters uses single and multiple Terfenol-D/(PMNT or PZT) transducers placed in a magnetic circuit formed by either rectangular or hollow magnets. In the next section, an FE model of magnetoelectricity is developed in order to investigate the magnetic field distribution for rectangular and hollow magnets' magnetic circuit designs.

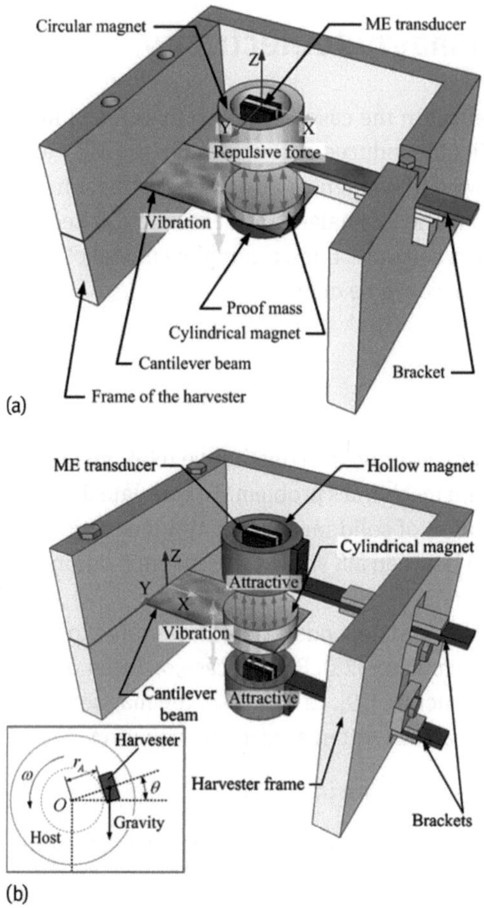

Fig. 5: Magnetoelectric energy harvesters based on a cantilever beam. (a) Single transducer reported by Dai et al., (b) single transducer reported by Zhang et al. and (c) multiple transducers reported by Dai et al. [26, 27].

Tab. 2: Summary of reviewed ME harvesters.

Magnetic circuit configuration	Ref.	Output power in mW	Type of excitation	
			Resonance frequency in Hz	Acceleration in g
Rectangular magnets	[23]	1.055	51	1
	[24]	1.9015	30.6	1
	[25]	1.44	26.2	1
	[25]	4.07	37	1.5
	[25]	3.95	29.4	2
	[25]	7.13	34.8	2.5
Hollow magnets	[26]	0.0688	25.4	0.2
	[27]	0.517	9.8	–

4 Finite element model of magnetoelectricity

In order to investigate in detail the ME effect in the case of the two mentioned magnetic circuit designs, two models to predict the induced deformation in the ME transducer for different relative positions between the magnets and the transducer in the two cases are developed by the FE method. The FE analyses presented in this section were carried out using Comsol Multiphysics software. First, a complete model of magnetostriction in the two Terfenol-D layers needs to be defined.

4.1 Modeling of magnetostriction

Comsol uses the generalized Hooke's law to describe the constitutive relationship between the stress and strain tensors in a solid mechanics problem. To simulate the elastic behavior of a MS material, a combination of solid mechanics and magnetic field modules is needed. To that end, a relation between MS strains and magnetization has to be developed.

Magnetostriction is an anisotropic effect and depends on the stress direction, because it is strongly related to the crystal symmetry [28]. For the case of Terfenol-D, it solidifies into the cubic C15 Laves phase structure [29]. In this case, the magnetostriction along the direction of magnetization can be written as given by equation (4),

$$\lambda = \lambda_{100} + 3(\lambda_{111} - \lambda_{100})(\alpha_1^2\alpha_2^2 + \alpha_2^2\alpha_3^2 + \alpha_3^2\alpha_1^2) \tag{4}$$

where λ_{100}, λ_{111} are the saturation magnetization along the directions $\langle 100 \rangle$ and $\langle 111 \rangle$, and α_i is the magnetization direction cosine.

By considering Terfenol-D as an isotropic material, equation (4) can be written as given by equation (5),

$$\lambda = \frac{3}{2}\lambda_s(\cos^2(\theta) - \frac{1}{3}) \tag{5}$$

where λ_s is the saturation magnetization and θ is the angle of magnetization.

Furthermore, the deformation due to magnetostriction along any direction can be considered as a nonlinear function of the direction of the magnetization as given by equation (6) [30],

$$\lambda = \frac{3}{2}\lambda_s\left(\left(\frac{M_i}{M_s}\right) - \frac{1}{3}\right) \tag{6}$$

where M_i and M_s are the magnetization along direction i and the saturation magnetization of the material, respectively. The negative term (–1/3) takes into account the magnetization required to align the individual domains within the MS material, and for this model it is ignored.

Thus, the magnetic effect in Magnetistriction Materials (MsM) can be incorporated by adding anisotropic initial strain, but the initial strain terms are proportional to the magnetization of the material along the corresponding direction.

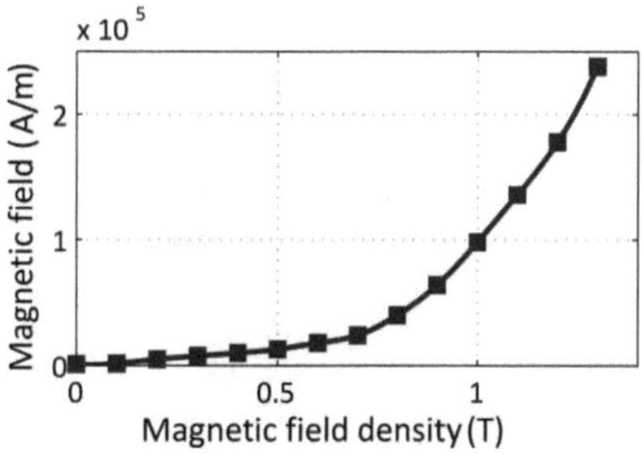

Fig. 6: *H*-*B*-curve of Terfenol-D.

The nonlinear magnetic behavior of Terfenol-D is modeled by using an HB curve to specify the magnetic constitutive relation in the MS layers. The curve used in the simulation model is shown in Figure 6. It can provide a value of the magnetic field H for a given magnetic flux density B in the material.

In the model under consideration, eddy currents are not taken into account. Thus, the electric conductivity of Terfenol-D is assumed to be equal to 0.

4.2 Modeling of adhesively bonded joints

In order to estimate the deformation transmitted to the PE layer, the adhesively bonded joints between the inner plate and the two outer MS layers have to be defined. Since the aim is to compare the performance of the ME transducer when placed in two different configurations of the magnetic circuit, mechanical losses in the bonded joints are neglected in the model under consideration, i.e., the bonding interface between the three layers are assumed to be ideal.

4.3 Simulation model

The simulation model of the two configurations is presented in Figure 7. The ME transducer consists of one PE plate between two Terfenol-D plates. All layers are of 12 mm length, 6 mm width and 1 mm thickness. In the case presented in Figure 7(a), the magnetic circuit is formed by two rectangular magnets (10 mm × 6 mm × 5 mm). In the second case, the ME transducer was placed inside the cavity of a hollow magnet (10 mm length, 10 mm outer diameter and 9 mm inner diameter), as shown in Figure 7(b). All magnets used were defined by a relative permeability of 1.09 and a remnant flux density parallel to z-axis. In both cases, an air domain is added to the model in order to eliminate the effect of the boundary condition outside of the modeling domain.

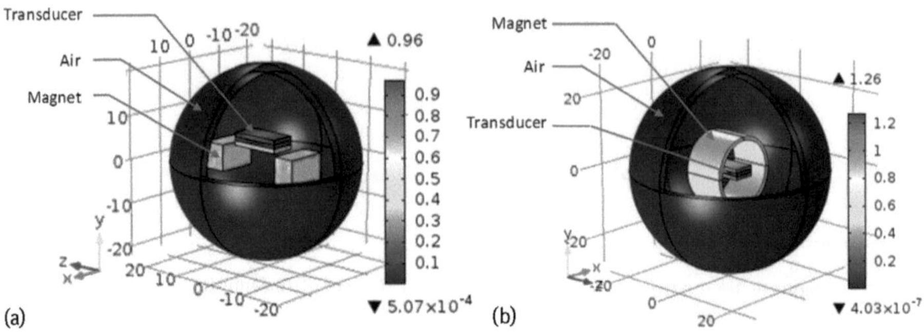

Fig. 7: Simulation model. (a) Magnetic circuit formed by two rectangular magnets. (b) Magnetic circuit formed by a hollow magnet.

5 Results and discussion

Figure 8 presents the simulation results. It can be seen in Figure 8(a) that the induced deformations in the two Terfenol-D layers are not identically the same due to the position of the magnets in the air gap of the two rectangular magnets. In this configuration, placing the transducer initially 2 to 3 mm above the magnets will induce more variation of deformation and then voltage, compared to other relative positions between the magnets and the laminate composite.

For the second case (Figure 8(b)), the deformation in the two Terfenol-D plates are similar due the presence of a symmetry axis in the case of a hollow magnet. In this case, the deformation transmitted to the PE layer also have similar variation. Since the ME transducer is able to generate more energy outcome by increasing the variation of the deformation in the PE layer, FE results prove that placing a part of the transducer (ideally half) inside the cavity of the hollow magnet presents the optimal initial position of transducer for this magnetic circuit configuration.

6 Conclusion

There is a continuing debate on the effective design for the magnetic circuit in ME converters, which can generate more energy at low vibration excitations. In this chapter, the basics of magnetoelectricity are described, and the main configurations for ME converters in the literature are discussed. It is reported that magnetic circuits formed by rectangular magnets can be more efficient than other magnetic circuit designs. A FE model is developed to initiate a model-based design method for ME converters to optimize their output power. Two magnetic circuit designs were investigated, the case of rectangular magnets and the case of a hollow magnet. Simulation

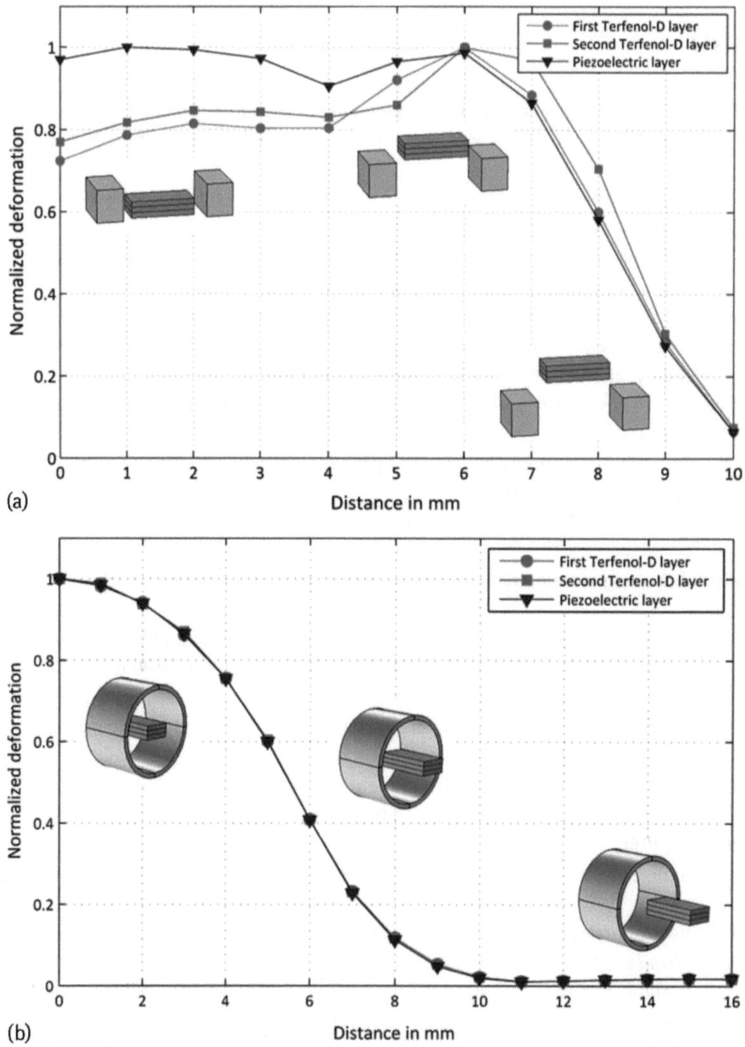

Fig. 8: Average deformation in the three layers versus the position of the transducer. (a) Magnetic circuit formed by two rectangular magnets. (b) Magnetic circuit formed by a hollow magnet.

results prove that placing the ME transducer in the cavity of two rectangular magnets with the same magnetization direction can induce more deformation to the transducer. However, the use of a hollow magnet has the advantage of generating a similar deformation in the three layers of the transducer, and then it can be more easily controlled and optimized in terms of the initial position of the transducer and other parameters such as magnet outer and the inner diameter of the magnet.

Bibliography

[1] X. Wang and A. Mortazawi. Medium wave energy scavenging for wireless structural health monitoring sensors. *IEEE Transactions on Microwave Theory and Techniques*, 62(4):1067–1073, 2014.

[2] F. Fei et al. Development of an indoor airflow energy harvesting system for building environment monitoring. *Energies*, 7(5):2985–3003, 2014.

[3] S. Naifar et al. Survey of electromagnetic and magnetoelectric vibration energy harvesters for low frequency excitation. *Measurement*, 106: 251–263, 2017.

[4] S. Naifar et al. Response analysis of a nonlinear magnetoelectric energy harvester under harmonic excitation. *The European Physical Journal Special Topics*, 224(14–15):2897–2907, 2015.

[5] J. Huang, R. C. O'Handley, and D. Bono. *New high-sensitivity hybrid magnetostrictive/electroactive magnetic field sensors.* Smart Structures and Materials. International Society for Optics and Photonics, 2003.

[6] L. Wang and F. G. Yuan. Vibration energy harvesting by magnetostrictive material. *Smart Materials and Structures*, 17(4):045009, 2008.

[7] L. D. Landau, E. M. Lifshitz, J. B. Sykes, J. S. Bell, and E. H. Dill. Electrodynamics of continuous media, 1961.

[8] M. Fiebig. Revival of the magnetoelectric effect. *Journal of Physics D: Applied Physics*, 38(8):R123, 2005.

[9] C.-S. Park, J. Evans, and S. Priya. Quantitative understanding of the elastic coupling in magnetoelectric laminate composites through the nonlinear polarization-magnetic (P-H) response. *Smart Materials and Structures*, 20(8):082001, 2011.

[10] N. Ortega et al. Multifunctional magnetoelectric materials for device applications. *Journal of Physics: Condensed Matter*, 27(50):504002, 2015.

[11] T. Kimura et al. Magnetic control of ferroelectric polarization. *nature*, 426(6962):55, 2003.

[12] I. E. Dzyaloshinskii. On the magneto-electrical effect in antiferromagnets. *Soviet Physics Jetp-Ussr*, 10(3):628–629, 1960.

[13] D. N. Astrov. The magnetoelectric effect in antiferromagnetics. *Sov. Phys. JETP*, 11(3):708–709, 1960.

[14] E. Ascher et al. Some Properties of Ferromagnetoelectric Nickel-Iodine Boracite, Ni3B7O13I. *Journal of Applied Physics*, 37(3):1404–1405, 1966.

[15] D. J. S. Santoro, R. P. and R. E. Newnham. Magnetic properties of LiCoPO4 and LiNiPO4. *Journal of Physics and Chemistry of Solids*, 27(6–7):1192–1193, 1966.

[16] S. Roy, B. Biswas, and S. B. Majumder. Investigations on flexible multiferroic composites. In S. M. Bose, S. N. Behera, and B. K. Roul, editors, *AIP Conference Proceedings*, volume 1063(1). AIP, 2008.

[17] K. H. Shin, M. Inoue, and K. I. Arai. Preparation and properties of elastically coupled electromagnetic elements with a bonding structure. *IEEE transactions on magnetics*, 34(4):1324–1326, 1998.

[18] F. Yang et al. Resonant magnetoelectric response of magnetostrictive/piezoelectric laminate composite in consideration of losses. *Sensors and Actuators A: Physical*, 141(1):129–135, 2008.

[19] H. Palneedi et al. Status and perspectives of multiferroic magnetoelectric composite materials and applications. *Actuators*, 5(1). Multidisciplinary Digital Publishing Institute, 2016.

[20] J. Ryu et al. Effect of the Magnetostrictive Layer on Magnetoelectric Properties in Lead Zirconate Titanate/Terfenol-D Laminate Composites. *Journal of the American Ceramic Society*, 84(12):2905–2908, 2001.

[21] Y. Wang et al. Enhanced magnetoelectric effect in longitudinal-transverse mode Terfenol-D/$Pb(Mg_{1/3}Nb_{2/3})O_3$-$PbTiO_3$ laminate composites with optimal crystal cut. *Journal of applied physics*, 103(12):124511, 2008.
[22] Z. Wang, W. Wang, and X. Luo. Enhancement of capacitive type magnetoimpedance effect in ring-type magnetoelectric transducers vibrator via size-dependent resonance frequency. *Sensors and Actuators A: Physical*, 247:234–238, 2016.
[23] X. Dai et al. Modeling, characterization and fabrication of vibration energy harvester using Terfenol-D/PZT/Terfenol-D composite transducer. *Sensors and Actuators A: Physical*, 156(2):350–358, 2009.
[24] Z. Zhang, X. Dai, and Y. Wang. An improved magnetoelectric vibration energy harvester for wireless sensors. In *Communication Technology (ICCT), 2012 IEEE 14th International Conference on*. IEEE, 2012.
[25] X. Dai et al. Energy harvesting from mechanical vibrations using multiple magnetostrictive/piezoelectric composite transducers. *Sensors and Actuators A: Physical*, 166(1):94–101, 2011.
[26] M. Li et al. *A magnetostrictive/piezoelectric laminate transducer based vibration energy harvester with resonance frequency tunability*. Sensors, 2011 IEEE. IEEE, 2011.
[27] M. Li et al. A resonant frequency self-tunable rotation energy harvester based on magnetoelectric transducer. *Sensors and Actuators A: Physical*, 194:16–24, 2013.
[28] S. Tumanski. *Handbook of magnetic measurements*. CRC Press, 2016.
[29] J. H. Alexander and O. J. Myers. Microstructure Properties and Strengthening Mechanisms of the AS4-3501-6 Polymeric Resin With Embedded Terfenol-D Particles, 2014. Strain (estimated linear) 800: 1200ppm.
[30] S. Chikazumi and C. D. Graham. *Physics of Ferromagnetism 2e. 94*. Oxford University Press on Demand, 2009.

Carlo Trigona, Sonia Bradai, Slim Naifar, Olfa Kanoun, and Salvatore Baglio

Nonlinear electromagnetic vibration converter with bistable RMSHI for power harvesting from ambient vibration

Abstract: Self powering for small electronic devices such as wearable sensors and wireless sensor networks has been in high demand in recent years. Therefore, much research has been conducted that aimed to harvest energy from ambient sources to power such systems. Due to the presence of mechanical vibration in almost all places, it is considered as the main interesting energy source. Dealing with vibration is also challenging due to its changing behavior relative to the vibration source (machines, human motions, trains, cars, etc.).

This work presents an overview on the importance and the challenges of nonlinear vibration energy harvesting systems to scavenge and store energy from a real mechanical vibration source. A nonlinear electromagnetic vibration converter managed through a nonlinear random mechanical switching harvesting on an inductor (RMSHI) is presented. In particular, the proposed electromagnetic converter is based on the use of the magnetic spring principle (a moving magnet), which ensures the ability of the device to harvest energy for larger bandwidths from real ambient sources compared to linear vibration converters. The system has been studied, realized and characterized.

Keywords: Nonlinear electromagnetic transducer, bistable RMSHI, mechanical switches, wideband vibrations, zero voltage threshold, nonlinear energy harvesting

1 Introduction

Harvesting energy from a real vibration source is challenging due to the energy sources available in the environment. As an example for an ambient variable vibration source, we can take human motion, which is variable relative to the movement variation, bridge vibration, train vibration, industrial machines, etc. Therefore, during the past

Carlo Trigona, Salvatore Baglio, D.I.E.E.I., Dipartimento di Ingegneria Elettrica Elettronica e Informatica, University of Catania, Viale Andrea Doria 6, 95125 Catania, Italy
Sonia Bradai, Slim Naifar, Olfa Kanoun, Technische Universität Chemnitz, Chair for Measurement and Sensor Technology, Reichenhainer Straße 70, 09126 Chemnitz, Germany
Sonia Bradai, Slim Naifar, Laboratory of Electromechanical Systems, National Engineering School of Sfax, University of Sfax, Route de Soukra km 4, 3038 Sfax, Tunisia

https://doi.org/10.1515/9783110445053-006

few years, more attention has been focused on smart architectures [1], which are also based on the nonlinear energy harvester being able to scavenge energy from random vibration sources.

Several solutions have been proposed to introduce or increase the nonlinearity in vibration transducers. In fact, nonlinearity can be introduced by the use of some materials that exhibit nonlinear behavior like the case of piezoelectric [2–5] or magnetoelectric materials [6–8]. Further, vibration converters based on the electromagnetic principle and the magnetic spring, where the magnet moves freely, are also promising solutions [9–11]. Other research focuses on the use of hybrid systems, which is the case of combining the piezoelectric and the electromagnetic principles [12–14]. In [14], the authors propose a nonlinear hybrid piezoelectric electromagnetic converter. The aim of this latter system is to reduce the resonant frequency and increase the frequency bandwidth in order to be able to scavenge energy from ambient sources. The system consists of the use of a cantilever beam within a pendulum architecture. The magnets are placed as the tip mass and a coil is placed on the system. A repulsive magnet is placed relative to the tip mass, which leads to the nonlinear behavior of the system. Another method has been developed that proposes the use of a bistable energy management [15] and the excitation of the converter in more than one direction [16], aiming to store more energy outcome from the vibration converter for a real random vibration source [17]. Some of the solutions treat only the nonlinearity of the device and use an active energy management to store the generated energy. In this work, a nonlinear electromagnetic converter with a low output voltage under the voltage of the threshold diodes, using a passive bistable-RMSHI energy management solution able to scavenge energy from real vibration sources is presented. For the nonlinear electromagnetic converter, the coil is fixed, and the magnetic spring principle is used to realize the repulsive force instead of the usually used mechanical springs. This leads to nonlinear behavior for the converter and reduces friction and aging effects and ensures a high reliability of the system.

The paper is organized as follows: Section 2 is devoted to the proposed nonlinear system to harvest energy from a real vibration source and the experimental setup to test the system. In Section 3, the results and the meteorological characterization of the conceived system are presented.

2 The proposed nonlinear vibration converter

2.1 Principle

Due to the importance of the nonlinearity for the energy harvesting system, in the following, a nonlinear electromagnetic (EM) vibration converter with a nonlinear passive energy management solution is proposed.

The EM converter is based on the use of a moving magnet oscillating due to the presence of two fixed magnets from both sides, placed in repulsive direction (see Figure 1(a)). A coil with a resistance of 22 Ω is placed surrounding the moving magnet. The total volume for the converter is 15.5 cm^3.

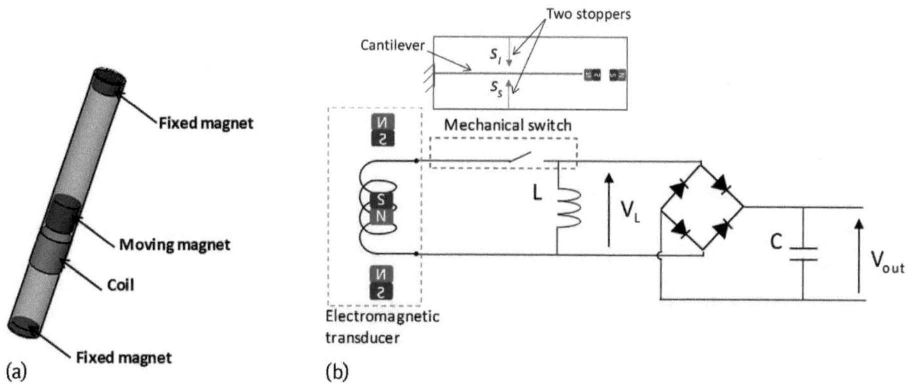

Fig. 1: (a) Proposed design for the EM converter, (b) schematic diagram of the energy management circuit.

Because of the reduced size of the converter, the output voltage level is too low (less than the bridge rectifier threshold 1.4 V), and a passive energy management solution is needed to store the generated energy. The proposed energy management circuit is based on the RMSHI [14] principle. A full bridge is used to rectify the output voltage. The bistable behavior in this case is due to the mechanical switch geometry, which consists of a brass beam where a permanent magnet is placed at the tip of the beam and a fixed magnet is placed in front in the repulsive direction, as shown in Figure 1(b) (details and modeling can be found in [18]).

The working principle of the mechanical switch is based on the use of two stoppers as electrical contacts. For an applied vibration, the cantilever beam will move between the two stable states defined by the stoppers, which leads to closing of the electrical circuit (Figure 1(b)). In this case, it is not required to used a power supply to control the stoppers. A capacitor with variable values of capacitance varying from 0.22 μF to 3.3 μF is used to store the rectified output. The energy management circuit has a size of 14 × 4 × 3 cm^3.

2.2 Experimental setup

The system is tested using the presented experimental setup (see Figure 2). It consists of an electrodynamic shaker to generate the applied excitation to the system, which is controlled via a laser sensor (Opto NCDT 1800). The applied vibration is defined

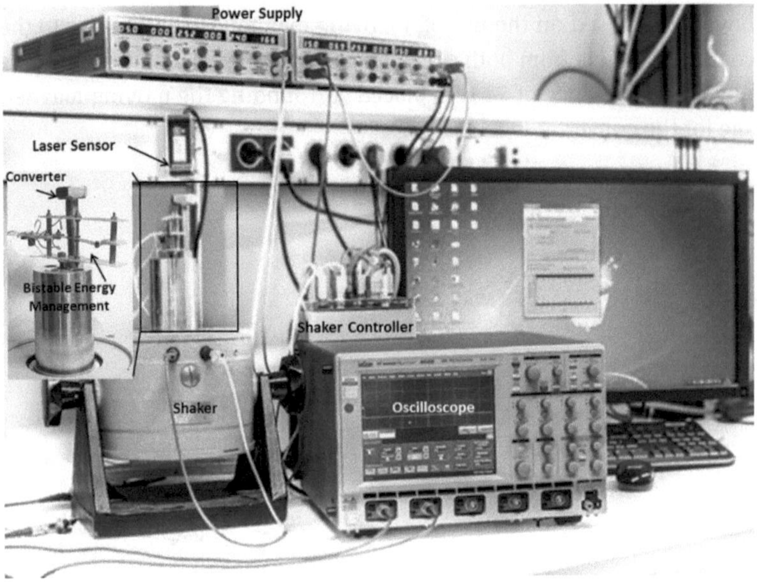

Fig. 2: Experimental setup.

via a shaker controller and a labVIEW program. A power supply and an oscilloscope are used for the laser sensor and the evaluation of the output voltage of the system, respectively (more details can be found in [19]).

3 Results

In this section, the converter behavior is characterized relative to different applied excitations and relative to the used energy management circuit.

As a first evaluation, the converter output is measured relative to an applied random signal generated by Matlab, characterized by a frequency between 1 Hz and 30 Hz and an applied sinusoidal signal characterized by a single frequency of 29 Hz. The same applied amplitude is ensured for both signals. Figure 3 presents V_{out}, the voltage output through the capacitor, which has a capacitance of 3.3 μF. As can be seen, the maximum voltage output is about 0.07 V, and it is reached in the case of a randomly applied excitation that is almost nine times higher compared to the generated voltage in the case of an applied excitation characterized by a single frequency (see Figure 3). This confirms that a nonlinear harvester vibration source ensures a better voltage output in the case of a random vibration source.

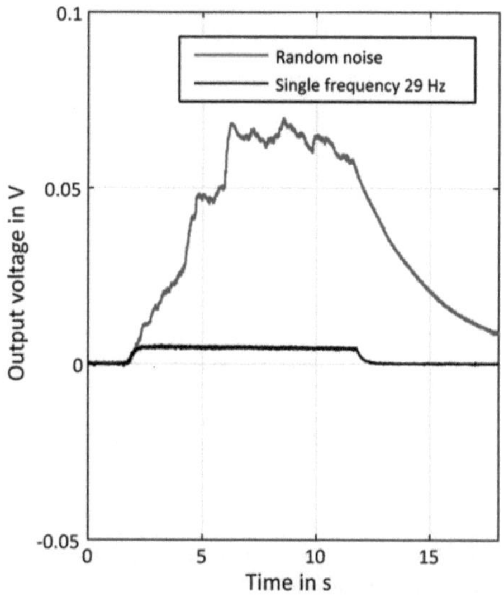

Fig. 3: Voltage output for applied random Matlab noise vibration and a sinusoidal excitation with a characterized frequency of 29 Hz.

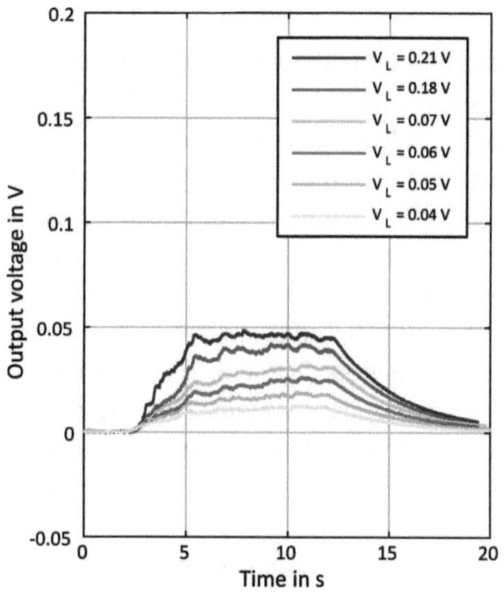

Fig. 4: Voltage output using a bridge rectifier with a mechanical switch for different applied accelerations.

Further, the system is evaluated for a random applied excitation signal with variable acceleration. The voltage output increases relative to the higher acceleration amplitude. V_L (Figure 4) presents the peak to peak voltage output from the converter, which varies from 0.04 V to 0.21 V (the values are under the bridge threshold), which corresponds to an applied root mean square (RMS) acceleration value varying from $8.18\,m/s^2$ to $12.25\,m/s^2$, respectively (see Figure 4).

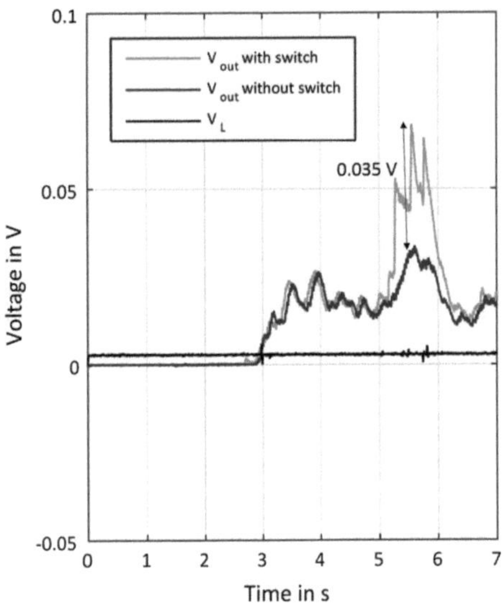

Fig. 5: Comparison of the output voltage with and without a mechanical switch.

In the following, the converter is evaluated for two different types of the energy management circuit, which consist of the presence of the (bistable) mechanical switch and without it (Figure 5). In this case, a capacitor value of 3.3 μF is used.

As it is shown in Figure 5, an improvement of the output voltage is reached only for a few seconds. In this case, the mechanical switch just moves for one of the positions up (S_I) or down (S_S) and is not able to move correctly because of the low applied excitation. This confirms that the use of a mechanical switch enables improving the RMS voltage outcome from about 0.034 V to about 0.069 V, since in this case, the mechanical switch moves correctly between the two positions S_I and S_S (Figure 1(b)).

Figure 6 shows the calibration diagram of the proposed system. The output voltage and the mechanical power are characterized for six different vibrations applied to the system. The mechanical power, in this case, varies from about $1.92 \cdot 10^{-5}$ W to about 6 $5.96 \cdot 10^{-5}$ W.

For each point, five measurements have been realized, and the mean value for the voltage output is calculated and presented in Figure 6. The red line presents the linear interpolation for the six measured points. The blue lines define the maximum uncertainty, which, in this case, is evaluated at one sigma that can be reached.

A maximum uncertainty of 0.0034 V is reached for a mechanical power and a mean voltage level of $5.96 \cdot 10^{-5}$ W and 0.039 V respectively. The minimum level of kinetic power able to overcome the energy barrier of the bistable system is about $1.92 \cdot 10^{-5}$ W, which represents the mechanical threshold.

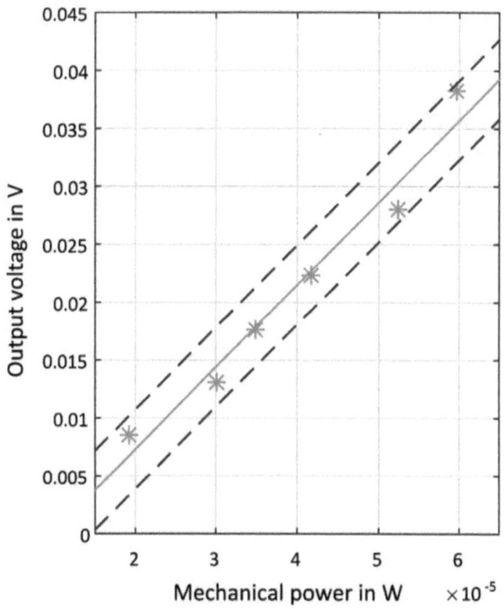

Fig. 6: Calibration diagram of the converter.

4 Conclusion

This work focuses on the nonlinear vibration energy harvesting systems. A nonlinear EM converter, based on a bistable RMSHI principle energy management circuit is developed and characterized. The system generates low output voltage less than 100 mV because of the low and random applied excitation, nevertheless, the system is able to store a small amount of energy within a passive energy management circuit. Work is in progress with a MEMS prototype of the bistable RMSHI to be integrated inside the nonlinear EM converter.

Bibliography

[1] B. Andò, S. Baglio, L. Malfa, A. Pistorio, and C. Trigona. A smart wireless sensor network for AAL. In *2011 IEEE International Workshop on Measurements and Networking Proceedings (M&N)*, 2011.
[2] P. Firoozy and M. P. S. EKhadem. Power enhancement of broadband piezoelectric energy harvesting using a proof mass and nonlinearities in curvature and inertia. *International Journal of Mechanical Sciences*, 133:227–239, 2017.
[3] T. Tan, Z. Yan, and W. Huang. Broadband design of hybrid piezoelectric energy harvester. *International Journal of Mechanical Sciences*, 516–526, 131–132, 2017.
[4] M. Rezaei and P. F. S. EKhadem. Broadband and tunable PZT energy harvesting utilizing local nonlinearity and tip mass effects. *118, 1–15*, 2017.
[5] B. Ando, S. Baglio, and C. Trigona. Autonomous sensors: From standard to advanced solutions [Instrumentation notes]. *IEEE Instrumentation & Measurement Magazine*, 13(3):33–37, 2010.

[6] X. Dai, Y. Wen, P. Li, J. Yang, and M. Li. Energy harvesting from mechanical vibrations using multiple magnetostrictive/piezoelectric composite transducers. *Sensors and Actuators*, A 166:94–101, 2011.
[7] X. Bai, Y. Wen, P. Li, J. Yang, X. Peng, and X. Yue. Multi-modal vibration energy harvesting utilizing spiral cantilever with magnetic coupling. *Sensors and Actuators A*, 209:78–86, 2014.
[8] M. Li, Y. Wen, P. Li, and J. Yang. A resonant frequency self-tunable rotation energy harvester based on magnetoelectric transducer. *Sensors and Actuators A*, 194:16–24, 2013.
[9] J. Leicht and Y. Manoli. A 2.6 μW–1.2 mW autonomous electromagnetic vibration energy harvester interface IC with conduction-angle-controlled MPPT and up to 95% efficiency. *IEEE Journal of Solid-State Circuits*, 52(9), 2017.
[10] S. Bradai, S. Naifar, and O. Kanoun. Finite element analysis of combined magnetoelectric-electrodynamic vibration energy converter. *J. of Physics Conference Series*, 660, Issue 1, 12, 115, 111–112, 2015.
[11] H. Ulusan, O. Zorlu, A. Muhtaroglu, and H. Kulah. Highly integrated 3 V supply electronics for electromagnetic energy harvesters with minimum 0.4 V_{peak} input. *IEEE Transactions on Industrial Electronics*, 64(7), 2017.
[12] A. Khaligh, P. Zeng, X. Wu, and Y. Xu. Hybrid energy harvesting topology for human-powered mobile electronics. In *34th Conf. IEEE Ind. Electron. Soc.*, pages 448–453, 2008.
[13] W. Yang and S. Towfighian. A hybrid nonlinear vibration energy harvester. *Mechanical Systems and Signal Processing*, 90:317–333, 2017.
[14] M. A. Karami, P. S. Varoto, and D. J. Inman. Experimental study of the nonlinear hybrid energy harvesting system. *Modal Analysis Topics*, 3:461–478, 2011.
[15] C. Trigona, N. Dumas, L. Latorre, B. Andò, S. Baglio, and P. Nouet. Exploiting benefits of a periodically-forced nonlinear oscillator for energy harvesting from ambient vibrations. *Procedia Engineering*, 25:819–822, 2011.
[16] M. Daqaq, C. Stabler, Y. Qaroush, and T. Seuaciuc-Osorio. Investigation of power harvesting via parametric excitations. *Journal of Intelligent Material Systems and Structures*, 20:545–557, 2009.
[17] S. Naifar, S. Bradai, T. Keutel, and O. Kanoun. Design of a vibration energy harvester by twin lateral magnetoelectric transducers. In *IEEE International Instrumentation and Measurement Technology Conference (I2MTC) Proceedings, Montevideo*, pages 1157–1162, 2014.
[18] F. Maiorca, F. Giusa, C. Trigona, B. Ando, A. R. Bulsara, and S. Baglio. Diode-less mechanical H-bridge rectifier for "zero threshold vibration energy harvesters. *Sensors and Actuators A: Physical*, 201:246–253, 2013.
[19] S. Bradai, S. Naifar, C. Trigona, S. Baglio, and O. Kanoun. Electromagnetic transducer with bistable-RMSHI for Power Harvesting from very weak kinetic sources. In *IEEE I2MTC-International Instrumentation and Measurement Technology Conference, Texas, USA*, 2018.

Krystian Łygas, Piotr Wolszczak, Paweł Stączek, and Grzegorz Litak

Energy harvesting from an oscillating vertical piezoelectric cantilever with clearance

Abstract: Vibration energy harvesting is proposed for a nonlinear electromechanical system. A mechanical resonator, which is realized as an inverted elastic pendulum excited by kinematic excitation, is characterized for a broad frequency band due to clearance and impact interactions. The moving frame induces pendulum beam deflection, and mechanical stress is converted into an electrical voltage by a piezoelectric patch. Finally, the output power is dissipated on the load resistor. The results of electric power and voltage outputs for various frequencies of kinematic excitations of the frame are shown. In the assumed working conditions, the system described transforms the input frame displacement frequency into the higher harmonics of beam deflection.

Keywords: Nonlinear vibration, impact, escape from the potential well, energy harvesting, piezoelectric, wavelets

1 Introduction

The energy harvesting research discipline focuses on building an alternative source for powering small devices and to recharge batteries. Especially, sensors or modern electronic applications, like IoT (Internet of Things), can be designed to work on harvested energy from ambient vibrations [1–3]. Preliminary results were obtained by using simple cantilever beams with piezoelectric, electrostatic, or electromagnetic coupling [4]. However, in such cases, to obtain the considerable power output, the excitation frequency was fixed to the resonance region of an energy harvesting resonator [4, 5]. Consequently, parameters of energy harvesting devices need to be tuned to the ambient energy conditions, including the frequency of ambient vibrations. A number of nonlinear devices have been proposed to overcome these limitations by broadening the frequencies ranges [3, 6–9].

Nonlinear systems, with hardening or softening characteristics, provide the inclinations of the resonance curve to lower and higher frequencies. For a large enough amplitude of excitation, nonlinearities lead to the occurrence of super and subharmonic

Krystian Łygas, Piotr Wolszczak, Paweł Stączek, Grzegorz Litak, Department of Automation, Lublin University of Technology, Nadbystrzycka 36, PL-20-618 Lublin, Poland
Grzegorz Litak, Department of Process Control, AGH University of Science and Technology, Mickiewicza 30, PL-30-059 Krakow, Poland
Grzegorz Litak, Laboratoire Génie Electrique et Ferroélectricité, Institut National des Sciences Appliquées de Lyon, 8 rue de la Physique, 69621 Villeurbanne cedex, France

https://doi.org/10.1515/9783110445053-007

resonances. Multiple solutions that appear in the nonlinear system provide various output amplitudes in the displacement of a mechanical resonator and, consequently, different power outputs [10]. The most promising solution for powering small electronic devices is the piezoelectric beam converting mechanical vibration energy into the electrical one, because of the relatively large factor of output power to volume [1].

In recent years, nonlinear mechanical resonators with double and single wells have been discussed [9]. Therefore, various systems with a piezoelectric beam and a double-well potential or similar systems with impacts have been studied in parallel. Furthermore, systems similar to an inverted pendulum with clamping to the frame have been discussed and tested [11–14]. On the other hand, systems with mechanical or magnetic impact interactions have also been studied experimentally and theoretically [15–17]. Among these studies, a bistable system with impacts using the magnetic field and mechanical stoppers was investigated [18]. This chapter presents a continuation of this direction. The bistable system is based on a piezoelectric beam, which can move as an inverted pendulum within the angle limits defined by mechanical stoppers.

2 Methods and experimental setup

An experimental stand is presented in Figure 1(a). Note that the bending inverted pendulum, which has the amplitude limiters, is a highly nonlinear system with bistability. The two stable positions (static, without excitation) are located at the edges of the bumpers. During motion, the contact between pendulum and limiters generates impacts. For better clarity, the beam contacted with an edge of one of the bumpers is depicted in Figure 1(b). In the experiment, the following parameters were used: mass of the elastic beam $M_b = 10.56$ g and the tip mass $M_t = 1.29$ g. The characteristic lengths are: the beam length – $l = 184$ mm; the breadth and thickness of the beam – $b = 17$ mm and $a = 3.5$ mm, respectively; the distance between the bumpers $h_1 = 7.8$ mm; the distance between the axis of rotation of the beam and the bumpers $h_2 = 19.3$ mm; the piezoelectric Macro Fiber Composite (MFC) M8514-P2 (with capacity of electrodes on the piezoelectric element – $C = 84.04$ nF).

The beam was a product of the 3D printer, made by SHS (selective heat sintering) printing. The beam's characteristic parameters are: material – SHS-COPA60 (mixture of polyamide and polycarbonate); tensile strength 11 MPa; Young's modulus 350 MPa; elongation at break 3–5%, density 0.9 g/cm^3. The anisotropy was relatively small. However, damping of this material had a unique multiple scale character, which gave an advantage of matching the mechanical and electrical impedance to broader a frequency band.

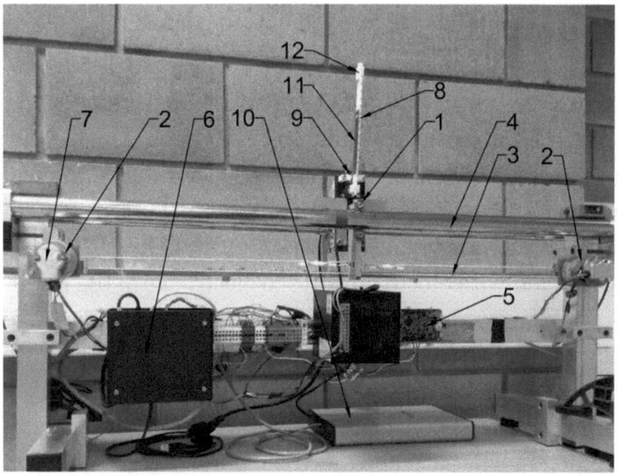

(a)

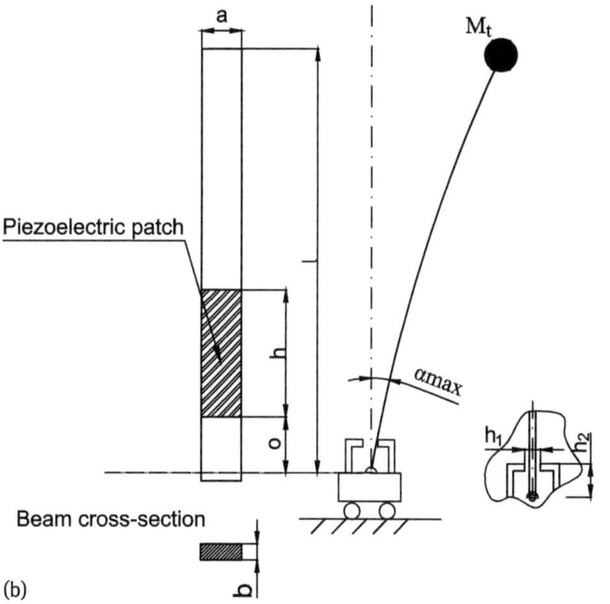

(b)

Fig. 1: Experimental setup. (a) The inverted elastic pendulum resonator with a piezoelectric stand: 1 – carriage of the inverted pendulum, 2 – DC servomotor, 3 – drive belt, 4 – guides, 5 – micro-controller, 6 – servo controller, 7 – incremental encoder, 8 – beam, 9 – bumpers, 10 – data acquisition card NI USB-6341, 11 – piezoelectric patch (No. MFC M-8514 P2), 12 – tip mass. (b) Characteristic dimensions of bending piezoelectric beam with angle limiters. Note that the piezoelectric was placed just above the impact zone to maximize the stress. Its length was fairly short to work in various modes shapes of the beam.

3 Results on power output

The presented experimental setup (Figure 1) was used to perform measurements on the electric power output with harmonic carriage excitation of the frequency fixed to 7 Hz and the amplitude at 9 mm, respectively. The aim was to consider the dynamical response of the system at a relatively low frequency limit where the current system can work successfully. The sampling frequency of voltage output was $f = 100$ kHz. The loading resistor, R, was connected with electrical wires to electrodes of the piezoelectric, varied to obtain the power output, P, and the efficiency curve. This curve satisfies approximately the relation known from the linear system, called impedance matching [4, 5], which in this case, would correspond to the maximum efficiency (see experimental point (b) in Figure 2, $R = 46.2$ kΩ). More generally, the optimum condition is realized at the point of equal mechanical and electrical damping. In this case, the nonlinear mechanical damping is realized by impacts and material deformations.

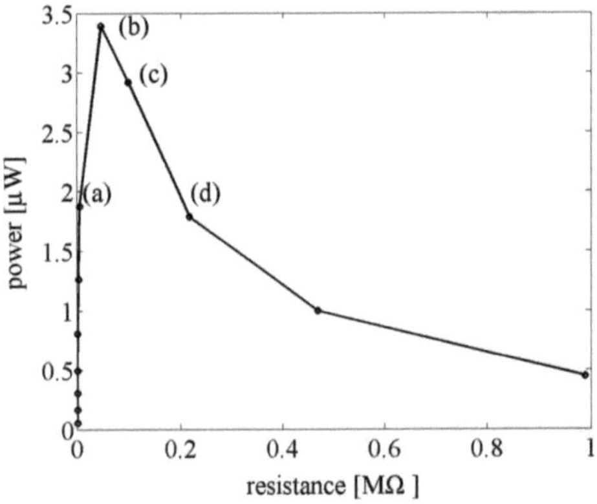

Fig. 2: Measured electric power output, P, versus load resistance, R. Cases (a)–(d) correspond to $R = 4610, 46{,}200, 98{,}600, 218{,}000$ Ω, respectively.

The detailed studies for the four chosen cases (a)–(d) are presented in Figure 3. The particular solutions and their evolution with a resistance load, R, in the transition through the maximum power output were obtained using different techniques.

4 Solution identification

In Figure 3, one can see the results for the described system during fixed frequency excitation and four different resistance load values R in increasing order. The main effect is a large increase of voltage output amplitude from case (a) to case (b). Then, further small increases are observed by going through the next cases (c) and (d).

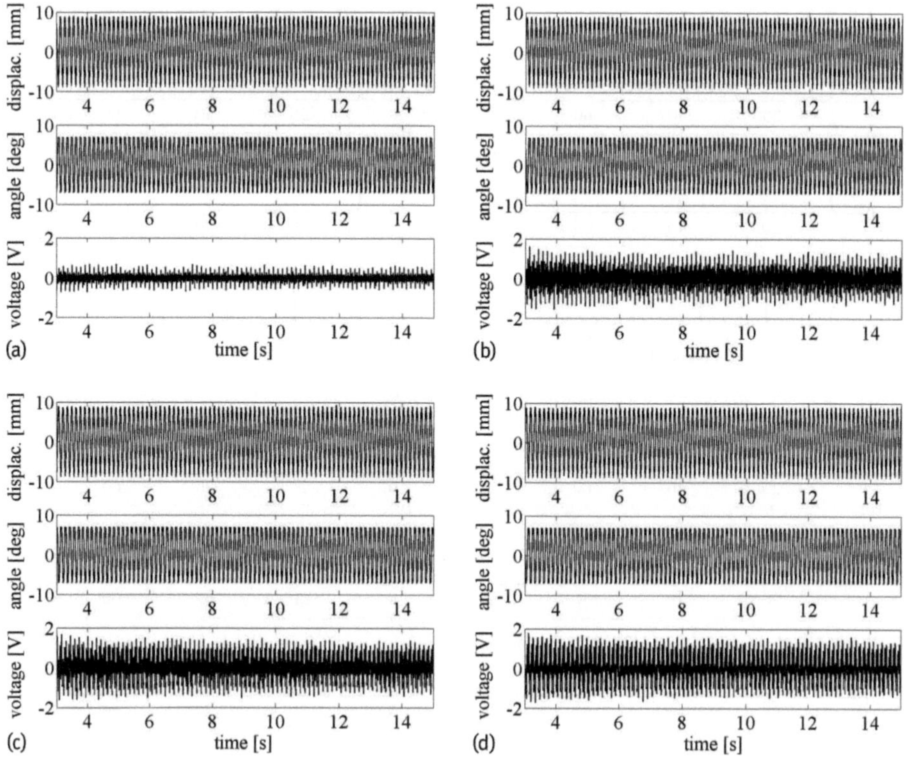

Fig. 3: Time series of the carriage excitation displacement, pendulum angle, and voltage output for the cases (a)–(d) from Figure 2, corresponding to $R = 4610, 46{,}200, 98{,}600, 218{,}000\ \Omega$, respectively.

The Fourier transform of the time series allows us to find more features of the results obtained. The results are presented in Figure 4. One can see that the carriage displacement indicates a single assumed excitation frequency of the range $f = 7$ Hz, while the beam angle has small contributions of odd higher harmonics (because of the cutoffs on the stoppers). Interestingly, the voltage Fourier spectra show not only considerable influence of odd higher harmonics but also the peak response shifted to the third harmonic component $f = 21$ Hz.

This could be related to the location of a piezoelectric patch above the impact zone (see Figure 1b). This tendency of the peak response frequency shift is stronger for small loading resistance, while for larger R contributions of the first and the third harmonics are of a similar level. On the other hand, for the case (a) of the lowest loading resistance, the fifth superharmonic is comparable to the third one. To a small extent, this is also true for the intermediate case (b).

Presumably, the higher harmonics appear as the response to impact interactions, where the position of the impact zone and the excited mode shape of the beam are important. Some more light can be shed by the voltage waveform with respect to the

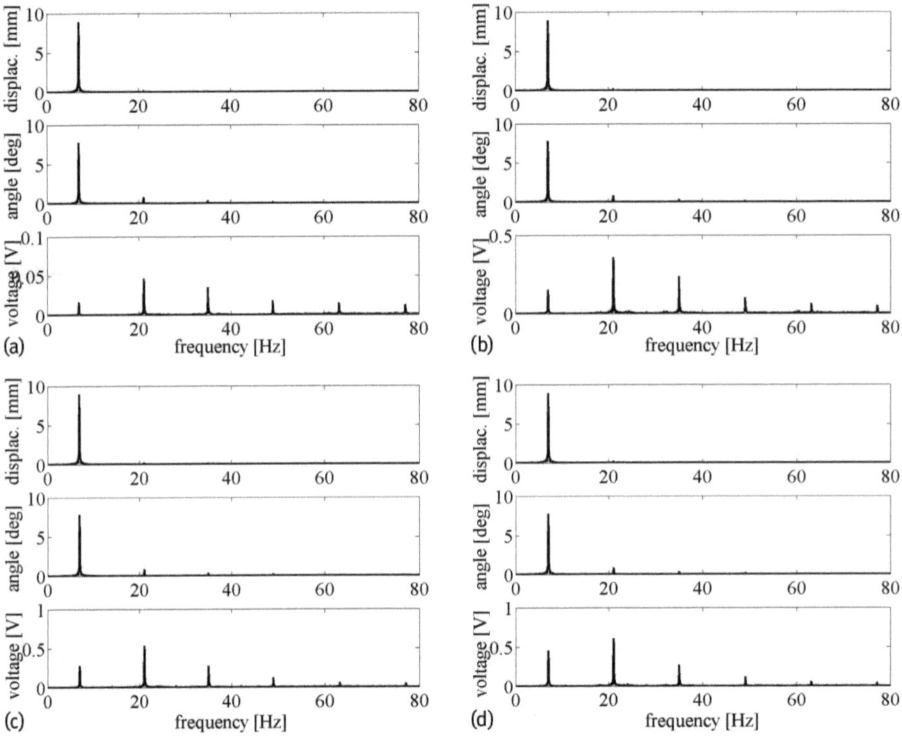

Fig. 4: Fourier transform of the carriage excitation displacement, pendulum angle and voltage output for cases (a)–(d) from Figure 2, corresponding to $R = 4610, 46{,}200, 98{,}600, 218{,}000\ \Omega$, respectively.

carriage displacement and beam hinged angle courses presented in Figure 5. One can see that the dominant period of the voltage course is three times smaller than the periods of excitation (displacement and angle) showing the third superharmonic solution. Apart from that, higher superharmonics are also visible from the Fourier spectra (Figure 4). The evolution of the characteristic third superharmonic solution is clearly visible in the phase diagrams (angle voltage) in Figure 6. One can see that the voltage oscillates in some quasi-periodic or other complex way. Fortunately, one can distinguish different patterns of responses. For instance, at the smallest R, the fluctuations (Figure 6(a)) of the voltage, V, are fairly small, and their magnitude increases only due to impact (because of bending after impact), and then it relaxes immediately.

In later cases (Figure 6(b)–(d)) more readable responses of the third superharmonic nature were observed. The better performance and visibility of the third superharmonic pattern could be also related to the growing relaxation time $\tau \sim RC$, where C is the capacity of the piezoelectric, with resistivity, R. Interestingly, in the later cases, the voltage output evolves in its response from a relatively symmetric waveform (Figure 6(c)) to a clearly asymmetric form (Figure 6(d)).

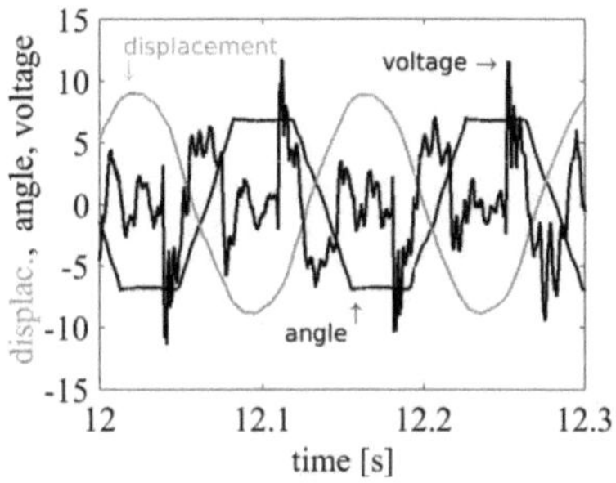

Fig. 5: Voltage output waveform with respect to the displacement carriage excitation and the beam's hinged angle courses (magnification of the case in Figure 3(b). Voltage is expressed in units [0.1 V], displacement in [mm], and the angle in [deg.], respectively.

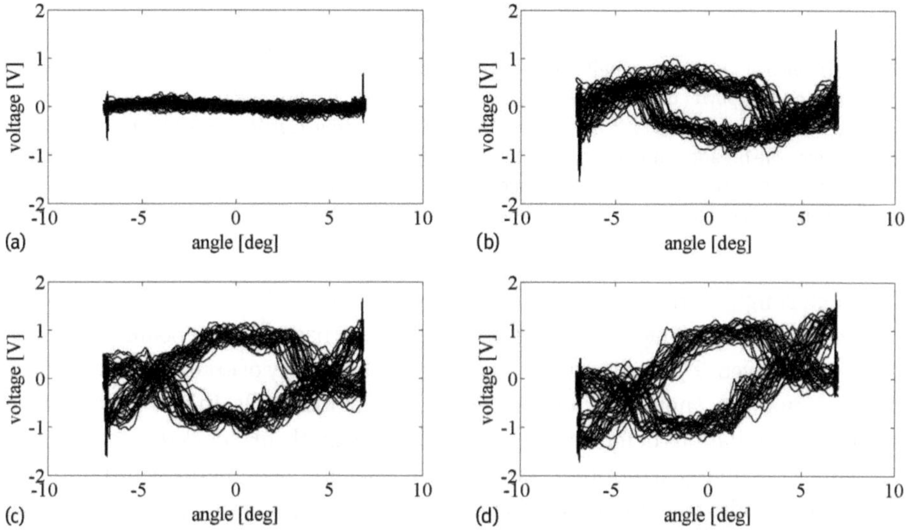

Fig. 6: Pendulum angle versus voltage-output phase diagrams for the cases (a)–(d) from Figure 2, corresponding to $R = 4610, 46{,}200, 98{,}600, 218{,}000\ \Omega$, respectively.

5 Frequency sweep: Evolution of the solution

A series of experiments, with different system parameters including beam sizes and piezoelectric types, were performed. Taking into account that the ambient frequency can vary in time, a sweep of the frequency was performed. The commercial piezoelectric No. MFC M-2814-P2, with dimensions 36×17 mm, was used. The results for the angular displacement of the beam and the voltage output are presented in Figure 7(a). Additionally, in Figure 7(b) the power output is shown. As the power output increases continuously with the rise of the excitation frequency, it is obvious that the experimental working conditions are below the resonance frequency. In the limit of the small amplitude of excitation, such a maximum would correspond to the natural frequency of a hinged beam.

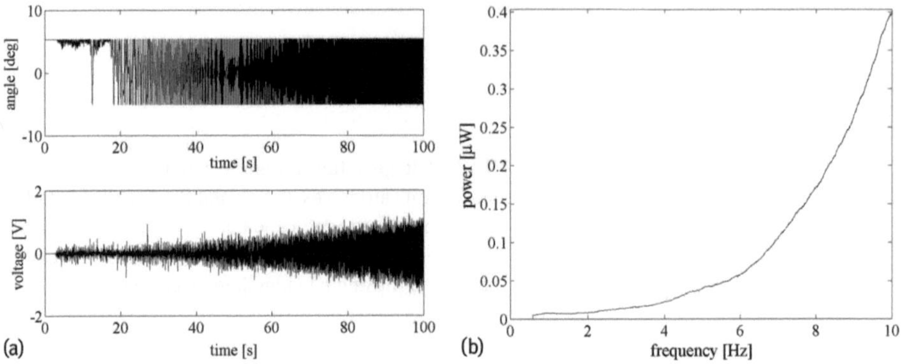

Fig. 7: (a) Beam angle (upper panel) and voltage (bottom panel) response in the frequency sweep. (b) Corresponding power output versus frequency. In this case, the excitation frequency was increased quasi-statically at a ratio of 1 Hz per 9.85 s while the frame (carriage) excitation amplitude was 8.9 mm. The breadth and thickness of the beam made of the same material were b = 17 mm and a = 4.2 mm, respectively. The load resistance was R = 1 MΩ. The tip mass M_t was the same as in the last section. The sampling frequency was intentionally reduced to f = 100 Hz.

Interestingly, during the excitation frequency increase (Figure 7(a)), single potential well oscillations (angle) and cross potential barrier passes for the beam's hinge angle were observed. This indicates that with a lower frequency of excitation, the system energy is not high enough to jump between the potential wells. In this region, some single well oscillations of the beam angle occur. For slightly higher frequency, the system jumps irregularly to the other potential well, showing a non-periodic response. Finally, the periodic response in the oscillation with a fairly large angular velocity was observed for large frequencies. Simultaneously, the extrema of the beam angle displacement are determined by the mechanical beam limiters. It should be noted that in the presented experiment the left-to-right symmetry was broken by an imperfect geometrical configuration. Presumably, the impacting limiters were not placed ideally. It is also worth noting that the voltage output (bottom panel in Figure 7(a)) is less periodic than the beam angle (upper panel in Figure 7(a)). This can be explained by the presence of higher harmonics in the beam deflection and impact phenomena. Consequently, an additional structure is visible in the voltage output. Note that a similar level of voltage fluctuations was also observed for a fixed frequency (Figures 5 and 6).

The following wavelet analysis (Figure 7) gives a better perspective of the experimental time series. The continuous wavelet transform (CWT) starts from the acceleration time series $x(t)(x$ = angle(t) or the voltage(t) [19–21]:

$$W_{s,n}(x) = \sum_{i=1}^{N} \frac{1}{s} \psi\left(\frac{i-n}{s}\right) \frac{(x(t_i) - \langle x \rangle)}{\sigma_x}, \qquad (1)$$

where $\psi(t)$ is the wavelet function, and $\langle x \rangle$ and σ_x are mean values and standard deviations of the measured value, respectively. The letters s and n denote the scale and

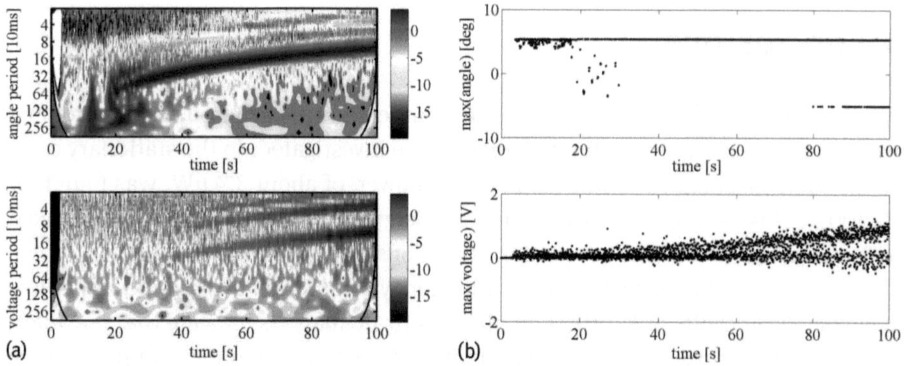

Fig. 8: (a) Angular and voltage output results of wavelet analysis for the cases from Figure 6(a). The colors red and blue represent the highest and lowest power levels, respectively, with the other colors denoting intermediate power levels in the logarithmic scale. (b) Peak values of angle and voltage, respectively.

the time indices. The square modulus of the CWT defines the wavelet power spectrum (WPS), P_W, of the corresponding time series $x(t)$:

$$P_W = |W_{s,n}|^2 \ . \tag{2}$$

The specific mother wavelet $\psi(t)$ of complex Morlet type, which was used for calculations, consists of a plane wave modulated by a Gaussian function:

$$\psi(\eta) = \pi^{-1/4} e^{i\theta_0 \eta} e^{-\eta^2/2} \ , \tag{3}$$

where θ_0 is the center frequency, also referred to as the order of the wavelet, and η is a renormalized time variable. The balancing of the time/frequency resolutions allows the parameter $\theta_0 = 6$, which defines six oscillations in the mother wavelet. An extended discussion on wavelets can be found in [19–22].

The corresponding beam angle and voltage time series were transformed by CWT and are presented in Figure 8(a) as the corresponding wavelet power spectra. One can easily distinguish the periodic and non-periodic intervals in the angular response. Furthermore, characteristic superharmonics are visible in the voltage response in analogy to the chosen excitation frequency cases studied by the Fourier transform (Figure 4). In Figure 8(b) the peak values (local maxima) of the angle and voltage are presented. For higher frequencies (later time), there is an additional lower angle branch (upper panel), which appeared because of the more complex beam deflection (Figure 8(b)). Simultaneously, the voltage output response splits.

6 Conclusions

In the experiment, the power and voltage outputs of the system based on the harmonic excitation of pendulum with limited angles were investigated. In the stationary case, for the same mechanical input, the maximal power, of about 3.4 μW, was found for 46,200 Ω. This corresponds to an optimum energy conversion efficiency for the fixed geometrical and mechanical excitation conditions.

This system has a potential application for systems with variable and rather low frequency. The results of Fourier and phase portrait analyses indicate that one interesting feature of this system is transforming the excitation frequency applied to the carriage into the third and fifth superharmonics (Figures 4 and 5). These results correspond to the choice of the elastic and damping properties of the mechanical resonator. Furthermore, they determine the maximal power output for the chosen excitation frequency and amplitude. The experiment with frequency sweep revealed more features of the system investigated. Especially, the appearance of the higher harmonics and jumps thorough the potential well are clearly visible on the wavelets in Figure 8(a), while the characteristic unperiodicity and bifurcations are presented in Figure 8(b) via maximal values of angle and voltage outputs.

It is worth mentioning that this structure with double wells and impacts has benefits from both types of nonlinear force characteristics: softening and hardening, leading to effective broadening of the transmission frequency band.

The above features can be very useful in energy harvesting applications to a moving train or a highway bridge, where the frequencies of ambient vibrational kinematic excitation are relatively small.

Acknowledgment: GL gratefully acknowledges the support of the Polish National Science Center under Grant No. 2012/05/B/ST8/00080.

Bibliography

[1] P. D. Mitcheson, E. M. Yeatman, G. K. Rao, A. S. Holmes, and T. C. Green. Energy harvesting from human and machine motion for wireless electronic devices. *Proceedings of the IEEE*, 96:1457–1486, 2008.

[2] G. Litak and E. Manoach. Dynamics of composite nonlinear systems and materials for engineering applications and energy harvesting–The role of nonlinear dynamics and complexity in new developments. *European Physical Journal Special Topics*, 222:1479–1482, 2013.

[3] G. Litak and E. H. E. Manoach. Nonlinear and multiscale dynamics of smart materials in energy harvesting. *European Physical Journal – Special Topics*, 224:2671–2673, 2015.

[4] S. P. Beeby, M. J. Tudor, and N. M. White. Energy harvesting vibration sources for microsystems applications. *Measurement Science and Technology*, 17:R175–R195, 2006.

[5] A. Koszewnik, P. Grzes, and W. Walendziuk. Mechanical and electrical impedance matching in a piezoelectric beam for energy harvesting. *European Phys. J-Spec. Topics*, 224:2719–2731, 2015.

[6] R. L. Harne and K. W. Wang. A review of the recent research on vibration energy harvesting via bistable systems. *Smart Mat. Struct.*, 22:023001, 2013.

[7] S. P. Pellegrini, N. Tolou, M. Schenk, and J. L. Herder. Bistable vibration energy harvesters: A review. *J. Intell. Mater. Syst. Struct.*, 24:1303–1312, 2013.

[8] J. Twiefel and H. Westermann. Survey on broadband techniques for vibration energy harvesting. *J. Intell. Mater. Syst. Struct.*, 24:1291–1302, 2013.

[9] M. F. Daqaq, A. E. Masana, R., D. D. Q. A., and D. D. On the role of nonlinearities in vibratory energy harvesting: a critical review and discussion. *Applied Mechanics Reviews*, 66:040801, 2014.

[10] A. Syta, G. Litak, M. I. Friswell, and S. Adhikari. Multiple solutions and corresponding power output of a nonlinear bistable piezoelectric energy harvester. *European Physical Journal B*, 89:99, 2016.

[11] M. I. Friswell, S. F. Ali, S. Adhikari, A. W. Lees, O. Bilgen, and G. Litak. Nonlinear piezoelectric vibration energy harvesting from a vertical cantilever beam with tip mass. *Journal of Intelligent Material Systems and Structures*, 23:1505–1521, 2012.

[12] G. Litak, M. I. Friswell, and S. Adhikari. Regular and chaotic vibration in a piezoelectric energy harvester. *Meccanica*, 51:1017–1025, 2016.

[13] G. Litak, A. Rysak, M. Borowiec, M. Scheffler, and J. Gier. Vertical beam modal response in a broadband energy harvester. *Proc IMechE Part K: Journal of Multi-body Dynamics*, 230:541–552, 2016.

[14] P. Harris, C. R. Bowen, H. A. Kim, and G. Litak. Dynamics of a vibrational energy harvester with a bistable beam: voltage response identification by multiscale entropy and "0-1" test. *European Physical Journal Plus*, 131:109, 2016.

[15] A. Rysak, M. Muller, M. Borowiec, J. Zubrzycki, G. Litak, A. Godlewska-Lech, and V. Wittstock. Broadband concept of energy harvesting in beam vibrating systems for powering sensors. *Advances in Science and Technology Research Journal*, 8:62–67, 2014.

[16] M. Borowiec, G. Litak, and S. Lenci. Noise effected energy harvesting in a beam with stopper. *International Journal of Structural Stability and Dynamics*, 14:1440020, 2014.

[17] S. Stoykov, E. Manoach, and G. Litak. Vibration energy harvesting by a Timoshenko beam model and piezoelectric transducer. *European Physical Journal – Special Topics*, 224:2755–2770, 2015.

[18] C.-B. Lan and W.-Y. Qin. Vibration energy harvesting from a piezoelectric bistable system with two symmetric stops. *Acta Physica Sinica*, 64:210501, 2015.

[19] P. Kumar and E. Foufoula-Georgiou. Wavelet analysis for geophysical applications. *Rev. Geophys.*, 35:385–412, 1997.

[20] C. Torrence and G. P. Compo. A practical guide to wavelet analysis. *Bull. Amer. Meteor. Soc.*, 79:61–78, 1998.

[21] G. Koszalka, J. Hunicz, A. Rysak, and G. Litak. Nonlinear vibration of semi-trailer suspension: wavelet and multiscale entropy-based approaches. *Archive of Applied Mechanics*, 86:761–768, 2016.

[22] P. Lonkwic, K. Lygas, P. Wolszczak, S. Molski, and G. Litak. Braking deceleration variability of progressive safety gears using statistical and wavelet analyses. *Measurment*, 110:90–97, 2017.

Sonia Bradai, Slim Naifar, Riadh Elleuch, and Olfa Kanoun

On hybridization of electromagnetic vibration converters

Abstract: Nowadays, harvesting energy from vibration is more and more challenging in terms of the maximum energy density that can be reached through the converter. The maximum energy density depends strongly on the applied excitation properties (frequency, acceleration), the principles used to convert the mechanical energy to electrical energy and the system design itself.

In this chapter, we focus on possible vibration converters dedicated to low vibration frequencies. Electromagnetic and magnetoelectric energy harvesting techniques deliver energy with relatively high efficiencies and are, therefore, the most promising principles, compared to piezoelectric, electrostatic and magnetostrictive principles. Furthermore, opting for hybrid solutions enables us to improve the limited energy outcome of a single converter. In this chapter, hybrid vibration converters based on the combination with the electromagnetic principle in macro scale are reviewed and compared.

Keywords: Vibration converter, energy harvesting, electromagnetic converter, magnetoelectric converter, low frequency vibration sources, magnetostrictive materials, composite, piezoelectric

1 Introduction

Powering systems from an ambient vibration source is challenging and interesting due to the availability as a source in several environments; indoor and outdoor as well as in different applications e.g., trains, cars, bridges, machines and also its energy density level, which is higher compared to other ambient sources. This mechanical energy is converted to an electrical energy using different principles, such as electrostatic, electromagnetic, magnetoelectric and piezoelectric principles. Improving the power density and the working frequency bandwidth of such devices can be realized only by the appropriate selection of the principle and design for the converter, taking into account the available vibration profiles.

Based on the state of the art, the use of classical electromagnetic, piezoelectric or magnetoelectric principles is not enough to reach the needed energy. Therefore, different approaches aiming to improve the energy density have been developed. Some of this research focuses on the improvement of the electronics for a better converter performance [1].

Sonia Bradai, Slim Naifar, Riadh Elleuch, National School of Engineers of Sfax, Tunisia
Sonia Bradai, Slim Naifar, Olfa Kanoun, Technische Universität Chemnitz, Germany

https://doi.org/10.1515/9783110445053-008

Another perspective is the improvement of the mechanical conversion to electrical energy. In [2], harvesters based on the same principle are connected together to increase their frequency bandwidth. This method is mostly used for the case of piezoelectric converters using multi cantilever beams with different sizes and tip masses to ensure a higher energy outcome and a converter working for a frequency bandwidth [3, 4]. The design of these systems is complex, and these converters are limited by a short lifetime because of the aging of the piezoelectric materials.

Another method is proposed to overcome some limitations of the use of multi-harvesters, which consists of the design of a hybrid converter using the same vibration source and using two principles to convert this mechanical energy to electrical energy. In this case, the proposed solutions differ relative to the selected combined principles and the design of the converter itself in order to combine both principles.

This chapter is organized in two main sections. First, the state of the art and a comparison of possible hybrid solutions using different principles is presented. Further, the modeling and design of a hybrid solution based on electromagnetic and magnetoelectric principles is developed.

2 Hybrid converters

Different combinations to harvest energy from vibration are being investigated in research. Opting for hybrid solutions aims to improve the performance of the converter according to the application needs. The main interest to combine two principles can be summarized as the increase in energy outcome, the increase in energy density, the decrease of the converter resonant working frequency, the improvement of the frequency bandwidth, ensuring better robustness and a long lifetime for the system. In the following, the main developed solutions are presented and compared.

Combined systems for vibration converters can be classified mainly into three types, which consist of the combination of piezoelectric and electromagnetic (EM) principles, EM combined with the magnetoelectric (ME) principle and EM combined with the magnetostrictive (MS) principle. Only converters excited along one direction for a frequency lower than 150 Hz are considered throughout this review.

2.1 Electromagnetic-piezoelectric vibration converter

The classical structure for a hybrid electromagnetic-piezoelectric is based on the use of a cantilever beam where the piezoelectric layer is fixed and at its end a magnetic sys-

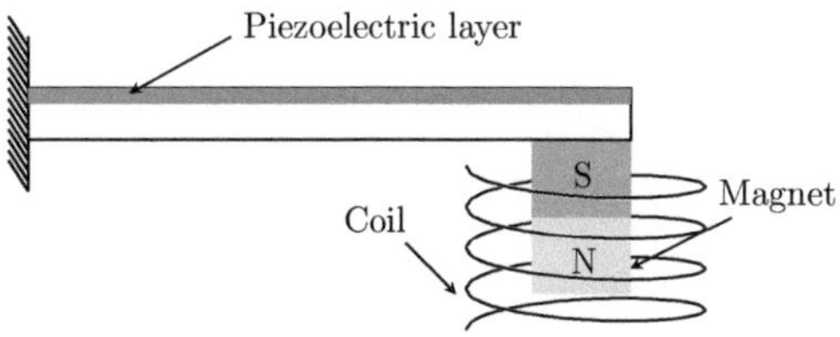

Fig. 1: General architecture of a hybrid electromagnetic-piezoelectric-based cantilever vibration converter.

tem ensuring the EM output is implemented (Figure 1). Mainly, the magnetic system is used as a tip mass to reduce the working resonant frequency, and different architectures for this part were investigated.

Various researchers investigated this design in order to reduce the resonant frequency of the piezoelectric vibration converter or to improve the frequency bandwidth of the system. The difference in the systems developed based on this structure is mainly the used cantilever beam material and the magnetic system architecture.

Based on the classical design for an EM–piezoelectric converter, [5] developed one based on the use of a piezoelectric layer placed along the cantilever beam and a magnet fixed at its end on the bottom, which is moving due to an applied excitation oscillating through a fixed coil (Figure 2). The converter is able to harvest 2.26 mW using a coil resistance of 1.5 Ω for an applied excitation of 23.3 Hz and an acceleration of 0.4 *g*.

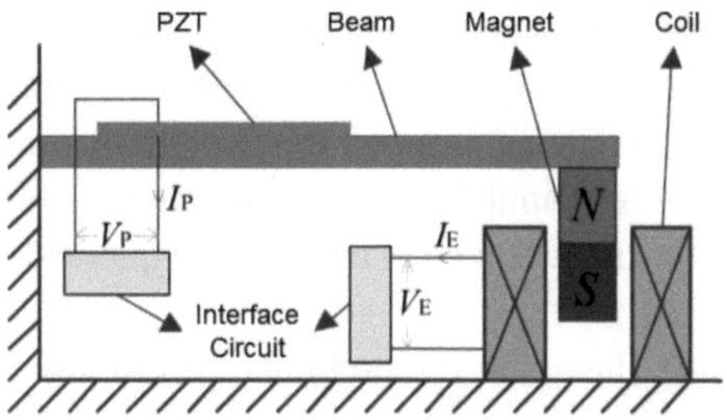

Fig. 2: The EM-piezoelectric converter-based cantilever beam proposed by [5].

Optimization of the magnetic system architecture has investigated by different researcher. In [6], an outer fixed magnet is placed in parallel to the magnet used at the end of the cantilever (Figure 3). In this case, the converter generates 0.127 mW/cm^3 for an applied excitation of 1 *g* and 49 Hz as acceleration and frequency, respectively. The authors proved that the proposed design enables an improvement of the working frequency bandwidth by 60% compared to a single linear EM or a single piezoelectric energy harvester due to the added magnet effect.

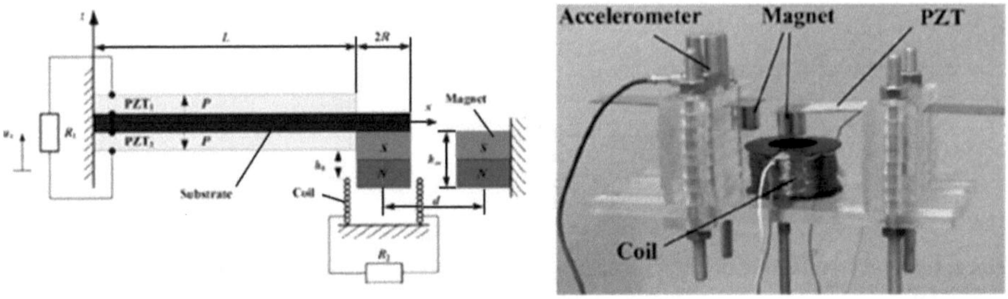

Fig. 3: Combined EM-piezoelectric-based cantilever beam with improved frequency bandwidth proposed by [6].

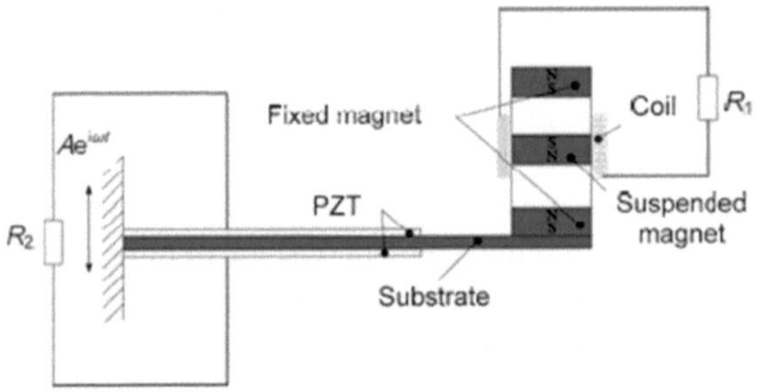

Fig. 4: Combined EM-piezoelectric bandwidth vibration converter-based cantilever beam developed by [7].

Another architecture based on the cantilever beam is commonly used and aims to widen the working frequency for the converter. It was developed by [7] and consists of placing not only one magnet at the end as a mass but a different magnetic system, which is based on the use of three magnets, one moving between the two others due to the presence of an excitation (Figure 4). For an applied acceleration of 0.5 *g*, the converter generates 8.46 mW and 19.9 mW for the two typical resonant frequencies 9 Hz and 16 Hz, respectively.

[8] proposed a nonlinear converter using the same EM architecture proposed by [7], which in this case, is placed in the middle of two clamped beams (Figure 5). This causes a decrease of the system's equivalent stiffness, which, in turn, leads to a decrease of the resonant frequency and an increase of the magnets displacement, hence, the applied stress to the piezoelectric layer; 0.44 mW is generated at a resonant frequency of 113.5 Hz, an applied acceleration of 0.2 *g* for a load resistance of 140 kΩ and 15.5 Ω for PE and the EM, respectively.

As mentioned, previously presented electromagnetic-piezoelectric converters are mainly based on the classical cantilever beam with different investigations. Another structure based on the pendulum principle was investigated by [9] in order to combine the EM and piezoelectric principles.

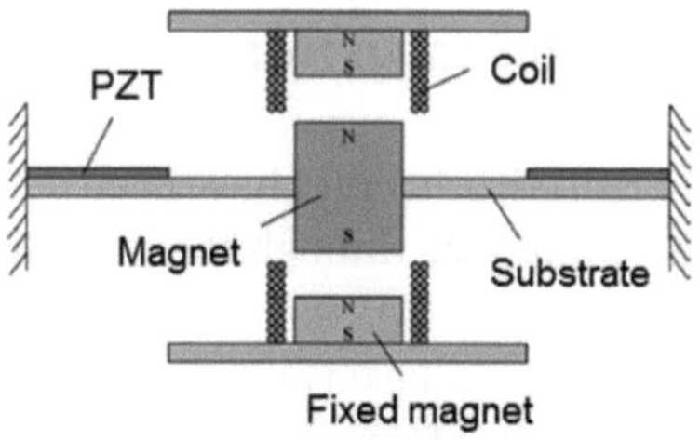

Fig. 5: The EM-piezoelectric-based cantilever beam with decreased equivalent stiffness for the converter proposed by [8].

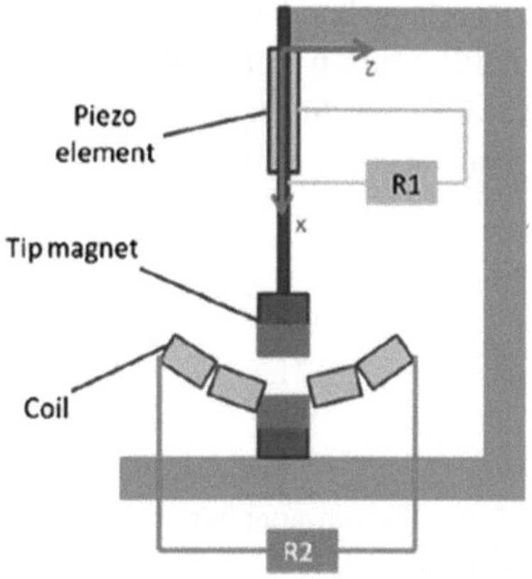

Fig. 6: The EM-piezoelectric-based pendulum developed by [9].

The pendulum is formed with an oscillating cantilever beam and a magnet (Figure 6). A second magnet is fixed on the bottom, which leads to the nonlinearity of the system and enabling one to increase the magnetic flux density passing through the coil. The system generates a power output in the range of 10^{-3} W and 10^{-5} W for the piezoelectric and EM generator, respectively, for an applied frequency in the range of 10 to 13 Hz [9]. Few investigations for such design are conducted because of the needed volume and the complexity of the implementation of such a device.

Recently, a different architecture was proposed combining the two principles. It consists of a spiral piezoelectric where a magnet is placed at its center (Figure 7). The coil is fixed surrounding the magnet. For a volume of $1.66\,\text{cm}^3$, $52\,\mu\text{W}$ is generated for an applied excitation of 21 Hz and 0.3 *g* of frequency and acceleration, respectively [10].

Fig. 7: The EM-piezoelectric vibration converter-based spiral piezoelectric geometry developed by [10].

2.2 Electromagnetic-magnetoelectric vibration converter

Recently, novel investigations for the combination of EM/ME principles have been developed. Only two designs are proposed for such combinations through the state of the art, which are presented in the following.

The first design developed by [11] is based on the cantilever beam architecture. At the cantilever's free end, an ME transducer is placed, which is formed with two magnetostrictive layers and one Lead zirconate titanate (PZT) layer bonded in between and surrounded by a fixed coil (Figure 8). The magnetic field variation is created due to the presence of a magnetic circuit placed surrounding the ME transducer. Optimization of the ME transducer and the magnetic circuit is taken into account for the converter.

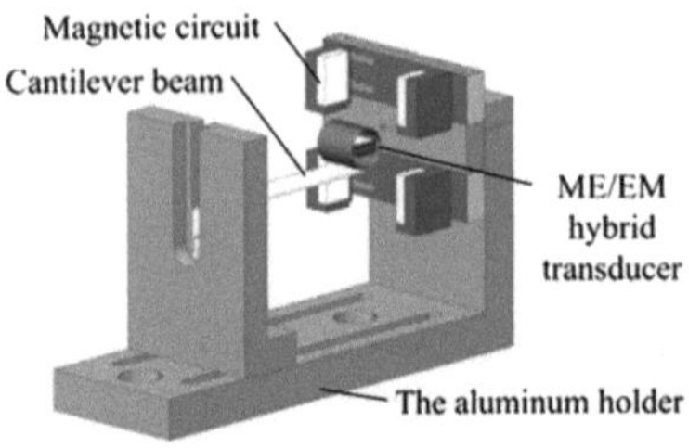

Fig. 8: The EM-ME vibration converter-based cantilever beam developed by [11].

For a resonant frequency equal to 25.7 Hz and an applied acceleration of 0.75 g, the system generates 40.84 mW within a total volume of 80 cm^3[1]. [11] confirmed a good improvement of the power output compared to a single EM or ME, but no quantitative comparison is mentioned. In this case, it is possible to tune the system resonant frequency through the coil and cantilever characteristics' change (size, mass, material,...).

The second architecture for an EM-ME principle combination was proposed by [12]. The system is formed mainly with a magnetic circuit, a magnetoelectric transducer and a coil. In this case, applied excitation is transferred to the converter using the magnetic spring principle (Figure 9). The authors proposed a compact system with a size of 20 cm^3.

In this case, the converter works for a frequency bandwidth due to the nonlinearity of the magnetic spring. The magnetic circuit, its position and the effect of combining both principles are investigated in this work. An improvement relative to a single EM converter is proved, but no details about energy outcome are presented.

1 The value is estimated based on the data mentioned in [11].

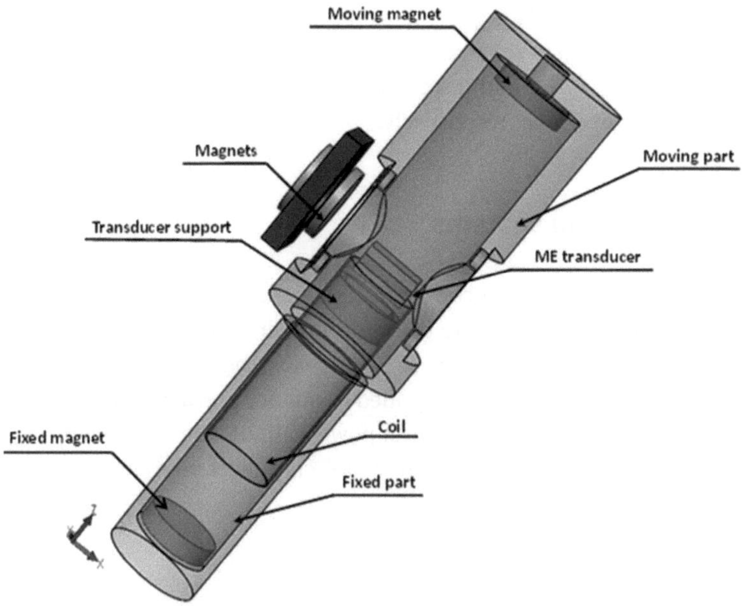

Fig. 9: The EM-ME vibration converter-based magnetic spring principle proposed by [12].

2.3 Electromagnetic-magnetostrictive vibration converter

Recent research has been devoted to the combination of MS and EM principles. [13] proposed one with the size of an AA battery (11 cm^3), oriented to railways, bridges and other applications characterized by an excitation frequency less than 20 Hz. The architecture is based on a moving magnet through a cylindrical tube surrounded by a fixed coil. For the MS principles, it is ensured by the placement of two rods on the top and bottom of the cylindrical tube, which are also surrounded by a coil (Figure 10). Due to the Villari effect, an energy outcome is generated through the coil in this case.

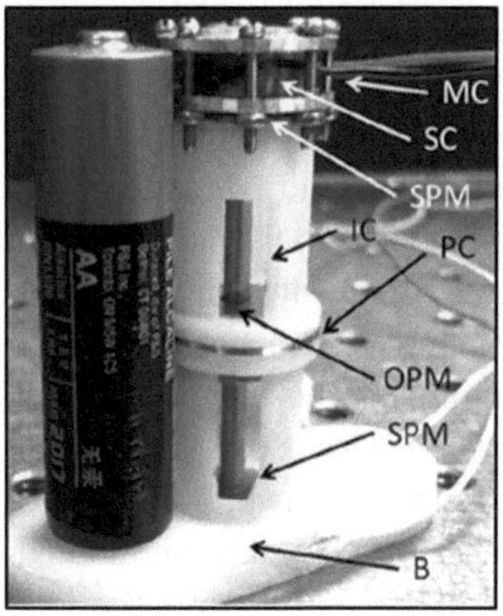

Fig. 10: The EM-MS vibration converter-based oscillating magnet proposed by [13].

For the MS principle, 0.0055 μW is generated for an excitation frequency equal to 14 Hz. For the EM principle, for an excitation frequency of 12.5 Hz, 11.5 Hz and 10 Hz and an acceleration of 0.9 *g*, 0.7 g and 0.4 g, an energy outcome equal to 5.3 mW, 2.57 mW and 0.27 mW is generated, respectively.

2.4 Comparison of hybrid systems

Different solutions for hybrid vibration converters are presented. In Table 1, the reviewed prototypes are compared relative to the important criterion for any vibration converter, which can be summarized on the used principles, the architecture, the energy outcome of the converter relative to the applied frequency and acceleration. As is shown, among all devices presented, the highest power output was achieved by [11], which is based on the EM-ME combined system-based cantilever beam as presented in Figure 9.

Tab. 1: Reviewed hybrid vibration converters.

Ref	Type	Architecture	Frequency (Hz)	Acceleration (g)	Energy outcome (mW)
[5]	EM/piezo	Cantilever	23.3	0.4	2.26
[6]	EM/piezo	Cantilever	49	1	1.63
[7]	EM/piezo	Cantilever	9	0.5	8.46
			16	0.5	19.9
[8]	EM/piezo	Two clamped beams	113.5	0.2	0.44
[10]	EM/piezo	Spiral piezoelectric	21	0.3	$52 \cdot 10^{-3}$
[11]	EM/ME	Cantilever	25.7	0.75	40.84
[13]	EM/MS	Moving magnet	12.5	0.9	5.3
			11.5	0.7	2.57
			10	0.4	0.27

To sum up, the main advantages for the EM-piezoelectric combination are principally the following: a decrease of the working frequency of the piezoelectric converter, an increase of the energy outcome of a single EM converter due to the possibility of a higher velocity for the system, which means higher magnet movement, and to broadband the working frequency range of a piezoelectric converter as well as the EM one. The main limitations of such EM-piezoelectric hybrid systems is the complexity to implement it, because of its size and the use of the cantilever beam architecture in most cases.

The EM-ME combination has the advantages of improving the energy density of the system, in this case compared to a single EM converter. This is due to the possibility of profiting form the effect of the magnetic field in both principles. In fact, for the same size, the ME principle can be implemented to a single EM converter through the placement of the ME transducer within the coil, which is the case for both existing EM-ME converters in the state of the art [11, 12]. In this case, the energy outcome can be generated for the same excitation and using twice the magnetic field variation without significantly increasing the system complexity. Such systems are more flexible for improving the working frequency bandwidth and have a better reliability.

For the EM-MS combination, through the MS principle, only a few microwatts are generated, which can be considered as a limited energy outcome. In this case, compared to a single EM, the converter improvement is limited to only the increase of the power output in the range of microwatts.

More important for the evaluation of vibration converters is to compare the energy density of the converters relative to the applied excitation. Figure 11 presents the energy density reached by the presented hybrid systems relative to the applied excitation, which, in this case, is limited to a working frequency and an acceleration less than 50 Hz and 1 *g*, respectively.

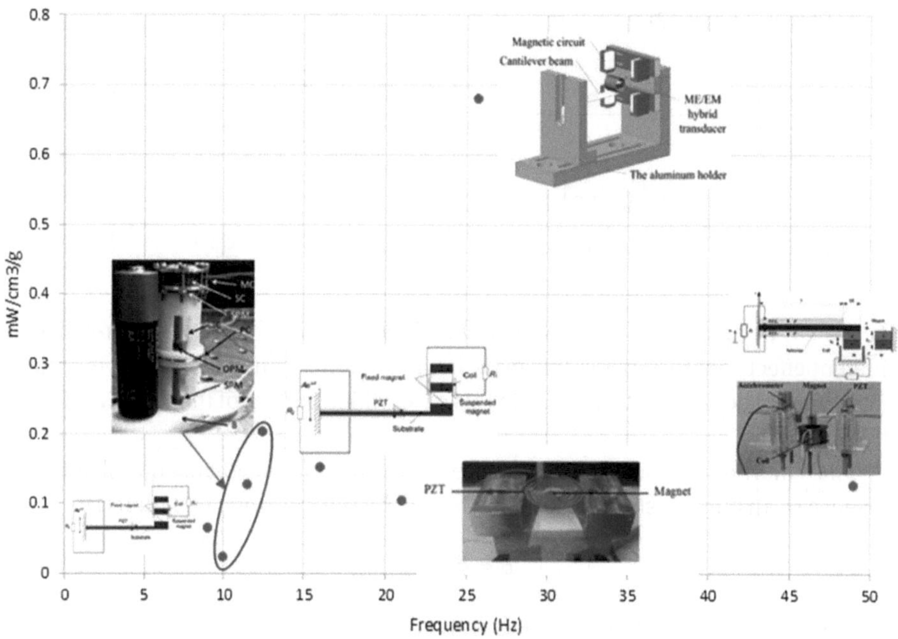

Fig. 11: Reviewed hybrid vibration converters for acceleration and frequency limited to 1 *g* and 50 Hz, respectively.

As has been shown, the energy outcome for EM-piezoelectric converters and EM-MS is limited to 0.2 mW/cm^3/g as the maximum. For the case of the EM-ME converter, it is 0.68 mW/cm^3/g, which can be considered as a relevant energy density level compared to other hybrid or single converters. The main reason leading to a higher energy density for the combination of the EM and ME principles compared to other hybrid systems is that for both principles, the energy outcome depends on the magnetic flux density variation, which is not the case of other combinations of principles. Based on the presented comparison, in the following, we propose focusing on the EM-ME combined converter design and properties.

3 Hybrid magnetoelectric-electromagnetic converters

For the EM-ME converter, due to the presence of magnetic field variation, electrical energy is generated through the coil due to the magnetic flux density passing through it, based on the principle of Faraday's law. Also, through the piezoelectric layer, a stress is applied due to the deformation of the MS layers, and energy is generated through the PZT layer (Figure 12).

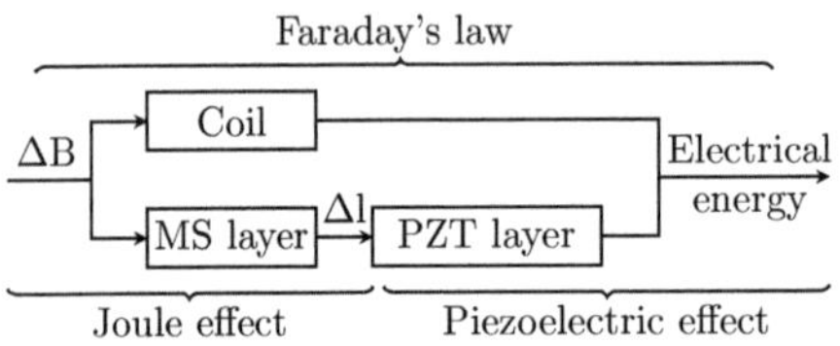

Fig. 12: EM and ME principles.

For the ME principle, MS material is used, which has a specific behavior relative to different effects characterizing this material. For a better converter design, two main effects should be taken into account in this case, which consist of the Villari and Joule effects.

The Villari effect consists of a change of the material magnetization due to the presence of an applied stress. It is characterized by the following equation (1):

$$B = d_\sigma + \mu^\sigma \tag{1}$$

where B, d, σ and μ^σ are the magnetic induction, the magnetostrictive constant, the stress change and the permeability at constant mechanical stress.

The Joule effect is the inverse of the Villari effect. In this case, due to the presence of a magnetic field, a change of the ferromagnetic material will occur. This change presents the magnetostriction of the material. It can be positive or negative relative to the material used and its behavior, and in both cases the sample volume remains constant. The main difference is that in the case of negative magnetostriction, the sample length decreases, and for positive magnetostriction, the sample thickness decreases.

The Joule effect is defined using the following equation (2):

$$S = c^H + d_\sigma H \tag{2}$$

where S, c^H and d_σ are the mechanical strain, the compliance coefficient at constant field strength H and the magnetostrictive constant at constant stress.

3.1 Design of an EM/ME vibration converter

Different parameters should be taken into account for the optimal design for an EM-ME converter. In this case, two differently generated outputs can be calculated.

First, the energy generated from the EM principle, which can be determined relative to Faraday's law by equation (3):

$$V_{\mathrm{EM}} = -N\frac{\mathrm{d}\Phi}{\mathrm{d}t} = -N\frac{\mathrm{d}\Phi}{\mathrm{d}z}\frac{\mathrm{d}z}{\mathrm{d}t} \tag{3}$$

where N, Φ and z are the number of coil windings, the magnetic field through the coil and the displacement of the magnet.

In this case, the output reached depends principally on the magnetic field variation passing through the coil and also the velocity of the moving part, which can be affected by the mass of the moving part and the excitation applied to the system.

For the energy generated through the ME principle, it is defined by equation (4):

$$V_{\mathrm{ME}} = \alpha_{\mathrm{v}}\Delta H \tag{4}$$

where ΔH is the variation of the magnetic field through the MS layers; α_{v} is the ME voltage coefficient and is defined by equation (5) [14]:

$$\alpha_{\mathrm{v}} = \frac{n(1-n)t_{\mathrm{c}}d_{33,\mathrm{m}}d_{31,\mathrm{p}}}{\epsilon_{33}\left[n(1-k_{31}^2)s_{11}^E + (1-n)s_{33}^H\right]} \tag{5}$$

where n, t_{c}, $d_{33,\mathrm{m}}$, $d_{31,\mathrm{p}}$, ϵ_{33}, s_{11}^E, k_{31}, and s_{33}^H are the thickness ratio of the composite, the total thickness of the composite, the piezomagnetic constant, the piezoelectric coefficient, the permittivity tensor, the piezoelectric elastic compliance at constant D, the piezoelectric electromechanical coupling coefficient and the longitudinal piezomagnetic elastic compliance at constant H, respectively.

3.2 Effect of combined EM-ME on the energy outcome

In this section, the effect of the combination of the EM-ME principles is studied by finite element analysis for a proposed combined model. The model is composed of a magnetic circuit (MC) presented by MC magnets (Figure 13). Between them, the

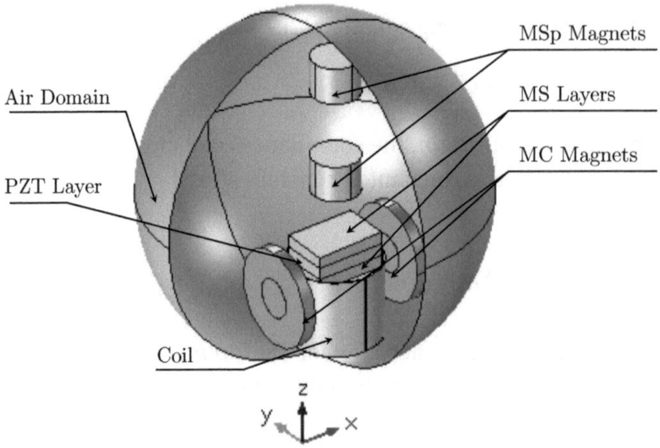

Fig. 13: Defined finite element for the combined converter.

ME transducer made of one PZT layer bonded by two MS layers is placed. To ensure the EM principle, a coil between the MC and the ME transducer is added. The magnetic spring (MSp) magnets as seen in Figure 14 are used to ensure the converter operation. An air domain enclosing all the components is added in order to reduce the time calculation.

In this case, NdFeB N42, copper and PZT-5H are defined from the data software for the magnets, coil and piezoelectric layer, respectively. For the MS material, Terfenol-D is used, and it is defined relative to its data sheet [15], since it is not available thorough the FEM software used. Based on the Villari effect, the MS layers are able to generate a magnetic flux density due to an applied stress. Therefore, it is very important to investigate the position of the coil, the magnetic circuit and the ME transducer relative to each other for an optimal converter output. In the following, the magnetic field variation relative to different parameters for a single and a combined system is evaluated. Figure 14 presents a comparison of the magnetic field level through the coil for single and combined EM.

Through this comparison, the same properties are maintained for the coil. As is shown, an increase of 125 Oe for the magnetic field is reached due to the presence of the ME transducer through the coil. This is due to the fact that once the MS layers are strained, a magnetization change occurs through the MS layers based on the Villari effect. This improvement will positively affect the voltage generated through the coil. The magnetic field level through the MS layers is improved in this case, compared to a single ME converter. This increase is not relevant, but it is proved that the presence of the coil does not negatively affect the ME transducer behavior.

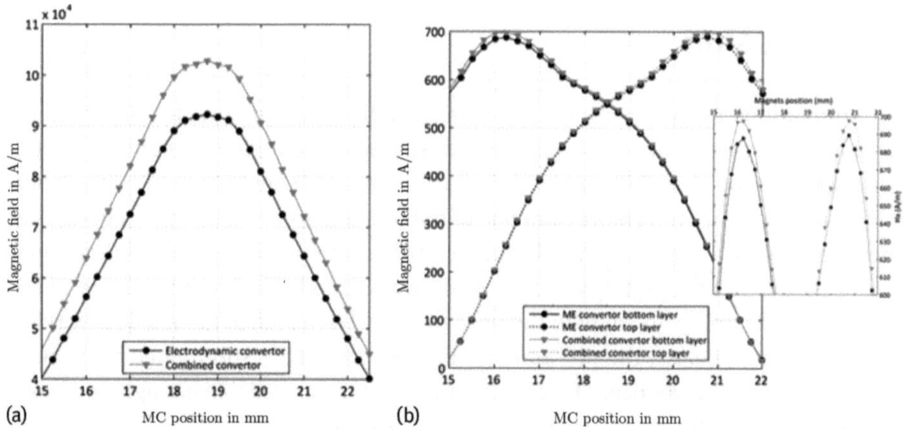

Fig. 14: Magnetic field level through (a) the coil for a single EM and combined EM-ME converter, (b) MS layers for a single ME and combined an EM-ME converter.

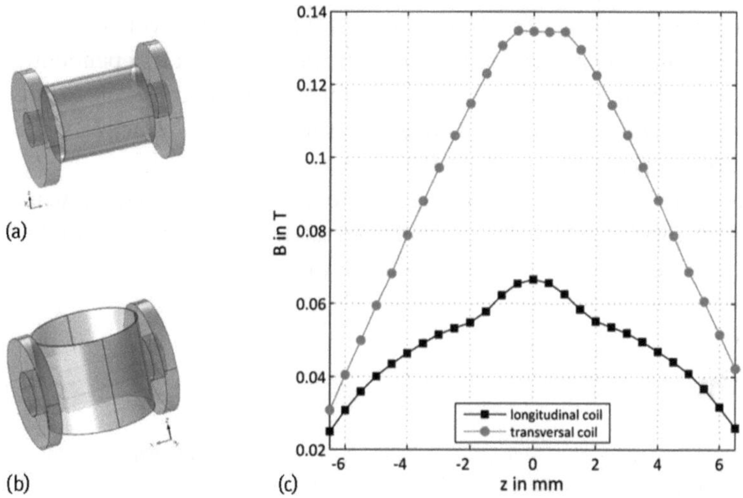

Fig. 15: Magnetic flux density for the longitudinal and transversal coil relative to its position along the *z*-direction.

In the following, the effect of the position of the coil direction relative to the ME transducer is studied. A comparison of transversal and longitudinal positions as shown in Figure 15 is investigated. The magnetic flux density level is increased from 0.065 T to 0.135 T in the case of the transversal coil.

4 Conclusion

Hybrid solutions are promising and have many challenges to improve the performance of vibration converter in terms of energy density, working frequency bandwidth, the reliability and the size of the converter. In this review, a comparison of hybrid solutions based on combining more than one of the principles used to convert mechanical energy to electrical energy using the same vibration source is conducted. Combination of the PE and EM principles is the most investigated solution in the state of the art due to its flexibility to adapt to the needed energy outcome or the needed working frequency. Nevertheless, the performance of these systems is still limited because of the use of the cantilever beam for the design, which is not reliable enough and has a limited lifetime in the case of powering wireless sensors.

For EM-ME converters, the positive effect of the ME transducer on the energy generated through the EM principle, which is not the case for the piezoelectric/EM converter, makes this solution more meaningful. Also, in this case, the lifetime of the converter is improved; the effect that one of the principles is not working will not lead to a defect of the complete system. Further, the two principles can work independently, as well as affecting each other positively. This is not the case for piezoelectric EM devices, where once the cantilever beam has been destroyed, neither principle works any longer, and no energy can be generated in this case.

To conclude, a combination of each of the principles together leads to some advantages and disadvantages, which can be optimized relative to the existing applied excitation, the available volume and the needed energy outcome.

Bibliography

[1] D. Zhu, S. Roberts, T. Mouille, M. J. Tudor, and S. P. Beeby. General model with experimental validation of electrical resonant frequency tuning of electromagnetic vibration energy harvesters. *J. Smart Materials and Structures*, 21(10):105039, 2012.

[2] Z. Dibin, J. Michael, and P. Stephen. Strategies for increasing the operating frequency range of vibration energy harvesters: a review. *Measurement Science and Technology*, 21(2):022001, 2010.

[3] H. Xue, Y. Hu, and Q. Wang. Broadband piezoelectric energy harvesting devices using multiple bimorphs with different operating frequencies. *IEEE Transactions on Ultrasonics, Ferroelectrics, and Frequency Control*, 55(9):2104–2108, 2008.

[4] S. M. Shahruz. Design of mechanical band-pass filters for energy scavenging. *J. of Sounds and Vibration*, 292(35):987–998, 2006.

[5] H. Xia, R. Chen, and L. Ren. Analysis of piezoelectric–electromagnetic hybrid vibration energyharvester under different electrical boundary conditions. *Sensors and Actuators A: Physical Journal*, 234:87–98, 2015.

[6] Z. L. Xu, X. B. Shan, R. J. Song, and T. Xie. Electromechanical modeling and experimental verification of nonlinear hybrid vibration energy harvester. In *2014 Joint IEEE International Sym-*

posium on Applications of Ferroelectrics, International Workshop on Acoustic Transduction Materials and Devices & Workshop on Piezoresponse Force Microscopy, 2014.

[7] X. Shan, S. Guan, Z. Liu, Z. Xu, and T. Xie. A new energy harvester using a piezoelectric and suspension electromagnetic mechanism. *Journal of Zhejiang University-SCIENCE A (Applied Physics & Engineering)*, 14(12):890–897, 2013.

[8] P. Li, S. Gao, H. Cai, and L. Wu. Theoretical analysis and experimental study for nonlinear hybrid piezoelectric and electromagnetic energy harvester. *Microsyst. Technol.*, 22:727–739, 2016.

[9] M. A. Karami and D. Inman. Equivalent damping and frequency change for linear and nonlinear hybrid vibrational energy harvesting systems. *Journal of Sounds and Vibration*, 330:5583–5597, 2011.

[10] M. Ibrahim. Design, modelling, and fabrication of a hybrid energy harvester. Master's thesis, Waterloo, Ontario, Canada, 2014.

[11] J. Qiu, Y. Wen, P. Li, and H. Chen. Design and optimization of a tunable magnetoelectric and electromagnetic hybrid vibration-based generator for wireless sensor networks. *IEEE Transactions on Magnetics*, 51(11), 2015.

[12] S. Bradai, S. Naifar, and O. Kanoun. Finite element analysis of combined magnetoelectric-electrodynamic vibration energy converter. *J. of Physics Conference Series*, 660(1):12, 111–112, 115, 2015.

[13] A. Marin. *Mechanical energy harvesting for powering distributed sensors and recharging storage systems*. PhD thesis, Blacksburg, VA, 2013.

[14] X. Dai, Y. Wena, P. Li, J. Yang, and G. Zhang. Modeling, characterization and fabrication of vibration energy harvester using Terfenol-D/PZT/Terfenol-D composite transducer. *Sensors and Actuators A: Physical*, 156:350–358, 2009.

[15] http://tdvib.com/terfenol-d/, Access online on 19.07.2017.

Andrzej Rysak, Marek Borowiec, Arkadiusz Syta, and Grzegorz Litak

Hybrid vibrational energy harvesting using piezoelectric and magnetostrictive transducers

Abstract: We study vibration energy harvesting in an electro-mechanical system with piezoelectric and magnetostrictive coil transducers. The mechanical resonator excited by harmonic excitation moves in two degrees of freedom. The output power is dissipated on a resistor load. We show the results of power output in two channels, piezoelectric and coil, for various frequencies. We observe a broadening of the total frequency transduction due to the hybrid structure of our harvester, which is sensitive to different modes.

Keywords: Vibration energy harvesting, piezoelectric, magnetostrictive

1 Introduction

Vibration energy harvesting is a fast developing branch of applications of electro-mechanical systems. The main idea is to provide powering to autonomous sensors and small portable devices [1–4]. Piezoelectrics and/or transducers are proposed because of the higher energy density. Beeby et al. [5] studied applications of piezoelectric materials for piezoelectric systems, while Yoo and Flatau [6] suggested a magnetostrictive galfenol beam, working in a bending mode high-efficiency energy harvester. The traditional approach used a single degree of freedom linear resonator, which limited the working conditions to a single frequency of ambient vibrations from the resonance region. To overcome fixed frequency working conditions, non-linear devices were proposed [7, 8]. In such cases, a number of non-linear effects were introduced, including inclination of the resonance curve, multiple solutions, sub and superharmonics. Alternative approaches were based on increasing the degrees of freedom by increasing the harvester feasibility to more frequencies [9], or hybrid systems sensitive to different coordinates such as the displacement and velocity of the mechanical resonator [10, 11].

In this chapter, we discuss a hybrid two degrees of freedom system based on the L-shaped beam. Such a beam was discussed earlier by Erturk at al. [12]. The degrees of

Andrzej Rysak, Marek Borowiec, Arkadiusz Syta, Grzegorz Litak, Faculty of Mechanical Engineering, Lublin University of Technology, Nadbystrzycka 36, PL-20-618 Lublin, Poland
Grzegorz Litak, Department of Process Control, AGH University of Science and Technology, Mickiewicza 30, PL-30-059 Krakow, Poland
Grzegorz Litak, Laboratoire Génie Electrique et Ferroélectricité, Institut National des Sciences Appliquées de Lyon, 8 rue de la Physique, 69621 Villeurbanne cedex, France

https://doi.org/10.1515/9783110445053-009

freedom are related to the L-shape, which is based on the two coupled beams with a minimum of two degrees of freedom in the lowest modes that are in-phase and out-of-phase modes with respect to the bending direction. An additional degree of freedom introduces an anti-resonance phenomenon into the power output.

2 Experimental setup and measurement results

The schematic picture of the experiment is presented in Figure 1(a). The system consists of the mechanical resonator (Figure 1(b)) with the piezoelectric and magnetostrictive transducers and electrical circuits with the resistance loads. Further parameters are shown in Table 1.

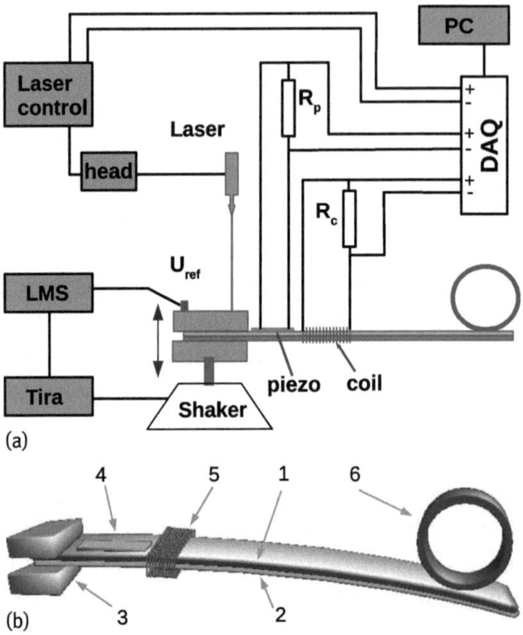

Fig. 1: (a) Schematics of the experimental setup consisting of the piezoelectric-magnetostrictive beam, electrical circuits, shaker, and excitation system with laser control and data acquisition system. (b) The resonating resonator excited vertically with an aluminium beam (1), a magnetostrictive layer (2), Galfenol – $Fe_{83}Ga_{17}$, a clamping system with the frame excited vertically (3), a piezoelectric element (4), an electrical coil with 320 turns (5) and a mass of the cylindrical shape (6) placed asymmetrically. Note that, effectively, our mechanical resonator based on glued beam and cylinder resembles an L-shaped system with two degrees of freedom. Optionally, there is an accelerometer at the end of the beam.

Tab. 1: Summary of parameters of the mechanical resonator and electrical circuit.

Description	Symbol and value
Aluminum beam:	
length	163 mm
width	20 mm
thickness	1.55 mm
Young's modulus	69 GPa
density of beam material	$\rho = 2.7\,\mathrm{g/cm^3}$
Piezoceramic layer:	
length	20 mm
width	6 mm
number of parallel fibers in the piezoceramic layer	$N = 10$
dimensions of the piezoceramic fibers (PZT)	11.5 mm × 0.24 mm × 0.26 mm
capacity of the piezoceramic layer	4.9 nF
Magnetostrictive layer (Galfenol):	
length	163 mm
width	20 mm
thickness	1.75 mm
Tip mass (steel tube):	
outer radius	20.2 mm
wall thickness	1.7 mm
height	15 mm
mass	9.7 g
Measurement:	
internal impedance of DAQ	1 GΩ
load of the piezoelectric circuit R_p	550 kΩ
load of the coil circuit R_c	48.3 Ω
acceleration amplitude	0.5 g

In the presented results, we fixed the forcing amplitude to $a = 0.5\,\mathrm{g}$ and performed sweeps of increasing frequencies. In Figure 2, we present the non-stationary time series and the corresponding input and output signals for the frame and beam acceleration and the voltage outputs and, consequently, the power outputs. Both of the voltage outputs (U_p and U_c in Figure 2) show not only the peaks associated with the resonance, at about $f_0 \approx 27.2\,\mathrm{Hz}$, but also lows identified as anti-resonance associated with the multi-degree of freedom or multi-modal characteristics of the system.

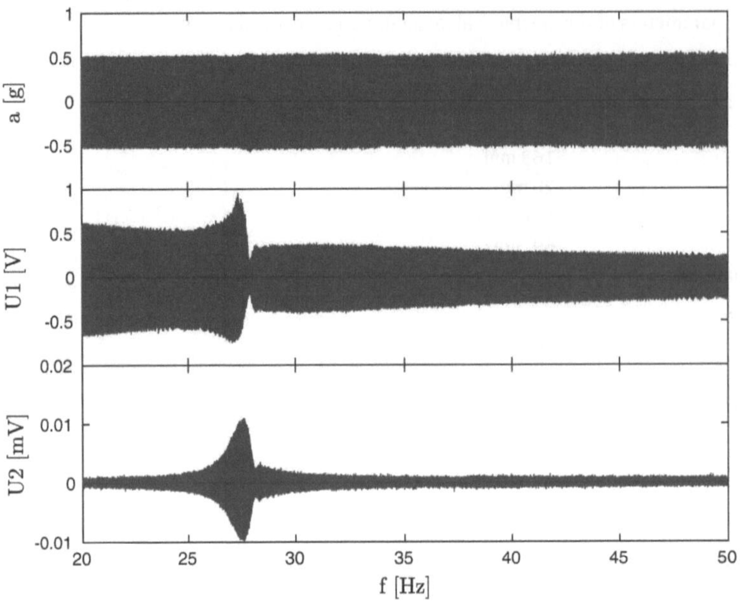

Fig. 2: Measured excitation frequency dependencies, f. Starting from the upper panel: Sweep time response of the excitation acceleration, a, then the time response of piezoelectric voltage, U_p, and the time response of the magnetostrictive coil voltage, U_c. Note that the time responses are presented in the frequency domain; there is a linear dependency for both the time and the frequency scales; $U1$ and $U2$ represent piezoelectric voltage U_p and coil voltage U_c.

Interestingly, the magnetostrictive output P_c (Figure 3(a)) dominates, showing the reflection of the beam amplitude in the resonance curve (Figure 3(b)). Its maximum value is ten times higher than the piezoelectric power output, P_p. This is partially due to the sizes of magnetostrictive and piezoelectric elements and different resistance loads R_p and R_c (see Table 1). Consequently, the anti-resonance, visible more clearly for P_p (for $f > f_0$) in Figure 3(a), is created by the phase change in the modal beam response. In the single degree of freedom system, amplitudes A_c and U_c change the sign of the response with respect to the driving acceleration. In our case, there is a more complex phase change of the cylinder (leading elongation contraction of the beam) and the beam (bending motion) relative motions, resulting in disturbance in a larger decrease in the piezoelectric power output. The smaller low on the magnetostrictive output is also visible. Note that the red curve is lowered drastically on the left-hand side of the resonance point ($f \leq f_0$). However, there is a fairly large difference between piezoelectric P_c and P_p (P_c), which is larger by one order of magnitude with respect to P_p.

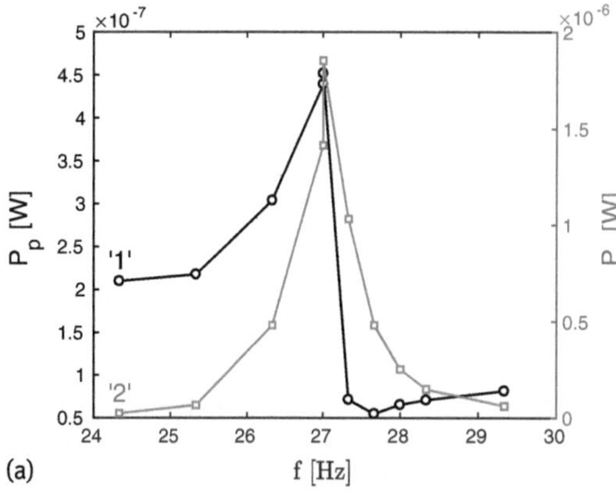

(a)

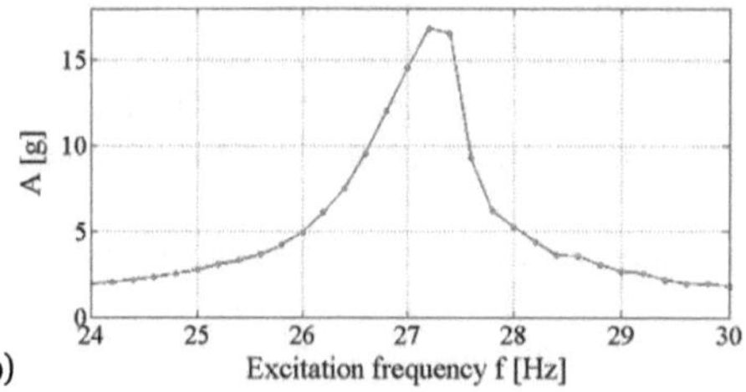

(b)

Fig. 3: (a) Power outputs versus frequency lines: '1' (black) and '2' (red) for U_p and U_c, respectively, (b) acceleration of the tip point A. The resonance frequency was estimated as $f_0 \approx 27.2$ Hz.

3 Conclusions

We have studied a hybrid piezoelectric-magnetostrictive energy harvester with two degrees of freedom. The results obtained show that the two transducers are sensitive to different modes of motion. We identified the influence of two different modes of the mechanical resonator leading to anti-resonance in the piezoelectric output. This is the main advantage of a hybrid system response where power outputs of the two approaches complete each other and increase the efficiency of energy harvesting. In our case, the magnetostrictive output was ten times higher compared to the piezoelectric one. In fact, this disturbance presumably had two sources. Firstly, the mode frequency dependence on the relative motion of beam and mass. The second, much weaker one, was related to the mechanical coupling of the piezoelectric and magnetostrictive layers. Obviously, the piezoelectric and magnetostrictive transducers can complement each other; however, to conclude that more precisely, we are preparing more systematic studies and system optimization with resistance loads. It is also worth adding that in our laboratory case study, the load factor at the end of the beam was very large at the resonance point (A = 15 g in Figure 3(b)). For any practical application it should be reduced.

Acknowledgment: A. R., A. S. and M. B. gratefully acknowledge the support of the Polish National Science Center under the grant Agreement No. DEC-2013/11/D/ST8/03308.

Bibliography

[1] P. D. Mitcheson, E. M. Yeatman, G. K. Rao, A. S. Holmes, and T. C. Green. Energy harvesting from human and machine motion for wireless electronic devices. *Proceedings of the IEEE*, 96:1457–1486, 2008.

[2] G. Litak and E. Manoach. Dynamics of composite nonlinear systems and materials for engineering applications and energy harvesting – The role of nonlinear dynamics and complexity in new developments. *European Physical Journal Special Topics*, 222:1479–1482, 2013.

[3] G. Litak, E. Manoach, and E. Halvorsen. Nonlinear and multiscale dynamics of smart materials in energy harvesting. *European Physical Journal – Special Topics*, 224:2671–2673, 2015.

[4] T. Yildirim, M. H. Ghayesh, W. Li, and G. Alici. A review on performance enhancement techniques for ambient vibration energy harvesters. *Renewable and Sustainable Energy Reviews*, 71:435–449, 2017.

[5] S. P. Beeby, M. J. Tudor, and N. M. White. Energy harvesting vibration sources for microsystems applications. *Measurement science and technology*, 17:R175–R195, 2006.

[6] J.-H. Yoo and A. B. Flatau. A bending-mode galfenol electric power harvester. *Journal of Intelligent Material Systems and Structures*, 23:647–654, 2012.

[7] M. F. Daqaq, A. E. Masana, R., D. D. Q. A., and D. D. On the role of nonlinearities in vibratory energy harvesting: a critical review and discussion. *Applied Mechanics Reviews*, 66:040801, 2014.

[8] M. I. Friswell, S. F. Ali, S. Adhikari, A. W. Lees, O. Bilgen, and G. Litak. Nonlinear piezoelectric vibration energy harvesting from a vertical cantilever beam with tip mass. *Journal of Intelligent Material Systems and Structures*, 23:1505–1521, 2012.

[9] S. M. Shahruz. Limits of performance of mechanical band-pass filters used in energy scavenging. *Journal of Sound and Vibration*, 293:449–461, 2006.

[10] V. R. Challa, M. G. Prasad, and F. Turner. A coupled piezoelectric–electromagnetic energy harvesting technique for achieving increased power output through damping matching. *Smart Mater. Struct.*, 18:095029, 2009.

[11] M. Lallart and I. D. J. Mechanical effect of combined piezoelectric and electromagnetic energy harvesting. In *Proc. IMAC (Jacksonville, USA)*, pages 261–272, 2011.

[12] A. Erturk, J. M. Renno, and D. J. Inman. Modeling of Piezoelectric Energy Harvesting from an L-shaped Beam-mass Structure with an Application to UAVs. *Journal of Intelligent Material Systems and Structures*, 20:529–544, 2009.

Part III: **Wireless energy transfer**

Elena Boshkovska, Nikola Zlatanov, Xiaoming Chen, Derrick Wing Kwan Ng, and Robert Schober

Beamforming design for secure SWIPT systems under a non-linear energy harvesting model

Abstract: Simultaneous wireless information and power transfer (SWIPT) is an appealing solution to extend the lifetime of wireless nodes and, hence, alleviate the energy bottleneck of energy-constrained wireless communication networks. SWIPT advocates the dual use of radio frequency signals for conveying information and energy concurrently, which introduces a paradigm shift in system design. This chapter focuses on the use of multiple antennas to improve the efficiency of wireless power transfer (WPT) and secure information transmission. In particular, our objective is to maximize the secrecy rate of a SWIPT system via beamforming. To this end, we formulate a non-convex optimization problem based on a practical nonlinear energy harvesting model. The problem formulation allows for the use of an energy signal to improve WPT efficiency and to provide communication security. The globally optimal solution of the design problem is obtained via a one-dimensional search and semidefinite programming (SDP) relaxation. Numerical results demonstrate that the proposed design can achieve a significant gain in secrecy rate compared to two baseline schemes.

Keywords: Beamforming, nonlinear energy harvesting, wireless information and power transfer, secure communication

1 Introduction

Wireless sensor networks serve as a key enabler of the Internet-of-Things (IoT) [1], which connects a tremendous number of sensors to computing systems to provide intelligent daily life services such as e-health, automated control, energy management (Smart City and Smart Grid), logistics, security control and safety management, etc. The European Commission has predicted that by 2020, there will be 50 to 100 billion devices connected to the Internet. In general, wireless sensors are small devices powered by batteries with limited energy storage capacity. Hence, the resulting limited lifetime of wireless sensor networks is foreseen to be a fundamental system performance bottleneck for the deployment of IoT. Conventionally, battery replacement pro-

Elena Boshkovska, Robert Schober, Friedrich-Alexander-University Erlangen-Nuremberg (FAU), Germany
Nikola Zlatanov, Monash University, Australia
Xiaoming Chen, Zhejiang University, P. R. China
Derrick Wing Kwan Ng, The University of New South Wales, Australia

https://doi.org/10.1515/9783110445053-010

vides an intermediate solution to energy depletion, which is suitable for networks with small numbers of devices. However, frequent replacement of batteries in IoT networks with massive numbers of sensors is costly and cumbersome. Besides, in some applications such as biomedical implant senors, it is almost impossible to replace sensor batteries once they are deployed. More importantly, battery replacement may create temporary service suspension, which is not possible in systems requiring stable communication services. Therefore, different promising approaches/technologies, such as energy efficiency optimization [2, 3], energy harvesting technology [4–7] and cooperative communications [8, 9] have been proposed to extend the lifetime of wireless communication networks. Energy harvesting is particularly appealing as it provides self-sustainability to wireless communication networks. Thereby, wireless communication devices are equipped with energy harvesting technology to collect energy from the environment. Solar, wind, tidal, biomass and geothermal are the major candidate renewable energy sources for generating electricity [4, 5]. Yet, these conventional natural energy sources are usually only available at specific locations, which limits the mobility of the devices. Besides, these sources are also climate dependent. Hence, the intermittent and uncontrollable nature of these natural energy sources makes it difficult to integrate conventional energy harvesting technology into wireless communication devices.

WPT has attracted much attention from both academia and industry [10–26], recently, as a building block for the IoT. The existing WPT technologies can be categorized into three classes: inductive coupling, magnetic resonant coupling and radio frequency (RF)-based WPT. In general, compared to the aforementioned conventional energy sources, WPT can provide regular and controllable energy supply. Also, if the locations of the energy harvesting nodes are fixed, the amounts of average harvested wireless power at the receivers are predictable, since the WPT efficiency depends mainly on distance. Inductive coupling and magnetic resonant coupling technologies rely on near-field electromagnetic (EM) waves. Hence, they cannot support the mobility of energy harvesting devices, due to the limited wireless charging distances and the required alignment of the EM-field with energy harvesting circuits. In contrast, RF-based WPT [10–15, 17–24, 26] exploits the far-field properties of EM waves to convey wireless energy. The use of RF for WPT in communication systems enables the possibility of simultaneous wireless information and power transfer (SWIPT), leading to a paradigm shift in system architectures and designs. In particular, SWIPT systems can exploit the broadcast nature of the wireless medium to facilitate one-to-many wireless charging and wireless communication, which may not be possible when near-field EM waves are employed.

In practice, RF-based wireless energy has to be transferred via a signal with high carrier frequency such that antennas with small size can be used for harvesting the power in portable devices. However, the associated path loss severely attenuates the signal, leading to a small amount of harvested energy at the receiver. For instance, for the short distance of 10 meters in free space, the attenuation of a wireless signal can be up to 50 dB for a carrier frequency of 915 MHz in the industrial, scientific and medical radio (ISM) frequency band. Hence, multiple antenna beamforming has been proposed to improve the efficiency of WPT [11–14]. With the spatial degrees of freedom offered by multiple antennas, one can focus information and energy beams, which improves the beamforming efficiency for information transfer and WPT. On the other hand, with the existing RF-based hardware circuit technology, the minimum required received power at an information receiver is −60 dBm, while that of an RF-based energy harvesting receiver (ER) is −10 dBm. In order to satisfy the sensitivity requirements, an RF-based ER has to be located closer to the transmitter than an the information receiver. Also, the transmitters can increase the energy of the information-carrying signal to facilitate RF energy harvesting at the receivers. However, these two common remedies may also increase the possibility of information leakage to eavesdroppers due to the associated higher signal power. Therefore, new quality of service (QoS) concerns regarding communication security and efficient WPT naturally arise in systems providing SWIPT services [21–26].

In fact, security is a crucial problem in wireless communication systems in general due to the broadcast nature of the wireless medium. Traditionally, communication security relies on cryptographic encryption algorithms employed at the application layer. Yet, these algorithms assume perfect secret key management and distribution, which may not be possible in future wireless IoT networks with a massive number of wireless sensor nodes. Alternatively, information-theoretic physical layer (PHY) security offers a complementary technology to cryptographic encryption [27–33]. The principle of PHY security is to exploit the physical characteristics of the wireless fading channel to ensure perfect secrecy of communication. In particular, it has been shown that in a wire-tap channel, a source and a destination can exchange perfectly secure information if the source-destination channel offers better conditions compared to the source-eavesdropper channel [27]. Hence, multiple-antenna technology has been proposed to ensure secure communication. Specifically, by exploiting the extra degrees of freedom introduced by multiple antennas, artificial noise is injected into the communication channel intentionally to impair the received signals at the eavesdropper. The notion of communication security for SWIPT systems has recently been pursued in [21, 22]. In particular, the dual use of energy signals for facilitating efficient WPT and providing communication security was proposed. However, the beamforming design for secure SWIPT systems in [21, 22] was based on a linear energy harvesting (EH) model, which does not capture the highly nonlinear characteristics of practical end-to-end WPT. In particular, existing beamforming schemes designed for a linear EH model may lead to severe resource allocation mismatches resulting in performance degradation in WPT and secure communications. Hence, in this chapter, we study beamforming designs enabling secure SWIPT based on a nonlinear EH model.

The remainder of this chapter is organized as follows. In Section 2, we introduce the adopted channel model and the RF EH model. Section 3 studies the beamforming design for guaranteeing secure communication in SWIPT systems and simulation results are provided in Section 4. In Section 5, we conclude with a brief summary of the chapter.

Notation

We use boldface capital and lower case letters to denote matrices and vectors, respectively. $\mathbf{A}^{\mathrm{H}}$, $\mathrm{Tr}(\mathbf{A})$, $\mathrm{Rank}(\mathbf{A})$, and $\det(\mathbf{A})$ represent the Hermitian transpose, trace, rank and determinant of matrix $\mathbf{A}$, respectively; $\mathbf{A} \succ \mathbf{0}$ and $\mathbf{A} \succeq \mathbf{0}$ indicate that $\mathbf{A}$ is a positive definite and a positive semidefinite matrix, respectively; $\lambda_{\max}(\mathbf{A})$ denotes the maximum eigenvalue of matrix $\mathbf{A}$; $\mathbf{I}_N$ is the $N \times N$ identity matrix; $\mathbb{C}^{N\times M}$ denotes the set of all $N \times M$ matrices with complex entries; $\mathbb{H}^N$ denotes the set of all $N \times N$ Hermitian matrices. The circularly symmetric complex Gaussian (CSCG) distribution is denoted by $\mathcal{CN}(\mathbf{m}, \mathbf{\Sigma})$ with mean vector $\mathbf{m}$ and covariance matrix $\mathbf{\Sigma}$; $\sim$ indicates "distributed as"; $\mathcal{E}\{\cdot\}$ denotes statistical expectation; $|\cdot|$ represents the absolute value of a complex scalar; $[x]^+$ stands for $\max\{0, x\}$.

2 Channel model

A downlink frequency flat fading channel is considered. We assume that the SWIPT system comprises a transmitter, an information receiver (IR) and J ER, cf. Figure 1. The transmitter is equipped with $N_{\mathrm{T}} \geq 1$ antennas. The IR is a single-antenna device, and each ER is equipped with $N_{\mathrm{R}} \geq 1$ receiver antennas to facilitate energy harvesting. In SWIPT systems, the signals intended for the IR are overheard by the ERs, since all receivers are in the range of service coverage. If the ERs are malicious, they may eavesdrop the information signal intended for IR. Hence, the ERs are potential eavesdroppers, which should be taken into account to provide secure communication. We assume that $N_{\mathrm{T}} > N_{\mathrm{R}}$ for the study of beamforming design. In each time slot, the re-

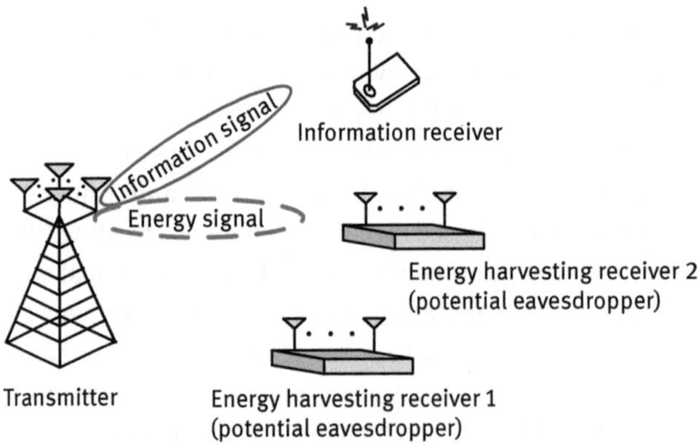

Fig. 1: A multiuser SWIPT model with an IR and $J = 2$ ERs. The transmitter emits both an information signal and an energy signal to facilitate secure communication for the IR and efficient WPT.

ceived signals at IR and ER $j \in \{1, \ldots, J\}$ are given by

$$y = \mathbf{h}^{\mathrm{H}}(\mathbf{w}s + \mathbf{w}_{\mathrm{E}}) + n \quad \text{and} \tag{1}$$

$$\mathbf{y}_{\mathrm{ER}_j} = \mathbf{G}_j^{\mathrm{H}}(\mathbf{w}s + \mathbf{w}_{\mathrm{E}}) + \mathbf{n}_{\mathrm{ER}_j}\,, \quad \forall j \in \{1, \ldots, J\}\,, \tag{2}$$

respectively, where $s \in \mathbb{C}$ and $\mathbf{w} \in \mathbb{C}^{N_{\mathrm{T}}\times 1}$ are the data symbol and the information beamforming vector, respectively. Without loss of generality, we assume that $\mathcal{E}\{|s|^2\} = 1$. $\mathbf{w}_{\mathrm{E}} \in \mathbb{C}^{N_{\mathrm{T}}\times 1}$ is an energy signal vector, which is a Gaussian pseudo-random sequence generated by the transmitter to facilitate efficient WPT and to guarantee communication security. In particular, $\mathbf{w}_{\mathrm{E}}$ is modeled as a complex Gaussian random vector with

$$\mathbf{w}_{\mathrm{E}} \sim \mathcal{CN}(\mathbf{0}, \mathbf{W}_{\mathrm{E}})\,, \tag{3}$$

where $\mathbf{W}_{\mathrm{E}} \in \mathbb{H}^{N_{\mathrm{T}}}$, $\mathbf{W}_{\mathrm{E}} \succeq \mathbf{0}$ denotes the covariance matrix of the pseudo-random energy signal. The channel vector between the transmitter and the IR is denoted by $\mathbf{h} \in \mathbb{C}^{N_{\mathrm{T}}\times 1}$ and the channel matrix between the transmitter and ER_j is denoted by $\mathbf{G}_j \in \mathbb{C}^{N_{\mathrm{T}}\times N_{\mathrm{R}}}$. $n \sim \mathcal{CN}(0, \sigma_{\mathrm{s}}^2)$ and $\mathbf{n}_{\mathrm{ER}_j} \sim \mathcal{CN}(\mathbf{0}, \sigma_{\mathrm{s}}^2\mathbf{I}_{N_{\mathrm{R}}})$ are the additive white Gaussian noises (AWGN) at the IR and ER_j, respectively, where σ_{s}^2 denotes the noise power at the receiver.

2.1 Nonlinear energy harvesting model

Figure 2 depicts the block diagram of an ER for SWIPT systems. In general, the RF-energy harvesting circuit consists of a bandpass filter and a rectifying circuit that converts the received RF power to direct current (DC) power. The total received RF power at ER_j is given by

$$P_{\mathrm{ER}_j} = \mathrm{Tr}\left((\mathbf{w}\mathbf{w}^{\mathrm{H}} + \mathbf{W}_{\mathrm{E}})\mathbf{G}_j\mathbf{G}_j^{\mathrm{H}}\right). \tag{4}$$

In the SWIPT literature [35–43], for simplicity, the total harvested power at ER_j is typically modeled by the linear equation:

$$\Phi_{\mathrm{ER}_j}^{\mathrm{Linear}} = \eta_j P_{\mathrm{ER}_j}\,, \tag{5}$$

where $0 \le \eta_j \le 1$ is the constant power conversion efficiency of ER_j. In other words, the total harvested power at the ER is linearly and directly proportional to the received

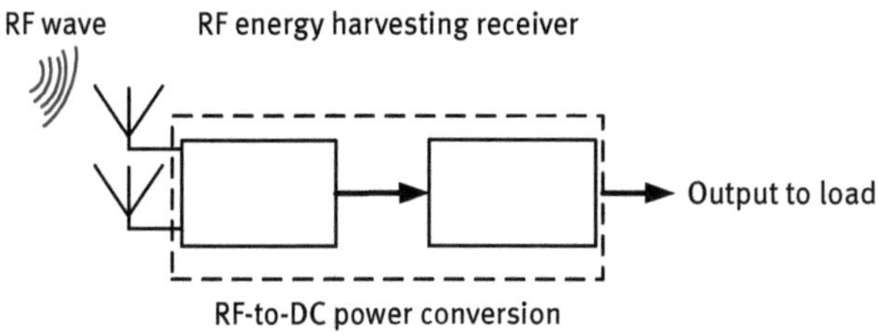

Fig. 2: Block diagram of an ER.

RF power. Although the linear model leads to simple beamforming designs, it has been shown in recent experiments that practical RF-based energy harvesting circuits [34, 44, 45] introduce various nonlinearities into the end-to-end WPT. In order to design a beamformer for practical secure SWIPT systems, we adopt the nonlinear parametric energy harvesting model from [26, 46]. Thus, the total harvested power at ER_j, Φ_{ER_j}, is modeled as:

$$\Phi_{\text{ER}_j} = \frac{[\Psi_{\text{ER}_j} - M_j\Omega_j]}{1-\Omega_j}, \quad \Omega_j = \frac{1}{1+\exp(a_j b_j)}, \tag{6}$$

$$\text{where} \quad \Psi_{\text{ER}_j} = \frac{M_j}{1+\exp\left(-a_j(P_{\text{ER}_j} - b_j)\right)} \tag{7}$$

is a logistic function that has the received RF power, P_{ER_j}, as its input; M_j is a constant denoting the maximum harvested power at ER_j when the EH circuit is saturated because of exceedingly large input power. Parameters a_j and b_j are constants that capture the joint effects of resistance, capacitance and circuit sensitivity. Specifically, a_j reflects the nonlinear charging rate with respect to the input power, P_{ER_j}, and b_j is related to the minimum turn-on voltage of the EH circuit. In practice, the parameters a_j, b_j and M_j of the model in (6) can be easily obtained using standard curve fitting algorithms for a given energy harvesting hardware circuit. In fact, it has been verified by experimental results that the proposed parametric nonlinear model is able to accurately capture the dynamics of the RF energy conversion efficiency and the joint effects of the nonlinear phenomena caused by hardware imperfections at different input power levels. In Figure 3, we show an example for the curve fitting for the nonlinear

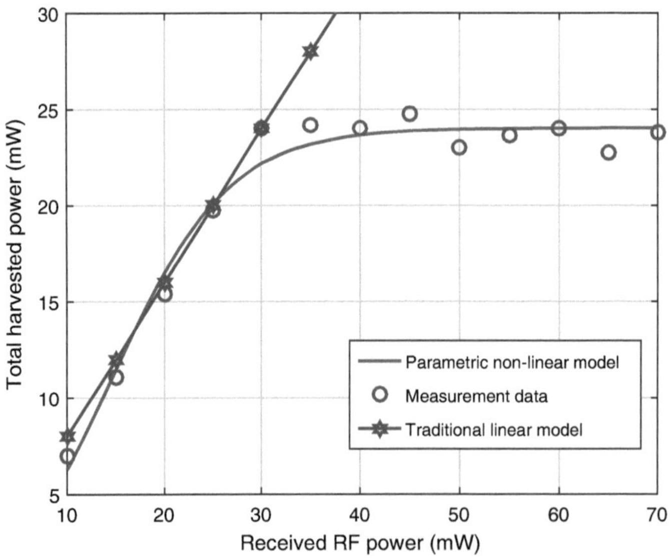

Fig. 3: A comparison between experimental data from [34], the harvested power for the nonlinear model in (6) and the linear energy harvesting model with $\eta_j = 0.8$ in (5).

EH model in (6) with parameters $M_j = 0.024$, $b_j = 0.014$ and $a_j = 150$ for ER_j. It can be observed that the parametric nonlinear model shows an excellent agreement with the experimental results provided in [34] for the wireless power harvested by a practical EH circuit. Figure 3 also illustrates the inability of the linear model in (5) to capture the nonlinear characteristics of practical EH circuits, especially in the high received RF power regime.

2.2 Achievable secrecy rate

Assuming perfect channel state information (CSI) at the IR for coherent signal detection, the achievable data rate (bit/s/Hz) between the transmitter and the IR is given by

$$R = \log_2\left(1 + \frac{\mathbf{w}^{\mathrm{H}}\mathbf{H}\mathbf{w}}{\sigma_{\mathrm{s}}^2}\right), \tag{8}$$

where $\mathbf{H} = \mathbf{h}\mathbf{h}^{\mathrm{H}}$. We note that since the energy signal is known to both the transmitter and the IR, the interference caused by the energy signal, i.e., $\mathrm{Tr}(\mathbf{H}\mathbf{W}_{\mathrm{E}})$ can be removed at the IR via successive interference cancellation (SIC) to improve the data rate.

In order to provide secure communication to the IR, all the ERs are treated as potential eavesdroppers and are taken into account for beamforming design. Besides, we focus on the worst case scenario. In particular, we assume noiseless ERs. Therefore, the capacity between the transmitter and ER_j for decoding the signal of the IR can be expressed as

$$\begin{aligned} R_{\mathrm{ER}_j} &= \log_2 \det(\mathbf{I}_{N_{\mathrm{R}}} + \mathbf{Q}_j^{-1}\mathbf{G}_j^{\mathrm{H}}\mathbf{w}\mathbf{w}^{\mathrm{H}}\mathbf{G}_j)\,, \\ \mathbf{Q}_j &= \mathbf{G}_j^{\mathrm{H}}\mathbf{W}_{\mathrm{E}}\mathbf{G}_j \succ \mathbf{0}\,, \end{aligned} \tag{9}$$

where $\mathbf{Q}_j$ is the interference covariance matrix for ER_j assuming the worst case for communication secrecy. We note that in contrast to the IR, the energy signal is not known to the ERs and appears random to them. Hence, the energy signal cannot be removed via interference cancellation techniques at the ERs. Thus, the achievable secrecy rate of the IR is given by [30, 33]

$$R_{\mathrm{sec}} = \left[R - \max_{\forall j}\{R_{\mathrm{ER}_j}\}\right]^{+}. \tag{10}$$

3 Problem formulation and solution

The considered system design objective is to maximize the secrecy rate at the IR while guaranteeing a minimum harvested power at each ER. The beamforming design is formulated as the following optimization problem:

Problem 1. Resource allocation for secrecy rate maximization

$$\begin{aligned} \underset{\mathbf{W}_\mathrm{E}\in\mathbb{H}^{N_\mathrm{T}},\mathbf{w}}{\text{maximize}} \quad & \log_2\left(1+\frac{\mathbf{w}^\mathrm{H}\mathbf{H}\mathbf{w}}{\sigma_\mathrm{s}^2}\right) - \max_{\forall j}\left\{R_{\mathrm{ER}_j}\right\} \\ \text{subject to} \quad & \text{C1:}\ \ \|\mathbf{w}\|_2^2 + \mathrm{Tr}(\mathbf{W}_\mathrm{E}) \le P_\mathrm{max}\,, \\ & \text{C2:}\ \ \Phi_{\mathrm{ER}_j} \ge P_{\mathrm{req}_j}^{\mathrm{min}}\,,\quad \forall j\in\{1,\dots,J\}\,, \\ & \text{C3:}\ \ \mathbf{W}_\mathrm{E} \succeq \mathbf{0}\,. \end{aligned} \tag{11}$$

Constants $P_{\max}$ and $P_{\mathrm{req}_j}^{\min}$ in constraints C1 and C2 are the maximum transmit power allowance and the minimum required harvested power at ER_j, respectively. Constraint C3 and $\mathbf{W}_\mathrm{E} \in \mathbb{H}^{N_\mathrm{T}}$ constrain matrix $\mathbf{W}_\mathrm{E}$ to be a positive semidefinite Hermitian matrix. We note that the objective function in (11) is equivalent to the achievable secrecy rate of the IR defined in (10), even though the $[\cdot]^+$ operator has been removed. This is because if the problem is feasible, the optimal beamforming design can always set $\mathbf{w} = \mathbf{0}$ to ensure a non-negative secrecy rate.

The objective function in (11) is a non-convex function due to the difference of two logarithmic functions. In particular, the log-det function in R_{ER_j} of the subtrahend of the objective function is non-convex. In order to obtain a globally optimal solution, we introduce several transformations in (11) in the following.

Optimization problem solution

In order to handle the non-convex objective function, we first introduce an auxiliary optimization variable τ and rewrite optimization problem (11) in the following equivalent form[1]:

1 In this chapter, “equivalent” means that both problem formulations lead to the same beamforming design.

Problem 2. Equivalent optimization problem

$$\begin{aligned} \underset{\mathbf{W}_\mathrm{E}\in\mathbb{H}^{N_\mathrm{T}},\mathbf{w},\tau}{\text{maximize}} \quad & \log_2\left(1+\frac{\mathbf{w}^\mathrm{H}\mathbf{H}\mathbf{w}}{\sigma_\mathrm{s}^2}\right)-\tau \qquad (12)\\ \text{subject to} \quad & \text{C1, C2, C3}\,,\\ & \text{C4: } \tau \geq \log_2\det(\mathbf{I}_{N_\mathrm{R}}+\mathbf{Q}_j^{-1}\mathbf{G}_j^\mathrm{H}\mathbf{w}\mathbf{w}^\mathrm{H}\mathbf{G}_j)\,,\ \forall j\,. \end{aligned}$$

We note that the objective function of the transformed problem is now jointly concave with respect to the optimization variables. Nevertheless, the new constraint C4 is non-convex. Hence, we solve the optimization problem in (12) for a fixed τ and obtain the corresponding beamforming design. Then, by adopting a one-dimensional search, we find the optimal value of the optimization problem and the corresponding τ[2].

Therefore, in the sequel, we focus on solving Problem 2 for a given τ. First, we further transform the considered problem into the following equivalent form:

Problem 3. Rank-constrained optimization problem

$$\begin{aligned} \underset{\mathbf{W}_\mathrm{E}\in\mathbb{H}^{N_\mathrm{T}},\mathbf{W},\beta_j,\delta}{\text{maximize}} \quad & \log_2\left(1+\frac{\delta}{\sigma_\mathrm{s}^2}\right)-\tau \qquad (13)\\ \text{subject to} \quad & \text{C1: } \mathrm{Tr}(\mathbf{W})+\mathrm{Tr}(\mathbf{W}_\mathrm{E}) \leq P_\mathrm{max}\,,\\ & \text{C2: } \frac{M_j}{1+\exp\Big(-a_j(\beta_j-b_j)\Big)} \geq P_{\mathrm{req}_j}^{\min}(1-\Omega_j)+M_j\Omega_j\,,\ \forall j\,,\\ & \text{C3: } \mathbf{W}_\mathrm{E}\succeq\mathbf{0}\,,\\ & \text{C4: } \tau \geq \log_2\det(\mathbf{I}_{N_\mathrm{R}}+\mathbf{Q}_j^{-1}\mathbf{G}_j^\mathrm{H}\mathbf{W}\mathbf{G}_j)\,,\ \forall j\,,\\ & \text{C5: } \delta \leq \mathrm{Tr}(\mathbf{H}\mathbf{W})\,,\\ & \text{C6: } \beta_j \leq \mathrm{Tr}\Big((\mathbf{W}+\mathbf{W}_\mathrm{E})\mathbf{G}_j\mathbf{G}_j^\mathrm{H}\Big)\,,\ \forall j\,,\\ & \text{C7: } \mathrm{Rank}(\mathbf{W}) \leq 1,\\ & \text{C8: } \mathbf{W}\succeq\mathbf{0}, \end{aligned}$$

where $\mathbf{W} = \mathbf{w}\mathbf{w}^\mathrm{H}$ is a new optimization variable matrix. Auxiliary optimization variables β_j and δ are introduced to simplify the analysis of the solution in the following. It can be observed that the non-convexity of the transformed problem arises from the

2 We note that the range of τ is $0 \leq \tau \leq \tau_\mathrm{max}$, where τ_max can be found by solving (13) with the right-hand side of constraint C4 set as zero.

log-det function in C4 and the combinatorial rank constraint C7. To circumvent the non-convexity, we first introduce the following proposition to handle constraint C4. Then, we handle the rank constraint C7 by recasting the considered problem as a convex optimization problem via SDP relaxation.

Proposition 1. *For $\tau > 0$, the following implication holds for constraint C4:*

$$C4 \Rightarrow \overline{C4}\colon \mathbf{G}_j^{\mathrm{H}}\mathbf{W}\mathbf{G}_j \preceq \alpha_{\mathrm{ER}}\mathbf{Q}_j\,, \quad \forall j\,, \tag{14}$$

where $\alpha_{\mathrm{ER}} = 2^{\tau} - 1$ is an auxiliary constant and $\overline{C4}$ is an linear matrix inequality (LMI) constraint. We note that constraints $\overline{C4}$ and C4 are equivalent, i.e., $\overline{C4} \Leftrightarrow$ C4, when $\mathrm{Rank}(\mathbf{W}) \leq 1$.

Proof. Please refer to Appendix 6.1 for the proof. □

Now, we apply Proposition 1 to Problem 3 by replacing constraint C4 with constraint $\overline{C4}$. Then, we adopt SDP relaxation [47] by removing constraint C7 which yields

Problem 4. Semidefinite relaxation of Problem 3

$$\begin{aligned}
\underset{\mathbf{W}_{\mathrm{E}}\in\mathbb{H}^{N_{\mathrm{T}}},\mathbf{W},\beta_j,\delta}{\text{maximize}} \quad & \log_2\left(1 + \frac{\delta}{\sigma_{\mathrm{S}}^2}\right) - \tau \\
\text{subject to} \quad & \text{C1: } \mathrm{Tr}(\mathbf{W}) + \mathrm{Tr}(\mathbf{W}_{\mathrm{E}}) \leq P_{\max}\,, \\
& \text{C2: } \frac{M_j}{1 + \exp\Big(- a_j(\beta_j - b_j)\Big)} \geq P_{\mathrm{req}_j}^{\min}(1 - \Omega_j) + M_j\Omega_j\,, \quad \forall j\,, \\
& \text{C3: } \mathbf{W}_{\mathrm{E}} \succeq \mathbf{0}\,, \\
& \overline{\text{C4}}\text{: } \mathbf{G}_j^{\mathrm{H}}\mathbf{W}\mathbf{G}_j \preceq \alpha_{\mathrm{ER}}\mathbf{G}_j^{\mathrm{H}}\mathbf{W}_{\mathrm{E}}\mathbf{G}_j\,, \quad \forall j\,, \\
& \text{C5: } \delta \leq \mathrm{Tr}(\mathbf{H}\mathbf{W})\,, \\
& \text{C6: } \beta_j \leq \mathrm{Tr}\Big((\mathbf{W} + \mathbf{W}_{\mathrm{E}})\mathbf{G}_j\mathbf{G}_j^{\mathrm{H}}\Big)\,, \quad \forall j \\
& \text{C7: } \cancel{\mathrm{Rank}(\mathbf{W}) \leq 1}\,, \\
& \text{C8: } \mathbf{W} \succeq \mathbf{0}\,.
\end{aligned} \tag{15}$$

As a result, the relaxed problem becomes a standard convex optimization problem and can be solved efficiently by numerical solvers such as CVX [48] via the interior point method. However, the relaxation may not be tight if $\mathrm{Rank}(\mathbf{W}) > 1$ occurs. Therefore, we reveal the tightness of the adopted SDP relaxation in (11) in the following theorem.

Theorem 1. *Let the optimal beamforming matrix and energy covariance matrix of* (15) *be* $\mathbf{W}^*$ *and* $\mathbf{W}_E^*$*, respectively. Assuming the considered problem is feasible for* $P_{max} > 0$*, then* $\text{Rank}(\mathbf{W}^*) \leq 1, \forall k$*, and* $\text{Rank}(\mathbf{W}_E^*) \leq 1$.

Proof. Please refer to Appendix 6.2. □

In other words, (11) can be solved optimally. Hence, information beamforming and energy beamforming are optimal for the maximization of the secrecy rate, despite the nonlinearity of the EH circuit.

4 Results

In this section, we evaluate the system performance of the proposed optimal resource allocation algorithm via simulations. We summarize the relevant simulation parameters in Table 1. We assume that the IR and the ERs are located 100 m and 5 m from the transmitter, respectively. For the nonlinear EH circuits, we set $M_j = 24$ mW, which corresponds to the maximum harvested power per ER. Besides, we adopt $a_j = 150$ and $b_j = 0.0014$. For the optimal beamforming design, we use 100 equally spaced intervals for quantizing the range of τ for facilitating the one-dimensional search. We solve the optimization problem in (2) and obtain the average system performance by averaging over different channel realizations. We note that the considered problem may be infeasible when the QoS requirements are stringent and/or the channels are in unfavorable conditions. In the simulation, we set the secrecy rate of the corresponding channel realizations to zero to account for the penalty incurred by an infeasible problem.

Tab. 1: Simulation parameters.

Carrier center frequency	**915 MHz**
Bandwidth	200 kHz
Transceiver antenna gain	10 dBi
Noise power σ^2	−95 dBm
Min. harvested power at each ER $P_{req_j}^{min}$	3 dBm
Transmitter-to-ER fading distribution	Rician with Rician factor 3 dB
Transmitter-to-IR fading distribution	Rayleigh

In Figure 4, we study the average secrecy rate versus the maximum transmit power at the transmitter, P_{max}, for different numbers of transmit antennas and beamforming schemes. As can be observed, the average secrecy rate increases with P_{max}. Indeed, with more transmit power available, the transmitter is able to increase the sig-

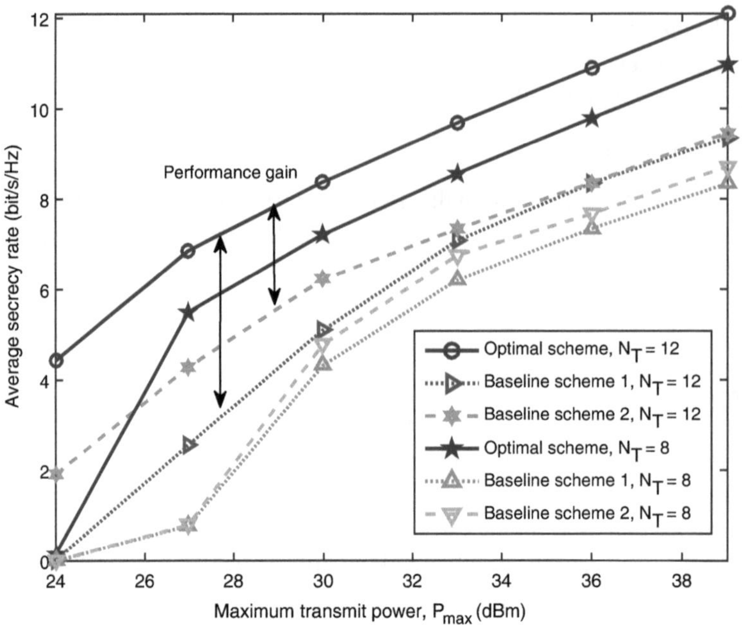

Fig. 4: Average secrecy rate (bit/s/Hz) versus the maximum available transmit power (dBm). There are $N_R = 3$ antennas equipped at each ER. The double sided arrows represent the performance gains due to the proposed optimization.

nal strength of the information signal. Besides, a higher power can also be allocated to the energy signal to degrade the channel quality of the ERs for information decoding. On the other hand, it can be seen that the achievable secrecy rate improves with the number of transmit antennas. In fact, the spatial degrees of freedom offered by extra transmit antennas facilitate a more flexible beamforming. In particular, the transmitter can steer the energy signal and the information signal towards the ERs more accurately to improve the efficiency of WPT. For comparison, we also show the performance of two baseline schemes. For baseline scheme 1, the beamforming is designed for the nonlinear EH model in (6). However, the power of the energy signal is set to zero and we solve the corresponding beamforming design problem in (15). For baseline 2, the existing linear EH model with $\eta_j = 1$, cf. (5), is adopted for beamforming design. Based on the linear EH model and assuming that at most half of the power is allocated to the information signal, we optimize $\mathbf{w}$ and $\mathbf{W}_E$ to maximize the secrecy rate subject to the constraints in (11). It can be observed that the proposed optimal algorithm designed for the nonlinear energy harvesting model provides a substantial performance gain compared to the two baseline schemes. In particular, baseline 1 obtains the worst system performance among all the schemes. In fact, transmitting an energy signal is necessary for achieving a high secrecy rate in SWIPT. On the other

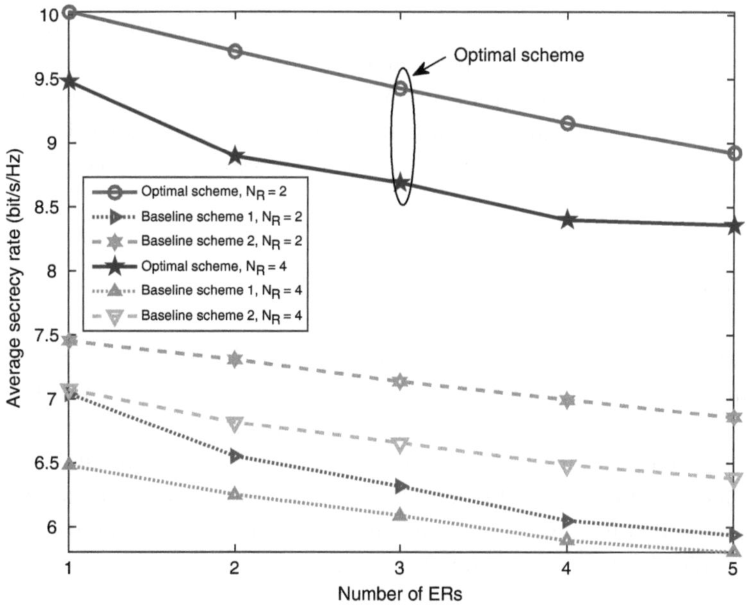

Fig. 5: Average system secrecy rate (bit/s/Hz) versus the number of ERs. There are $N_T = 8$ antennas equipped at the transmitter.

hand, although baseline scheme 2 also employs an energy signal, it may suffer from severe mismatches in resource allocation since it does not account for the nonlinear nature of the energy harvesting circuits.

Figure 5 shows the average secrecy rate versus the number of ERs for different beamforming schemes and different numbers of receive antennas equipped at each ER. The maximum transmit power at the transmitter is $P_{max} = 33$ dBm. It can be observed that the average system secrecy rate, i.e., R_{sec}, is a non-increasing function with respect to the number of ERs for the following two reasons. First, the minimum required harvested power constraint in C2 becomes more stringent when more ERs are in the system. In particular, the transmitter is forced to focus some of the energy of the information signal towards the ERs in order to satisfy the constraints. Second, there are more potential eavesdroppers present in the system resulting in a higher potential for information leakage. Thus, a higher amount of transmit power has to be allocated to the energy signal for interfering the ERs to guarantee communication secrecy. Hence, less power can be allocated to the desired signal. Also, it can be observed that the average secrecy rate decreases with the number of antennas equipped at each ER, N_R. In fact, the signal reception capability of the ERs improves with N_R. On the one hand, the ERs can harvest the wireless energy more efficiently, which makes constraint C2 less stringent. However, on the other hand, it also makes the considered system more vulnerable to eavesdropping. Therefore, there is a non-trivial tradeoff between the secrecy rate and the number of receive antennas equipped at ERs in the considered system. We also compare the performance of the proposed optimal beamforming scheme with the two baseline schemes. As expected, the optimal scheme outperforms the baseline schemes. This is because the proposed optimal scheme is able to fully exploit the available degrees of freedom for efficient beamforming design.

5 Conclusions

In this chapter, beamforming design for secure SWIPT was studied, which is of fundamental importance for wireless sensor networks facilitating IoT. The design was formulated as a non-convex optimization problem for the maximization of the secrecy rate of the IR. The problem formulation took into account a minimum required power transfer to practical nonlinear ERs. The optimization problem was solved by a one-dimensional search and SDP relaxation. Numerical results showed the potential gains in secrecy rate enabled by the proposed optimization.

Acknowledgment: This work was supported in part by the AvH Professorship Program of the Alexander von Humboldt Foundation. D. W. K. Ng is supported under the Australian Research Council's Discovery Early Career Researcher Award funding scheme (project number DE170100137).

6 Appendix

6.1 Proof of proposition 1

We start the proof by rewriting constraint C4 as

$$\text{C4:}\ \ \log_2 \det(\mathbf{I}_{N_\mathrm{R}} + \mathbf{Q}_j^{-1}\mathbf{G}_j^\mathrm{H}\mathbf{W}\mathbf{G}_j) \le \tau \tag{16}$$

$$\iff \det(\mathbf{I}_{N_\mathrm{R}} + \mathbf{Q}_j^{-1/2}\mathbf{G}_j^\mathrm{H}\mathbf{W}\mathbf{G}_j\mathbf{Q}_j^{-1/2}) \le 1 + \alpha_\mathrm{ER}\,. \tag{17}$$

Then, we propose a lower bound on the left-hand side of (17) by introducing the following lemma.

Lemma 1. *For any square matrix* $\mathbf{A} \succeq \mathbf{0}$*, we have the following inequality [49]:*

$$\det(\mathbf{I} + \mathbf{A}) \ge 1 + \mathrm{Tr}(\mathbf{A})\,, \tag{18}$$

where the equality holds if and only if $\mathrm{Rank}(\mathbf{A}) \le 1$.

Exploiting Lemma 1, the left-hand side of (17) is bounded below by

$$\begin{aligned}&\det(\mathbf{I}_{N_\mathrm{R}} + \mathbf{Q}_j^{-1/2}\mathbf{G}_j^\mathrm{H}\mathbf{W}\mathbf{G}_j\mathbf{Q}_j^{-1/2})\\ &\geq 1 + \mathrm{Tr}(\mathbf{Q}_j^{-1/2}\mathbf{G}_j^\mathrm{H}\mathbf{W}\mathbf{G}_j\mathbf{Q}_j^{-1/2})\,. \end{aligned} \tag{19}$$

Subsequently, by combining Equations (16), (17) and (19), we have the following implications:

$$\begin{aligned}&(16) \Longleftrightarrow (17)\\ \Longrightarrow\;& \mathrm{Tr}(\mathbf{Q}_j^{-1/2}\mathbf{G}_j^\mathrm{H}\mathbf{W}\mathbf{G}_j\mathbf{Q}_j^{-1/2}) \leq \alpha_\mathrm{ER} && (20a)\\ \Longrightarrow\;& \lambda_{\max}(\mathbf{Q}_j^{-1/2}\mathbf{G}_j^\mathrm{H}\mathbf{W}\mathbf{G}_j\mathbf{Q}_j^{-1/2}) \leq \alpha_\mathrm{ER} && (20b)\\ \Longleftrightarrow\;& \mathbf{Q}_j^{-1/2}\mathbf{G}_j^\mathrm{H}\mathbf{W}\mathbf{G}_j\mathbf{Q}_j^{-1/2} \preceq \alpha_\mathrm{ER}\mathbf{I}_{N_\mathrm{R}} && (20c)\\ \Longleftrightarrow\;& \mathbf{G}_j^\mathrm{H}\mathbf{W}\mathbf{G}_j \preceq \alpha_\mathrm{ER}\mathbf{Q}_j\,. && (20d)\end{aligned}$$

We note that Equations (16) and (20d) are equivalent, i.e., C4 $\Leftrightarrow$ $\overline{\text{C4}}$, when $\mathrm{Rank}(\mathbf{W}) \leq 1$. □

6.2 Proof of theorem 1

We follow a similar approach as in [47] to prove Theorem 1. We note that Problem 4 is jointly convex with respect to the optimization variables. Besides, it can be verified that the problem satisfies Slater's constraint qualification and thus has zero duality gap. Therefore, to reveal the structure of $\mathbf{W}$ and $\mathbf{W}_\mathrm{E}$, we consider the Lagrangian of Problem 4, which is given by:

$$\begin{aligned} L = &-\lambda\Big(\mathrm{Tr}(\mathbf{W}) + \mathrm{Tr}(\mathbf{W}_\mathrm{E}) - P_{\max}\Big) - \psi(\delta - \mathrm{Tr}(\mathbf{H}\mathbf{W}))\\ &-\sum_{j=1}^{J}\theta_j\Big(\beta_j - \mathrm{Tr}\big((\mathbf{W}+\mathbf{W}_\mathrm{E})\mathbf{G}_j\mathbf{G}_j^\mathrm{H}\big)\Big) + \mathrm{Tr}(\mathbf{W}\mathbf{Y})\\ &+\sum_{j=1}^{J}\mathrm{Tr}\Big(\mathbf{G}_j^\mathrm{H}(\alpha_\mathrm{ER}\mathbf{W}_\mathrm{E} - \mathbf{W})\mathbf{G}_j\mathbf{D}_{\overline{\text{C4}}_j}\Big) + \mathrm{Tr}(\mathbf{W}_\mathrm{E}\mathbf{Z}) + \mathbf{\Delta}\,, \end{aligned} \tag{21}$$

where $\lambda \geq 0, \mathbf{Z} \succeq \mathbf{0}, \mathbf{D}_{\overline{\text{C4}}_j} \succeq \mathbf{0}, \forall j \in \{1,\ldots,J\}, \psi \geq 0, \theta_j \geq 0$, and $\mathbf{Y} \succeq \mathbf{0}$ are the dual variables for constraints C1, C3, C4, C5, C6, and C8, respectively. Besides, $\mathbf{\Delta}$ is a collection of variables and constants that are not relevant to the proof. For notational convenience, we denote the optimal primal and dual variables of the SDP relaxed version in (13) by the corresponding variables with an asterisk superscript in the following.

Now, we focus on those Karush–Kuhn–Tucker (KKT) conditions, which are needed for the proof:

$$\mathbf{Y}^*, \mathbf{Z}^*, \mathbf{D}^*_{\overline{C4}_j} \succeq \mathbf{0}, \quad \lambda^*, \psi^*, \theta_j^* \geq 0, \tag{22a}$$

$$\mathbf{Y}^*\mathbf{W}^* = \mathbf{0}, \quad \mathbf{Z}^*\mathbf{W}_{\mathrm{E}}^* = \mathbf{0}, \tag{22b}$$

$$\mathbf{Y}^* = \lambda^*\mathbf{I}_{N_{\mathrm{T}}} - \Xi, \tag{22c}$$

$$\Xi = \psi_j\mathbf{H} + \sum_{j=1}^{J} \theta_j^*\mathbf{G}_j\mathbf{G}_j^{\mathrm{H}} - \sum_{j=1}^{J} \mathbf{G}_j\mathbf{D}^*_{\overline{C4}_j}\mathbf{G}_j^{\mathrm{H}} \tag{22d}$$

$$\mathbf{Z}^* = \lambda^*\mathbf{I}_{N_{\mathrm{T}}} - \sum_{j=1}^{J} \theta_j^*\mathbf{G}_j\mathbf{G}_j^{\mathrm{H}} - \sum_{j=1}^{J} \mathbf{G}_j\mathbf{D}^*_{\overline{C4}_j}\mathbf{G}_j^{\mathrm{H}}\alpha_{\mathrm{ER}}. \tag{22e}$$

From (22b), we know that the columns of $\mathbf{W}^*$ lie in the null space of $\mathbf{Y}^*$. In order to reveal the rank of $\mathbf{W}^*$, we investigate the structure of $\mathbf{Y}^*$. First, it can be shown that $\lambda^* > 0$, since constraint C1 is active for the optimal solution. Then, we show that $\mathrm{Rank}(\mathbf{W}) \leq 1$ in the following two cases. For the first case, if Ξ is a negative definite matrix, then from (22c), $\mathbf{Y}^*$ becomes a full-rank and positive definite matrix. By (22b), $\mathbf{W}^*$ is forced to be the zero matrix with $\mathrm{Rank}(\mathbf{W}^*) = 0$. For the second case, we focus on $\Xi \succeq \mathbf{0}$. Since matrix $\mathbf{Y}^* = \lambda^*\mathbf{I}_{N_{\mathrm{T}}} - \Xi$ is positive semidefinite, the following inequality holds:

$$\lambda^* \geq \lambda_{\Xi}^{\max} \geq 0, \tag{23}$$

where $\lambda_{\Xi}^{\max}$ is the maximum eigenvalue of matrix Ξ. From (22c), if $\lambda^* > \lambda_{\Xi}^{\max}$, matrix $\mathbf{Y}^*$ will become a positive definite matrix with full rank. This will again yield the zero solution, $\mathbf{W}^* = \mathbf{0}$, with $\mathrm{Rank}(\mathbf{W}^*) = 0$. On the other hand, if $\lambda^* = \lambda_{\Xi}^{\max}$, then, in order to have a bounded optimal dual solution, it follows that the null space of $\mathbf{Y}^*$ is spanned by vector $\mathbf{u}_{\Xi,\max} \in \mathbb{C}^{N_{\mathrm{T}}\times 1}$, which is the unit-norm eigenvector of Ξ associated with eigenvalue $\lambda_{\Xi}^{\max}$. As a result, the optimal beamforming matrix $\mathbf{W}^*$ has to be a rank-one matrix and is given by

$$\mathbf{W}^* = \upsilon\mathbf{u}_{\Xi,\max}\mathbf{u}_{\Xi,\max}^{\mathrm{H}}, \tag{24}$$

where υ is a parameter such that the power consumption satisfies constraint C1.

On the other hand, for revealing the structure of $\mathbf{Z}^*$, we focus on (22e). Define an auxiliary variable matrix $\mathbf{B} = \sum_{j=1}^{J} \theta_j\mathbf{G}_j\mathbf{G}_j^{\mathrm{H}} + \sum_{j=1}^{J} \mathbf{G}_j\mathbf{D}_{\overline{C4}_j}\mathbf{G}_j^{\mathrm{H}}\alpha_{\mathrm{ER}} \succeq \mathbf{0}$ and the corresponding maximum eigenvalue as $\lambda_{\mathbf{B}}^{\max}$. Since $\mathbf{Z}^* \succeq \mathbf{0}$, we have $\lambda^* \geq \lambda_{\mathbf{B}}^{\max} \geq 0$. If $\lambda^* = \lambda_{\mathbf{B}}^{\max}$, then $\mathrm{Rank}(\mathbf{Z}^*) = N_{\mathrm{T}} - 1$ and $\mathrm{Rank}(\mathbf{W}_{\mathrm{E}}^*) = 1$. If $\lambda^* > \lambda_{\mathbf{B}}^{\max}$, then $\mathrm{Rank}(\mathbf{Z}^*) = N_{\mathrm{T}}$ and $\mathrm{Rank}(\mathbf{W}_{\mathrm{E}}^*) = 0$. Therefore, $\mathrm{Rank}(\mathbf{W}_{\mathrm{E}}^*) \leq 1$, and at most one energy beam is required to achieve optimality. □

Bibliography

[1] M. Zorzi, A. Gluhak, S. Lange, and A. Bassi. From today's INTRAnet of things to a Future INTERnet of Things: a Wireless- and Mobility-Related View. *IEEE Wireless Commun.*, 17:44–51, 2010.

[2] D. W. K. Ng, E. S. Lo, and R. Schober. Energy-Efficient Resource Allocation in Multi-Cell OFDMA Systems with Limited Backhaul Capacity. *IEEE Trans. Wireless Commun.*, 11:3618–3631, 2012.

[3] D. W. K. Ng, Y. Wu, and R. Schober. Power Efficient Resource Allocation for Full-Duplex Radio Distributed Antenna Networks. *IEEE Trans. Wireless Commun.*, 15(4):2896–2911, 2016.

[4] D. W. K. Ng, E. S. Lo, and R. Schober. Energy-Efficient Resource Allocation in OFDMA Systems with Hybrid Energy Harvesting Base Station. *IEEE Trans. Wireless Commun.*, 12:3412–3427, 2013.

[5] I. Ahmed, A. Ikhlef, D. W. K. Ng, and R. Schober. Power Allocation for an Energy Harvesting Transmitter with Hybrid Energy Sources. *IEEE Trans. Wireless Commun.*, 12:6255–6267, 2013.

[6] V. W. S. Wong, R. Schober, D. W. K. Ng, and L.-C. Wang. *Key Technologies for 5G Wireless Systems*. Cambridge University Press, 2017.

[7] Q. Wu, G. Y. Li, W. Chen, D. W. K. Ng, and R. Schober. An Overview of Sustainable Green 5G Networks. *accepted for publication, IEEE Wireless Commun.*, Mar. 2017.

[8] I. Hammerstrom and A. Wittneben. Power Allocation Schemes for Amplify-and-Forward MIMO-OFDM Relay Links. *Transactions on wireless communications*, 6:2798–2802, 2007.

[9] D. W. K. Ng, E. S. Lo, and R. Schober. Dynamic Resource Allocation in MIMO-OFDMA Systems with Full-Duplex and Hybrid Relaying. *IEEE Trans. Commun.*, 60:1291–1304, 2012.

[10] P. Grover and A. Sahai. Shannon Meets Tesla: Wireless Information and Power Transfer. In *Proc. IEEE Intern. Sympos. on Inf. Theory*, pages 2363–2367, Jun. 2010.

[11] I. Krikidis, S. Timotheou, S. Nikolaou, G. Zheng, D. W. K. Ng, and R. Schober. Simultaneous Wireless Information and Power Transfer in Modern Communication Systems. *IEEE Commun. Mag.*, 52(11):104–110, 2014.

[12] Z. Ding, C. Zhong, D. W. K. Ng, M. Peng, H. A. Suraweera, R. Schober, and H. V. Poor. Application of Smart Antenna Technologies in Simultaneous Wireless Information and Power Transfer. *IEEE Commun. Mag.*, 53(4):86–93, 2015.

[13] X. Chen, Z. Zhang, H.-H. Chen, and H. Zhang. Enhancing Wireless Information and Power Transfer by Exploiting Multi-Antenna Techniques. *IEEE Commun. Mag.*, 4:133–141, 2015.

[14] X. Chen, D. W. K. Ng, and H.-H. Chen. Secrecy Wireless Information and Power Transfer: Challenges and Opportunities. *IEEE Commun. Mag.*, 2016.

[15] Q. Wu, M. Tao, D. Ng, W. Chen, and R. Schober. Energy-Efficient Resource Allocation for Wireless Powered Communication Networks. *IEEE Trans. Wireless Commun.*, 15:2312–2327, 2016.

[16] Q. Wu, W. Chen, D. W. K. Ng, J. Li, and R. Schober. User-Centric Energy Efficiency Maximization for Wireless Powered Communications. *IEEE Trans. Wireless Commun.*, 15(10):6898–6912, Oct., 2016.

[17] X. Chen, X. Wang, and X. Chen. Energy-Efficient Optimization for Wireless Information and Power Transfer in Large-Scale MIMO Systems Employing Energy Beamforming. *IEEE Wireless Commun. Lett.*, 2:1–4, 2013.

[18] D. W. K. Ng, E. S. Lo, and R. Schober. Wireless Information and Power Transfer: Energy Efficiency Optimization in OFDMA Systems. *IEEE Trans. Wireless Commun.*, 12:6352–6370, 2013.

[19] R. Zhang and C. K. Ho. MIMO Broadcasting for Simultaneous Wireless Information and Power Transfer. *IEEE Trans. Wireless Commun.*, 12:1989–2001, 2013.

[20] S. Leng, D. W. K. Ng, N. Zlatanov, and R. Schober. Multi-Objective Resource Allocation in Full-Duplex SWIPT Systems. In *Proc. IEEE Intern. Commun. Conf.*, 2016.

[21] D. W. K. Ng, E. S. Lo, and R. Schober. Robust Beamforming for Secure Communication in Systems with Wireless Information and Power Transfer. *IEEE Trans. Wireless Commun.*, 13:4599–4615, 2014.
[22] D. W. K. Ng and R. Schober. Secure and Green SWIPT in Distributed Antenna Networks with Limited Backhaul Capacity. *IEEE Trans. Wireless Commun.*, 14(9):5082–5097, 2015.
[23] M. Khandaker and K.-K. Wong. Robust Secrecy Beamforming With Energy-Harvesting Eavesdroppers. *IEEE Wireless Commun. Lett.*, 4:10–13, 2015.
[24] Q. Wu, W. Chen, and J. Li. Wireless Powered Communications With Initial Energy: QoS Guaranteed Energy-Efficient Resource Allocation. *IEEE Wireless Commun. Lett.*, 19, 2015.
[25] N. Zlatanov, Z. Hadzi-Velkov, and D. W. K. Ng. Asymptotically Optimal Power Allocation for Wireless Powered Communication Network with Non-orthogonal Multiple Access. Springer International Publishing, 2016. [Online]. doi:10.1007/978-3-319-46810-5_10.
[26] E. Boshkovska, D. Ng, N. Zlatanov, and R. Schober. Practical Non-Linear Energy Harvesting Model and Resource Allocation for SWIPT Systems. *IEEE Commun. Lett.*, 19:2082–2085, 2015.
[27] A. D. Wyner. *The Wire-Tap Channel*. Tech. Rep., Oct., 1975.
[28] X. Chen, D. W. K. Ng, W. Gerstacker, and H. H. Chen. A Survey on Multiple-Antenna Techniques for Physical Layer Security. *IEEE Commun. Surveys Tuts., PP*, 99:1–1, 2016.
[29] J. Zhu, R. Schober, and V. Bhargava. Secure Transmission in Multicell Massive MIMO Systems. *IEEE Trans. Wireless Commun.*, 13:4766–4781, 2014.
[30] S. Goel and R. Negi. Guaranteeing Secrecy using Artificial Noise. *IEEE Trans. Wireless Commun.*, 7:2180–2189, 2008.
[31] H. M. Wang, C. Wang, D. Ng, M. Lee, and J. Xiao. Artificial Noise Assisted Secure Transmission for Distributed Antenna Systems. *IEEE Trans. Signal Process., PP*, 99:1–1, 2016.
[32] J. Chen, X. Chen, W. H. Gerstacker, and D. W. K. Ng. Resource Allocation for a Massive MIMO Relay Aided Secure Communication. *IEEE Trans. on Inf. Forensics and Security*, 11(8):1700–1711, 2016.
[33] D. W. K. Ng, E. S. Lo, and R. Schober. Efficient Resource Allocation for Secure OFDMA Systems. *IEEE Trans. Veh. Technol.*, 61:2572–2585, 2012.
[34] J. Guo and X. Zhu. An Improved Analytical Model for RF-DC Conversion Efficiency in Microwave Rectifiers. In *IEEE MTT-S Int. Microw. Symp. Dig.*, pages 1–3, Jun. 2012.
[35] X. Zhou, R. Zhang, and C. K. Ho. Wireless Information and Power Transfer: Architecture Design and Rate-Energy Tradeoff. In *Proc. IEEE Global Telecommun. Conf.*, Dec. 2012.
[36] D. W. K. Ng, E. S. Lo, and R. Schober. Energy-Efficient Resource Allocation in Multiuser OFDM Systems with Wireless Information and Power Transfer. In *Proc. IEEE Wireless Commun. and Netw. Conf.*, 2013.
[37] S. Leng, D. W. K. Ng, and R. Schober. Power Efficient and Secure Multiuser Communication Systems with Wireless Information and Power Transfer. In *Proc. IEEE Intern. Commun. Conf.*, Jun. 2014.
[38] D. W. K. Ng, L. Xiang, and R. Schober. Multi-Objective Beamforming for Secure Communication in Systems with Wireless Information and Power Transfer. In *Proc. IEEE Personal, Indoor and Mobile Radio Commun. Sympos.*, Sep. 2013.
[39] D. W. K. Ng, R. Schober, and H. Alnuweiri. Secure Layered Transmission in Multicast Systems With Wireless Information and Power Transfer. In *Proc. IEEE Intern. Commun. Conf.*, pages 5389–5395, Jun. 2014.
[40] D. W. K. Ng and R. Schober. Resource Allocation for Coordinated Multipoint Networks With Wireless Information and Power Transfer. In *Proc. IEEE Global Telecommun. Conf.*, pages 4281–4287, Dec. 2014.

[41] M. Chynonova, R. Morsi, D. W. K. Ng, and R. Schober. Optimal Multiuser Scheduling Schemes for Simultaneous Wireless Information and Power Transfer. In *23rd European Signal Process. Conf. (EUSIPCO)*, Aug. 2015.
[42] Q. Wu, M. Tao, D. W. K. Ng, W. Chen, and R. Schober. Energy-Efficient Transmission for Wireless Powered Multiuser Communication Networks. In *Proc. IEEE Intern. Commun. Conf.*, Jun. 2015.
[43] D. Ng and R. Schober. Max-Min Fair Wireless Energy Transfer for Secure Multiuser Communication Systems. In *IEEE Inf. Theory Workshop (ITW)*, pages 326–330, Nov. 2014.
[44] C. Valenta and G. Durgin. Harvesting Wireless Power: Survey of Energy-Harvester Conversion Efficiency in Far-Field, Wireless Power Transfer Systems. *IEEE Microw. Mag.*, 15:108–120, 2014.
[45] T. Le, K. Mayaram, and T. Fiez. Efficient Far-Field Radio Frequency Energy Harvesting for Passively Powered Sensor Networks. *IEEE J. Solid-State Circuits*, 43:1287–1302, 2008.
[46] E. Boshkovska. Practical Non-Linear Energy Harvesting Model and Resource Allocation in SWIPT Systems. Master's thesis, University of Erlangen-Nuremberg., 2015. [Online]. Available: http://arxiv.org/abs/1602.00833.
[47] E. Boshkovska, D. W. K. Ng, N. Zlatanov, A. Koelpin, and R. Schober. *Robust Resource Allocation for MIMO Wireless Powered Communication Networks Based on a Non-linear EH Model.* IEEE Trans. Commun, 2017, accepted for publication.
[48] M. Grant and S. Boyd. CVX: Matlab Software for Disciplined Convex Programming, version 2.0 Beta, 2013. [Online] https://cvxr.com/cvx.
[49] Q. Li and W. K. Ma. Spatically Selective Artificial-Noise Aided Transmit Optimization for MISO Multi-Eves Secrecy Rate Maximization. *IEEE Trans. Signal Process.*, 61:2704–2717, 2013.

Issam Chaour, Ahmed Fakhfakh, and Olfa Kanoun

Radio frequency power transfer for wireless sensors in indoor applications

Abstract: Radio frequency (RF) energy transmission and harvesting techniques have recently become interesting alternative methods to support the new generation of wireless sensor network (WSN). This developing technology enables proactive energy replenishment for wireless control and monitoring applications. With RF power transfer long distances can be overbridged. It is possible to harvest energy from ambient RF energy sources and to convert it into usable electrical form. This chapter presents an investigation of the feasibility of the use of RF energy transfer to overcome the lack of different power sources.

Keywords: Radio frequency energy transmission, RF-DC converter, rectifier, power efficiency, voltage multiplier, Villard topology, Dickson topology

1 Introduction

Listen and control systems with a wireless access point build the main foundation for the improvement of wireless sensor node (WSN) architectures. In addition to the energy management system optimization, WSN needs to increase life cycle duration of the battery and reduce the maintenance intervals [1]. Limited energy supply and subsequent battery replacement are common issues, especially in industrial plants. One solution is energy harvesting, where the ambient energy is converted into usable direct current (DC) form to charge WSN batteries or to run free battery sensors [2]. The main function of energy management is to provide the available energy to the operating node and reduce conversion losses. Using energy harvesting enables energy autonomy. [3]. In some applications, energy harvesting is not able to provide a reliable and efficient energy supply because of the lack of ambient energy such as solar, wind and vibration [4]. To overcome the lack of different energy sources, RF power transmission helps to transfer energy for relatively long distances. In addition, energy harvesting from available RF sources, operating at different frequency bands, enables the supply of low-power systems. RF energy transfer techniques have recently become interesting alternative methods to power the next generation of wire-

Issam Chaour, Olfa Kanoun, Chair of Measurement and Sensor Technology, Technische Universität Chemnitz, 09126 Chemnitz, Germany
Issam Chaour, National Engineering School of Sfax, University of Sfax, Sfax 3038, Tunisia
Issam Chaour, Ahmed Fakhfakh, Laboratory of Technologies for Smart Systems (LT2S), Digital Research Center of Sfax (CRNS), Tunisia

https://doi.org/10.1515/9783110445053-011

less networks. This chapter investigates the feasibility of recharging WSNs by RF power transmission and harvesting energy from RF ambient sources.

2 RF to DC conversion and storage

RF energy harvesting from dedicated transmitters or from ambient RF sources presents an effective alternative to supply WSNs. Demands of low power consumption WSN maintains the ability to use RF power transfer for microwatt power levels. RF energy harvesting allows the handover of the necessary power to the interrogated WSN located far from the transmitter base station. As presented in Figure 1, the transmitted RF energy is collected by a receiver antenna. The rectifier transforms the RF wave to a DC power useful for charging typical storage elements for low power or running battery-free devices. The main goal of an RF harvester is to load high overall power conversion efficiency. This can be attributed to the interface optimization between the antenna and the rectifier circuit (rectenna). The passive RF harvester needs a good reactive matching circuit, in order to minimize signal reflections and losses [5].

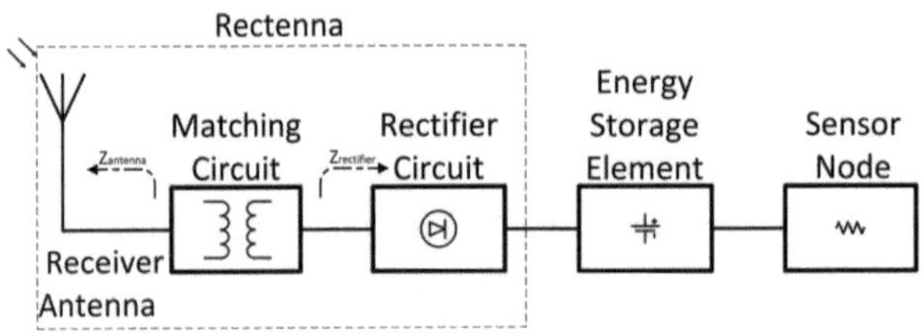

Fig. 1: System architecture of a WSN supplied by RF energy.

Multiple factors can influence the received power P_r, such as the RF source density, the distance between transmitter and receiver antennas, antennas performances and the operating frequency [6]. The radiated power on the receiver antenna placed at a distance d from the transmitter antenna can be expressed by Friis' theorem defined in equation (1) [7]:

$$P_r = P_t G_t G_r \left(\frac{c}{4\pi f}\right)^2 \left(\frac{1}{d}\right)^n e^{-\alpha R} \tag{1}$$

where P_t is the transmitted power, G_t and G_r are, respectively, the gains of receiver and transmitter antennas, c is the speed of light and f is the RF signal frequency; α is the effective decay coefficient in air and is equal to 0.001; n denotes the path loss exponent and is equal to 2 in free space. As shown in Figure 2, P_r depends on multiple factors, such as the transmitted power by the RF source, the performance of the receiver and the transmitter antennas, the operating frequency and the distance between the transmitter and the receiver antennas [8].

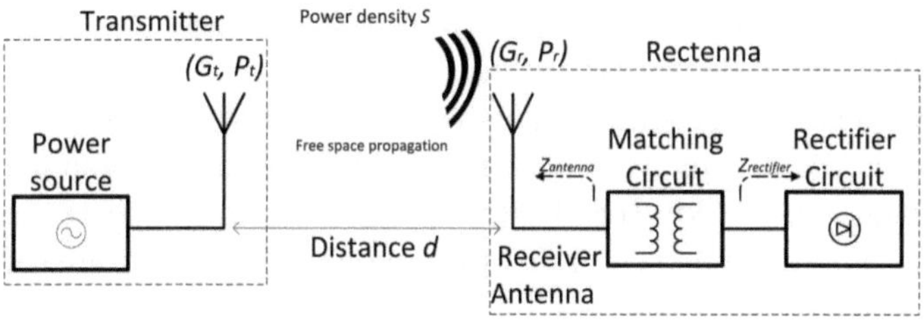

Fig. 2: Far-field RF power transmission for free space propagation.

The path loss P_L, which is expressed in dB, represents the signal strength losses for a propagated RF wave in free space along line of sight direct path [9]. It is expressed in dB and is given by equation (2). Figure 3 shows the effect of the distance on free space path loss for different frequencies:

$$P_L = 10 \log_{10} \left(\frac{4\pi f}{c} \right) \tag{2}$$

The available power at the receiver node is attenuated by 6.02 dB path loss for every doubled distance from the transmitter source. For example, at 915 MHz, the free space path loss for 2 m distance is equal to 37.69 dB, for 4 m distance is equal to 43.71 dB and for 8 m distance is equal to 49.73 dB. It is clear that the path loss increases as much

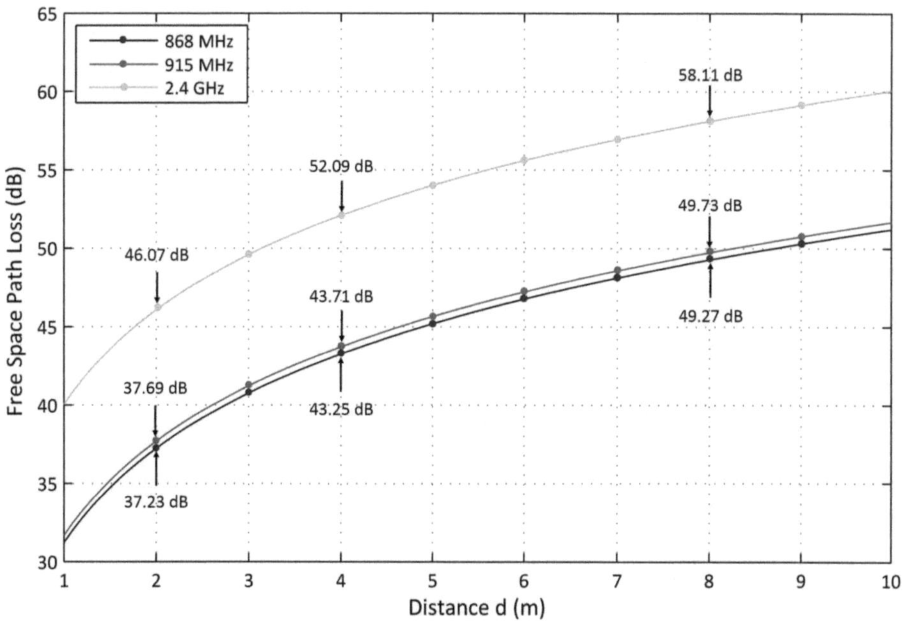

Fig. 3: Distance and frequency influence on the free space path losses.

as the distance increases. Thus, the path loss in free space increases from 2 to 4 m and from 4 to 8 m by 6.02 dB when the distance is double.

The power density S defines the distribution of the transmitted power over a free space propagation. The power density S at the receiving side is calculated from the effective isotropic radiated power EIRP. The RF signal power density S [11], over the transmitter antenna surrounding sphere, is given by equation (3). The sphere area of the antenna radiation is $4\pi d^2$,

$$S = \frac{P_t G_t}{4\pi d^2} = \frac{\text{EIRP}}{4\pi d^2} \tag{3}$$

Many rules specify RF power transmission limits in terms of effective radiated power ERP or EIRP. ERP and EIRP are defined in linear terms as the product of the antenna gain and the input power to the antenna. The antenna gain for ERP is relative to a half-

Tab. 1: Frequency ranges of RFID systems [10].

Frequency range	Description	Transmission power
865.0–865.6 MHz	Ultra-high frequency (UHF) RFID, listen before talk, backscatter coupling, channel spacing: 200 kHz	100 mW ERP Europe
865.6–867.6 MHz	UHF RFID, listen before talk, backscatter coupling, channel spacing: 200 kHz	2 W ERP Europe
867.6–868 MHz	UHF SRD, backscatter coupling, channel spacing: 200 kHz	500 mW ERP Europe
902–928 MHz	UHF SRD, backscatter coupling	4 W EIRP spread spectrum, USA/Canada
2.400–2.483 GHz	Super-high frequency (SHF), backscatter coupling	4 W spread spectrum, USA/Canada
2.446–2.454 GHz	SHF RFID automatic vehicle identification	<0.5 W EIRP outdoor Europe
2.446–2.454 GHz	less then 15% duty cycle, FHSS techniques, should be used	>500 mW 4 W EIRP indoor Europe
5.725–5.875 GHz	SHF, backscatter coupling	4 W USA/Canada, 500 mW Europe

wave dipole antenna, whereas the EIRP is expressed relative to the isotropic antenna gain. ERP is related mathematically to EIRP as expressed in equation (4),

$$\text{ERP} = \text{EIRP} - 2.15\,\text{dB} \tag{4}$$

Since special regulations exist for RF power transmission, it makes sense to use the free license frequency bands or industrial, scientific and medical bands (ISM). They should be classified as either non-specific short-range devices (SRD), wideband data transmission systems or radio frequency identification (RFID) applications [11, 12]. For example, the 867.6–868 MHz band is one of the RFID frequency ranges used in ultra-high frequency SRD applications. For this band, it is allowed to transfer until 500 mW ERP in Europe with 200 kHz of coupling channel spacing. Table 1 shows the standard RF ranges used for RFID and the maximum transmission power.

A lot of calculations and analyses are required to study the efficiency and performance of the RF power transmission. After this approximate study, it is necessary to note that with increase of the frequency, the received power decreases and, thus, the yield power decreases.

3 Reflection coefficient and matching circuit

RF power transfer requires sophisticated circuits for conversion and storage of the available RF ambient energy on the receiver side [4]. This can be reached by the interface optimization between the antenna, and typical RF-DC rectifier circuit. The main aim is to reach a high overall efficiency by minimizing discontinuities and signal reflections [5]. The reflection coefficient is defined by the scattering parameter S_{11} and can be used to evaluate the reflected power from the rectifier to the receiver antenna. When the antenna is coupled to the rectifier circuit, as shown in Figure 4, S_{11} is given by equation (5) [13]. Z_{antenna} and $Z_{\text{rectifier}}$ are the internal impedances of the receiver antenna and the rectifier,

$$S_{11} = \frac{Z_{\text{rectifier}} - Z_{\text{antenna}}}{Z_{\text{rectifier}} + Z_{\text{antenna}}} \tag{5}$$

where S_{11} is a complex quantity. It also determines the return loss L_r expressed in dB and given by equation (6):

$$L_r = -20\log(|S_{11}|) \tag{6}$$

The return loss is a metric of the reflected power, which is not transmitted to the rectifier and, therefore, returned to the antenna. For that, a reactive matching circuit connects the antenna to the rectifier circuit under optimized operating conditions [6]. The matching network ensures the transfer of the maximum captured power by the receiver antenna to the rectifier and reduces the reflected power at the operating frequency. It is formed by different connection configurations of inductors and capacitors [14].

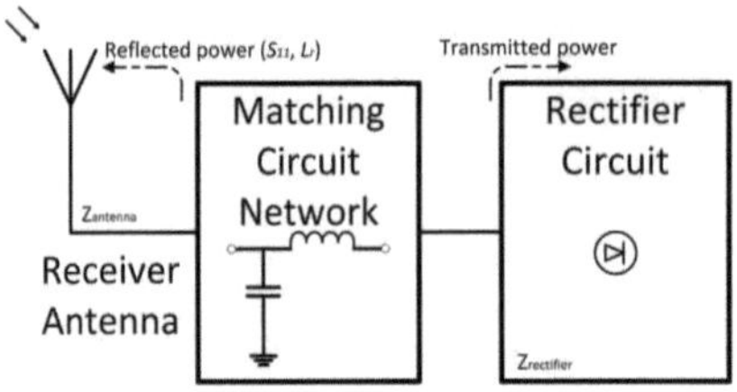

Fig. 4: Return loss and impedance transformation network.

4 RF-DC rectifier circuits for RF power transfer

For RF power transfer applications, the main task to be fulfilled after the RF to DC conversion, is to maximize the amount of DC power and to reduce the power losses caused by the rectifier circuit. Many literature references show that when the incident RF power is low, the rectifier circuit efficiency is also low. So, it is recommended to improve the rectifier circuit for low RF incident power [15]. Power conversion efficiency (PCE) or RF to DC conversion efficiency, is an important rectification metric for RF power transfer [16] and is given by equation (7):

$$\text{PCE} = \frac{\text{DC output power}}{\text{incident RF power} - \text{reflected power}} \tag{7}$$

Many RF-DC rectifier structures, suitable for RF power transfer, are mentioned in the state of the art [17–21]. The basic element for an RF-DC rectifier circuit is the Schottky diode, which is characterized by a low drop voltage and a high operating frequency.

4.1 Schottky diode

A Schottky diode is based on a contact between a metal and a semiconductor, which is doped n or p. The use of the metal with the doped semiconductor increases the switching frequency. Electrons flow across the metal and the semiconductor junction. The potential difference created between the semiconductor and the metal, defines the zero bias Schottky barrier. Schottky diodes are generally used for small signal detector applications because of their low turn on voltage [22].

Schottky diodes are modeled according to the non-linear property of the curve behavior $I_D(V_D)$, shown in Figure 5 [18]; I_D denotes the diode current and V_D denotes the diode forward voltage. The $I_D(V_D)$ characteristic of a Schottky diode defines three main regions. The Schottky diode conducts in the reverse condition for an input voltage V less than the reverse breakdown voltage V_{bv}. Between the reverse breakdown voltage V_{bv} and the turn on threshold voltage V_Y, the Schottky diode is blocked, and a low level leakage current flows across the diode. For an input voltage V higher than the turn on threshold voltage V_Y, the Schottky diode is supposed to forward current that is proportional to the voltage.

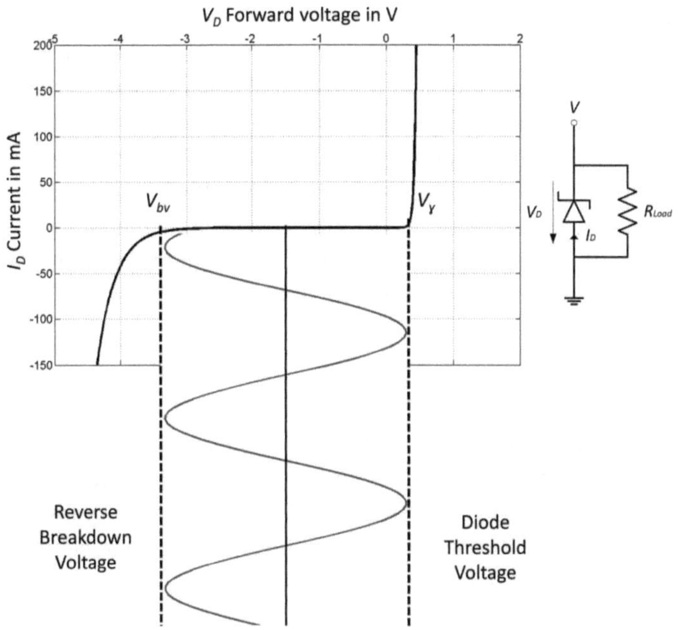

Fig. 5: A typical Schottky diode current voltage characteristic [23].

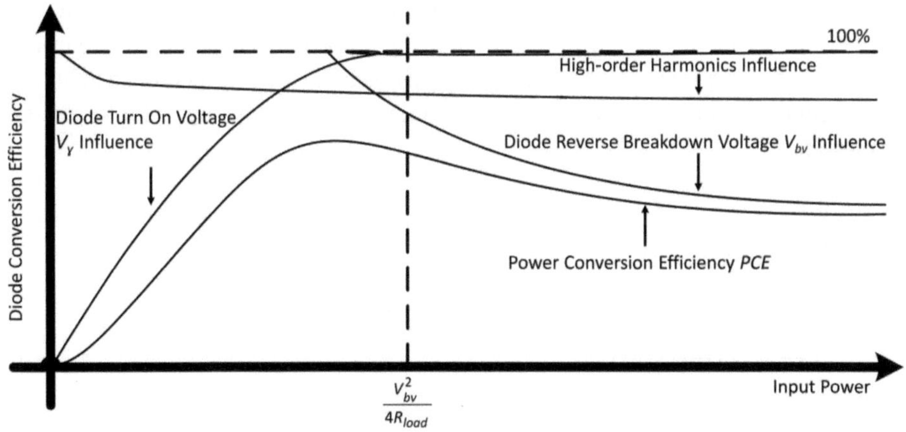

Fig. 6: Input power, reverse breakdown voltage V_{bv} and turn on threshold voltage V_γ influences on the Schottky diode power conversion efficiency [24].

Figure 6 presents the influence of the input power, the reverse breakdown voltage V_{bv} and the turn on threshold voltage V_γ on the Schottky diode power conversion efficiency PCE. For a low input power, the turn on voltage V_γ limits the Schottky diode PCE. This limitation is caused by the non-sufficient incident power to overcome the barrier V_γ. The Schottky diode PCE increases as the input power increases, until a restriction level generated by the higher-order harmonics. The higher-order harmonics is caused by the diode non-linearity. The diode PCE gradually decreases when the voltage across the Schottky diode surpasses the half reverse breakdown voltage V_{bv} [24]. The reverse breakdown influence becomes leading when the input power is higher than a critical power level P_c, which is expressed in equation (8). R_{Load} denotes the DC resistance connected to the Schottky diode. At the end, when the input power increases, the Schottky diode provides a constant DC output power and, therefore, the diode PCE decreases. The optimal Schottky diode PCE is related to the turn on the threshold voltage V_γ, the high-order harmonics and the reverse breakdown V_{bv},

$$P_c = \frac{V_{bv}^2}{4R_{Load}} \tag{8}$$

Figure 7 illustrates the non-linear model for a Schottky diode. A packaged Schottky diode is modeled as a non-linear junction resistance R_j in parallel with a non-linear junction capacitance C_j; R_S is the bulk resistance, L_S and C_P are, respectively, the packaging inductance and capacitance [25]. The non-linear current characteristic is defined by the Richardson expression [26–29] in equation (9),

$$I = I_S \left[\exp\left(\frac{qV_j}{\eta_0 KT}\right) - 1\right] \tag{9}$$

I_S denotes the reverse saturation current of the Schottky diode; V_j is the voltage losses of the Schottky diode rectifying junction; K is the Boltzmann constant; T denotes the temperature and q is the electron charge; η_0 denotes the ideality factor of a Schottky diode. Schottky diodes are mostly used in common rectifier topologies for RF energy harvesting.

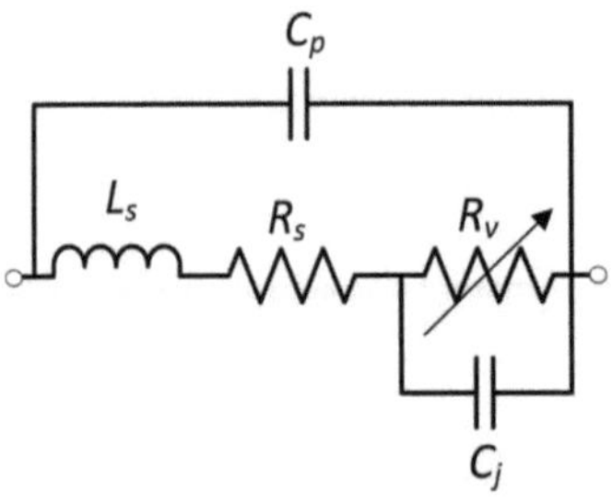

Fig. 7: Equivalent model of the Schottky diode [25].

4.2 Topologies for RF to DC rectifiers

A rectenna, based on one Schottky diode connected to the receiver antenna, reaches higher efficiency. However, the output DC voltage achieved is not enough to drive the

WSN or to charge its battery [30–32]. For RF energy transfer, charge pump topologies are used most. Charge pumps topologies not only convert the RF wave to DC voltage, but also boost the output voltage. The following sections present an overview on typical used rectifier topologies.

4.2.1 Single Schottky diode half-wave rectifier

The half-wave rectifier circuit for RF energy harvesting is based on one single Schottky diode. It rectifies the RF wave from the positive half cycle and blocks the negative one. A smoothing capacitance C_1 is connected in series to the diode in order to filter the output voltage across the resistive load R_{Load}. Generally, the half-wave rectifier circuit produces low voltage and high conversion efficiency [30]. Figure 8 presents the topology of a half-wave rectifier circuit for RF energy harvesting [31–33].

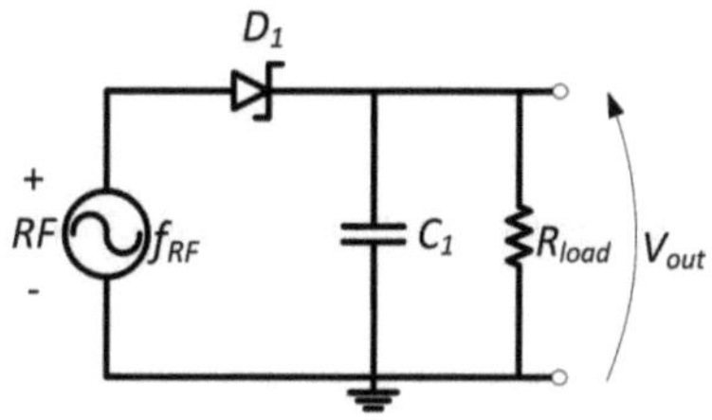

Fig. 8: Half-wave rectifier circuit for RF energy harvesting.

4.2.2 Single-shunt rectifier circuit

As presented in Figure 9, a single-shunt rectifier circuit is a simple topology of a Schottky diode followed by a quarter-wave transformer. The rectifier output voltage is filtered by the capacitance C_1. The single-shunt rectifier circuit reaches high efficiency [31, 34, 35]. It is characterized by low parasitic series losses. Moreover, the use of a quarter-wave transformer reduces the reflected wave and improves the converted power. In general, the output voltage of single-shunt rectifier circuits is low. Nevertheless, other topologies are used to boost the output voltage.

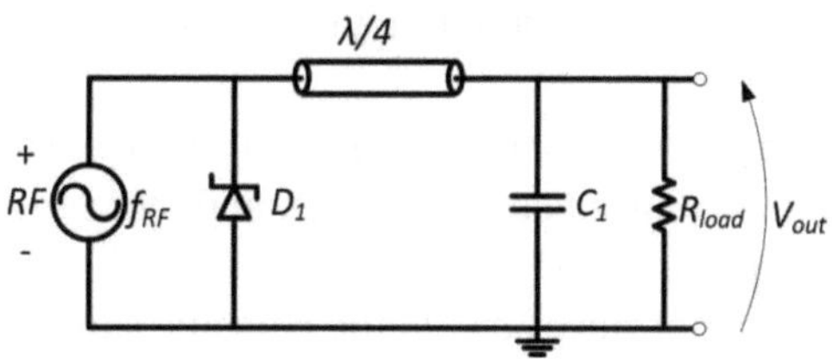

Fig. 9: Single shunt-rectifier circuit.

4.2.3 Voltage multiplier circuit

The voltage multiplier structure is a full-wave RF to DC rectifier circuit. Two configurations are arranged in a cascade using Schottky diodes to provide a passive voltage offset before rectification [36]. The conventional voltage multiplier rectifier forms a peak rectified by D_2 and C_2, while a voltage clamp is formed by C_1 and D_1.

Figure 10 shows the voltage multiplier circuit. The circuit can be also called a voltage doubler. The RF input signal is rectified during the positive phase. The stored charge on the input capacitor C_1 during the negative phase is transmitted to the output capacitor C_2 during the live phase of the RF input signal. Thus, the voltage on C_2 is roughly two times the peak voltages of the RF source minus two times the turn on voltage of the diodes [37].

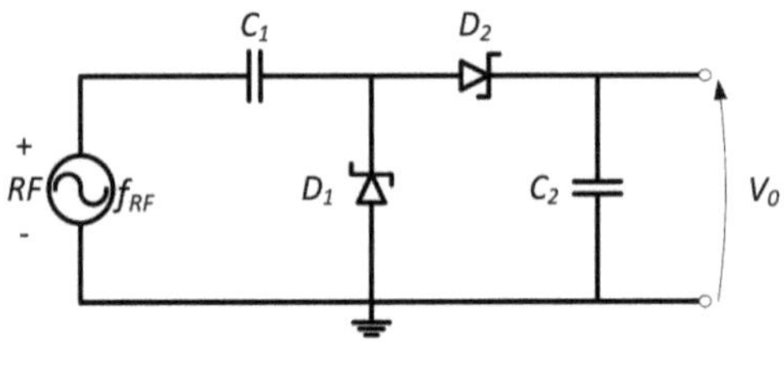

Fig. 10: One-stage voltage multiplier circuit.

4.2.4 Multiple-stage voltage multiplier circuit

The one-stage voltage multiplier can be extended to n stages in cascade to achieve a higher DC output voltage level. Succeeding stages in cascade provides more voltage than using only one stage. The rectifier output voltage V_{out} is expressed by equation (10), where V_{O} is the one-stage open-circuit voltage, R_{O} denotes the internal resistance of the circuit and R_{Load} is a resistance load,

$$V_{\text{out}} = \frac{nV_{\text{O}}R_{\text{Load}}}{nR_{\text{O}} + R_{\text{Load}}} \tag{10}$$

In Villard topology, stages are connected in series and behave similarly to many batteries in cascade. The output voltage of Villard topology is multiplied by the number of stages. Figure 11 illustrates a two-stage Villard voltage multiplier circuit. Each Villard stage acts as a passive voltage booster and provides a DC offset voltage for the next stage.

The number of stages used in the rectifier has a major influence on the output voltage circuit. One rectifier stage may yield to an unused output voltage, and too many Villard stages increase the series' reactive impedance and reduce the PCE of the circuit. Practical constraints limit the optimal number of stages due to the parasitic capacitance effect of each stage. To overcome these problems, Dickson proposed the circuit presented in Figure 12 [38]. It is characterized by an effective voltage multiplication

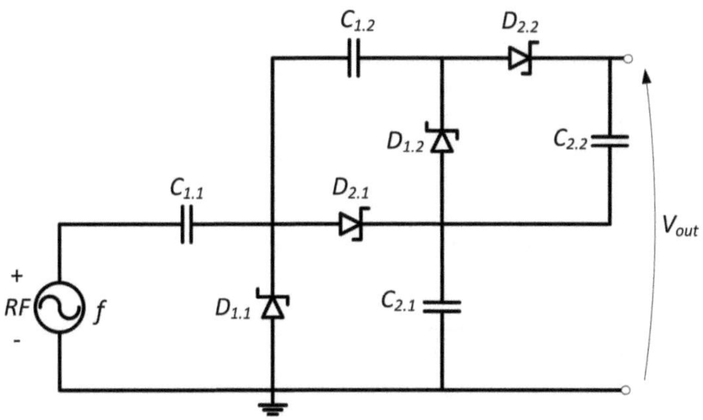

Fig. 11: Two-stage Villard voltage multiplier circuit.

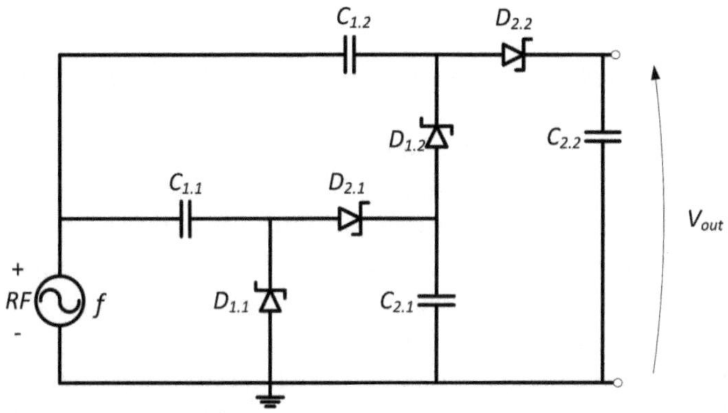

Fig. 12: Two-stage Dickson voltage multiplier circuit.

with relatively low parasitic effects [39]. Diode nodes in Dickson topology are coupled via parallel capacitors, which reduces the parasitic capacitance of the circuit. A comparison between the two architecture topologies was made in [40]. For low voltages, the two topologies offer similar performances.

Figure 13 illustrates a modified Greinacher charge pump topology for a multi-stage voltage multiplier circuit. The architecture of the modified Greinacher circuit is a symmetric, which limits the reflected harmonic. Furthermore, the circuit is based on parallel capacitors to reduce the input impedance of the circuit and losses [41, 42].

During the positive wave, the couples (C_1, D_1) forward the voltage up to the rectifier elements (C_2, D_2), which rectify the half-wave to DC voltage. The capacitor C_3 charges through D_3. During the negative wave, the couples (C_3, D_3) forward the voltage up to the rectifier elements (C_4, D_4), which rectify the half-wave to DC voltage. The capacitor C_1 charges through D_1. When the circuit reaches the equilibrium state, it provides a constant output power that is useful for the load.

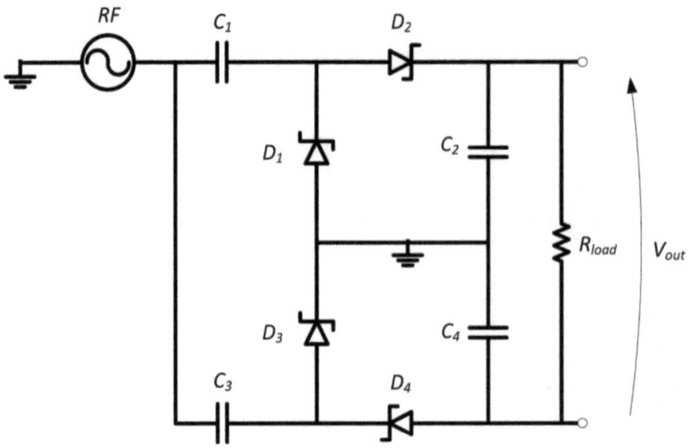

Fig. 13: Modified Greinacher rectifier.

4.3 RF to DC rectifier based on a diode-connected CMOS transistor

Another rectifier element that can be used for an RF to DC conversion, is the diode connected transistor. It offers a solution to reduce the threshold voltage based on a standard complementary metal oxide semiconductor (CMOS) process. The diode-connected CMOS transistor offers a solution for Schottky diode limitations by implementing integrated circuit (IC) rectifiers. The high costs and the complex processing are the main disadvantages of the RF to DC rectifier circuit based on a diode-connected CMOS transistor. As presented in Figure 14, the diode behavior can be achieved by connecting a P-type (PMOS) or a N-type (NMOS) transistor drain and gate together, which maintains the saturation of the transistor in the forward bias condition [43].

When V_{GS}, the potential between the transistor gate and source is lower than the transistor's threshold voltage V_{th}; the transistor is blocked and a low leakage current flows through the transistor. When V_{GS} is higher than V_{th}, a current flows through the diode-connected transistor from the drain to the source terminal.

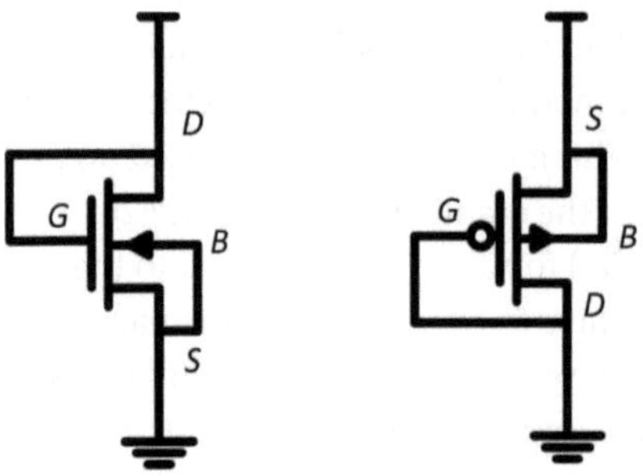

Fig. 14: Diode-connected (a) NMOS transistor and (b) PMOS transistor.

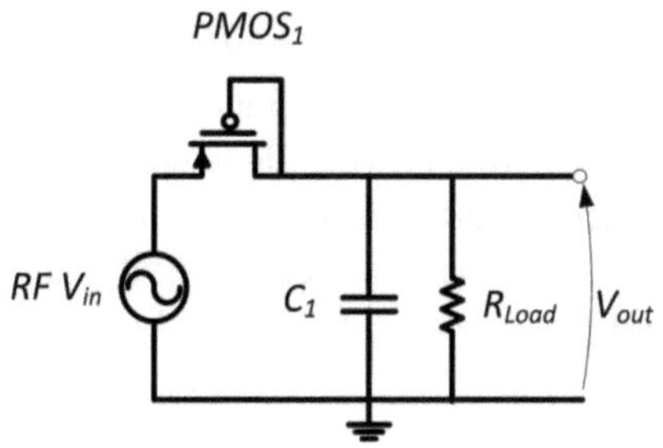

Fig. 15: Half-wave rectifier circuit based on a diode-connected PMOS.

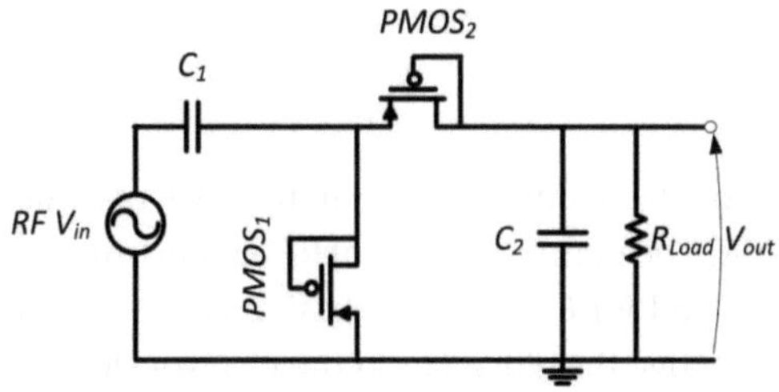

Fig. 16: Voltage multiplier circuit based on a diode-connected PMOS.

During the forward bias, the transistor losses are caused by its internal on-resistance, whereas the losses are mainly generated by the leakage current over the reverse-bias process [44].

For a typical process of CMOS transistor design, the PMOS transistor is named for its lower threshold voltage compared to the NMOS transistor [45]. Therefore, the diode-connected PMOS is frequently used for low-power consumption circuit and RF wave-rectifying applications [45]. Figure 15 presents a design for half-wave rectifier based on a diode-connected PMOS transistor. Figure 16 presents the voltage multiplier circuit design based on a diode-connected PMOS transistor [46].

The diode-connected CMOS transistor generates a voltage drop caused by the transistor threshold voltage V_{th}. Many solutions and techniques have been developed to reduce the transistor drop voltage. One solution is the PMOS floating gate drop voltage compensation of Figure 17 [47]. The idea of this technique consists of connecting

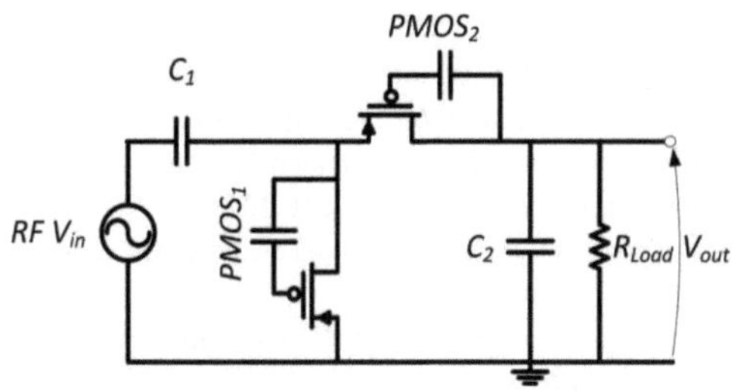

Fig. 17: Voltage multiplier circuit based on a diode-connected floating gate PMOS.

the transistor drain and gate terminals via a capacitor. First, the capacitor charges until a voltage level closed to V_{th} is achieved. The charged capacitor represents a series voltage source, which limits the drop voltage generated by the diode-connected PMOS transistor. The use of capacitance can produce losses due to the leakage current. The evolution of the technology used and the IC design process has encouraged many research groups to investigate and propose solutions for CMOS transistor threshold cancellation and reduce losses [17, 44, 48–50].

5 Conclusion

With RF energy transfer, it is possible to provide an output power in the level of some microwatts useful to supply low-power WSN. The main challenge of RF energy transfer is the low level of the received power and the RF-DC power conversion efficiency. In this chapter, we reported on methods to improve the efficiency at the receiver side. In order to be able to convert RF energy even from low-density ambient RF sources, it is necessary to optimize the system design. An overview on the RF energy transfer and its important metrics is presented in this chapter. The structure of the RF-DC conversion and storage is detailed. Different rectifier topologies for a low-input power to improve the voltage multiplication are discussed. The techniques proposed for enhancing the PCE and minimizing the rectifying losses are detailed.

Bibliography

[1] S. Kawasaki. The Green Energy Harvesting Winds by the RF/Microwave Power Transmission. In *Proceedings of the 2013 Wireless Power Transfer, Perugia, Italy, 15–16 May 2013*, pages 111–114, 2013.

[2] A. C. Patel, M. P. Vaghela, H. Bajwa, and P. K. Patra. Power Harvesting for Low Power Wireless Sensor Network. In *Proceedings of the Antennas and Propagation Conference, Loughborough, UK, 16–17 November 2009*, pages 633–636, 2009.

[3] W. Zhao, K. Choi, S. Bauman, T. Salter, D. A. Lowy, M. Peckerar, and M. K. Khandani. An Energy Harvesting System Surveyed for a Variety of Unattended Electronic Applications. *Solid State Electron*, 79:233–237, 2013.

[4] J. Masuch, M. Delgado-Restituto, D. Milosevic, and P. Baltus. Co-Integration of an RF Energy Harvester into a 2.4 GHz Transceiver. *IEEE J. Solid State Circuits*, 48:1565–1574, 2013.

[5] H. Shadmehr, F. Grimaccia, G. Gruosso, M. Mussetta, and R. Zich. Optimized Antenna for Low UHF Range Wireless Power Transfer. *Int. J. Commun. Antenna Propag.*, 3:21, 2013.

[6] A. Dolgov, R. Zane, and Z. Popovic. Power Management System for Online Low Power RF Energy Harvesting Optimization. *IEEE Trans. Circuits Syst.*, 57:1802–1811, 2010.

[7] M. Nariman, F. Shirinfar, A. P. Toda, S. Pamarti, A. Rofougaran, and F. A. Flaviis. Compact 60-GHz Wireless Power Transfer System. *IEEE Trans. Microwave Theory Tech.*, 64:2664–2677, 2016.

[8] U. Olgun, C. Chen, and J. Volakis. Investigation of Rectenna Array Configurations for Enhanced RF Power Harvesting. *IEEE Antennas Wirel. Propag. Lett.*, 10:262–265, 2011.

[9] J. D. Parsons. *Fundamentals of VHF and UHF Propagation*. John Wiley and Sons, Ltd, 2000.

[10] Federal Communications Commission. *Code of Federal Regulations*. Federal Communications Commission: Washington, DC, USA, 2002.

[11] ERC/REC 70-03. Recommendation Adopted by the Frequency Management, Regulatory Affairs and Spectrum Engineering Working Groups. Available online: http://www.arcep.fr/fileadmin/reprise/dossiers/frequences/ERC-REC-70-03E-version02.PDF (accessed on 3 March 2017).

[12] H. Visser and R. M. Vullers. RF Energy Harvesting and Transport for Wireless Sensor Network Applications: Principles and Requirements. *Proc. IEEE*, 101:1410–1423, 2013.

[13] J. Orfanidis. *Electromagnetic Waves and Antennas*, chapter 14: S-Parameters, pages 664–708. 2016.

[14] J. Rahola. Power Waves and Conjugate Matching. *IEEE Transactions on Circuits And Systems—I*, 55(1):92–96, 2008.

[15] H. Nishimoto, Y. Kawahara, and T. Asami. Prototype Implementation of Ambient RF Energy Harvesting Wireless Sensor Networks. In *Proceedings of the IEEE Sensor Conference, Waikoloa, HI, USA, 1–4 November 2010*, pages 1282–1287, 2010.

[16] J. Curty, N. Joehl, C. Dehollain, and M. Declercq. Remotely powered addressable UHF RFID integrated system. *IEEE J. Solid State Circuits*, 40:2193–2202, 2005.

[17] F. Yuan and N. Soltani. Design Techniques for Power Harvesting of Passive Wireless Microsensors, Circuits and Systems. In *MWSCAS 2008. 51st Midwest Symposium, 10–13 August 2008*, pages 289–293, 2008.

[18] C. R. Valenta and G. D. Durgin. Harvesting Wireless Power. *IEEE Microwave magazine*, pages 108–120, June 2014.

[19] H. J. Visser. Indoor Wireless RF Energy Transfer for Powering Wireless Sensors. *Radioengineering*, 21:963–973, 2012.

[20] S. S. Chouhan, M. Nurmi, and K. Halonen. Efficiency enhanced voltage multiplier circuit for RF energy harvesting. *Microelectronics Journal, Volume*, 48:95–102, 2016.

[21] M. Ghovanloo and K. Najafi. Fully Integrated Wideband High-Current Rectifiers for Inductively Powered Devices. *IEEE Journal of Solid-State Circuits*, 39:1976–1984, 2004.

[22] H. Watson. *Microwave semiconductor devices and their circuit applications*. Schottky-barrier Devices, 1969.

[23] C. R. Valenta. *Microwave-Energy Harvesting At 5.8 Ghz For Passive Devices*. PhD thesis, Georgia Institute of Technology, August 2014.

[24] T. Yoo and K. Chang. Theoretical and Experimental Development of 10 and 35 GHz Rectennas. *IEEE Transactions on Microwave Theory And Techniques*, 1259–1266, 1992.

[25] B. L. Sharma. *Metal-Semiconductor Schottky Barrier Junctions and Their Applications*. Springer US, 1984.

[26] C. T. Sah. P/N and Other Junction Diodes. In *Fundamentals of solid-state electronics*. World scientific publishing Co. Pte. Ltd., 1993.

[27] D. A. Neamen. *Semiconductor physics and device: basic principles*. McGraw-Hill Book, 3rd edition, 2002.

[28] D. Chattopadhyay. *Electronics (Fundamentals and Applications)*. New Age International (P) Limited, 2006.

[29] S. M. Sze and K. K. Ng. *Physics of Semiconductor Devices*. Wiley Interscience publication, Wiley-Interscience, 2007.

[30] M. K. Kazimierczuk and D. Czarkowski. *Resonant Power Converters*. New York, John Wiley and Sons, 1995.

[31] H. Takhedmit, L. Cirio, B. Merabet, and O. Picon. A 2.45-GHz dual-diode rectenna and rectenna arrays for wireless remote supply applications. *International Journal of Microwave and Wireless Technologies, June*, 2011.
[32] A. Douyere and F. A. J. D. Lan Sun Luk. High efficiency microwave rectenna circuit: modeling and design. *Electronics Letters*, 44:1409–1410, 2008.
[33] J. A. G. Akkermans, G. J. N. D. M. C. van Beurden, and H. J. Visser. Analytical models for low-power rectenna design. *IEEE Antennas Wirel. Propag. Lett.*, 4:187–190, 2005.
[34] Y. J. Ren and K. Chang. 5.8-GHz circularly polarized dual-diode rectenna and rectenna array for microwave power transmission. *IEEE Transactions on Microwave Theory and Techniques*, 54:1495–1502, 2006.
[35] B. Strassner and K. Chang. 5.8-GHz circularly polarized rectifying antenna for wireless microwave power transmission. *IEEE Transactions on Microwave Theory and Techniques*, 50:1870–1876, 2002.
[36] C. H. P. Lorenz, S. Hemour, and K. Wu. Physical Mechanism and Theoretical Foundation of Ambient RF Power Harvesting Using Zero-Bias Diodes. *IEEE Transaction Microwave Theory Technique*, 64:2146–2158, 2016.
[37] H. Jabbar, Y. Song, and T. Jeong. RF energy harvesting system and circuits for charging of mobile devices. *IEEE Transactions on Consumer Electronics*, 56:247–253, 2010.
[38] J. F. Dickson. On-chip high-voltage generation in MNOS integrated circuits using an improved voltage multiplier technique. *IEEE J. Solid State Circuits*, 11:374–378, 1976.
[39] T. Tanzawa and T. Tanaka. A dynamic analysis of the Dickson charge pump circuit. *IEEE J. Solid State Circuits*, 32:1231–1240, 2013.
[40] H. Yan, J. M. Montero, A. Akhnoukh, and J. N. Burghartz. An integration scheme for RF power harvesting. In *Proceedings of the STW Annual Workshop on Semiconductor Advances for Future Electronics and Sensors, Utrecht, The Netherlands, 17–18 November 2005*, pages 17–18, 2005.
[41] J. P. Curty, M. Declercq, C. Dehollain, and N. Joehl. *Design and Optimization of Passive UHF RFID Systems*, chapter Analysis of the Modified-Greinacher Rectifier, pages 17–35. Springer, 2017.
[42] J. Curty, N. Joehl, F. Krummenacher, C. Dehollain, and M. Declercq. A model for μ-power rectifier analysis and design. *IEEE Transactions on Circuits and Systems*, 52:2771–2779, 2005.
[43] P. E. Allen and D. R. Holberg. *CMOS Analog Circuit Design*. Oxford University Press, New York, 2nd edition, 2002.
[44] K. Kotani. High-efficiency differential-drive CMOS rectifier for UHF RFIDs. *IEEE J. Solid-State Circuits*, 44:3011–3018, 2009.
[45] W. Wang, H. Wong, and Y. Han. A high-efficiency full-wave CMOS rectifying charge pump for RF energy harvesting applications. *Micro-electronics Journal*, 46:1447–1452, 2015.
[46] D. S. Liu, F. B. Li, X. C. Zou, Y. Liu, X. M. Hui, and X. F. Tao. New Analysis and Design of a RF Rectifier for RFID and Implantable Devices. *Sensors*, 24, 2011.
[47] T. Le, K. Mayaram, and T. Fiez. Efficient far-field radio frequency energy harvesting for passively powered sensor networks. *IEEE Journal of Solid-State Circuits*, 43:1287–1302, 2008.
[48] T. Umeda. A 950-MHz rectifier circuit for sensor network tags with 10-m distance. *IEEE J. Solid-State Circuits*, 41:35–41, 2006.
[49] H. Nakamoto. A passive UHF RF identification CMOS tag IC using ferroelectric RAM in 0.35-μm technology. *IEEE J. Solid-State Circuits*, 42:101–110, 2007.
[50] K. Kotani and T. Ito. High efficiency CMOS rectifier circuit with self-Vth-cancellation and power regulation functions for UHF RFIDs. In *IEEE Asian Solid-State Circuits Conf.*, pages 119–122, Nov. 2007.

Mohammad Haerinia

Modeling and simulation of inductive-based wireless power transmission systems

Abstract: This chapter studies an inductive-based wireless power transfer system for low-power applications at short distances. The transferring power system has been modeled, simulated and analyzed via finite element method. A wireless power transfer system includes important parts such as coil, core and driver. In this chapter, the important parts of an inductive power transfer system have been analyzed. Receiving and transmitting printed spiral coils are designed in an optimized procedure. The experimental results were in a good agreement with the simulation results. Moreover, based on the performed modulation and simulation the use of the pot core as the receiving core is proposed. It is concluded that this type of core can improve magnetic flux density in the receiving side. Different geometries of coils for transmitting side have been modeled and simulated. An electromagnetic analysis has been done; the experimental result was in a good agreement with the simulation result. This work presents an efficient perspective to coil design.

Keywords: Inductive power transfer, finite element method, modeled, simulated, experimental result

1 Introduction

Wireless power transfer systems have attracted great attention recently [1]. Power transfer has been electrified for environmental, energy-related and other reasons for many years [2]. There are several schemes for wireless power transmission as inductive, capacitive, laser, microwave, lighting applications, etc. Induction power transfer (IPT) is the most popular and has been widely studied in the last decades [3, 4]. Inductive power transfer is the wireless transformation of electrical power over distances. Some applications of wireless power transfer include wireless charging and powering implanted electronic devices, medical devices, mobile electronics, implantable medical devices and powered radio way electric vehicles [5–8]. Inductive power transfer (IPT) systems have been widely used over the past decade for transmitting tens to hundreds of watts [9]. This technology has proved to be essential [10]. This method has been known as a technique for delivering small amounts of energy to remote devices, however, there are other power transfer methods including infrared and radio frequency (RF) methods, which all have their own limitations. The restriction regard-

Mohammad Haerinia, College of Engineering and Mines, University of North Dakota, Grand Forks, USA.

https://doi.org/10.1515/9783110445053-012

ing infrared technology is requiring a direct line of sight between the receiver and the transmitter, and RF technology has a low power limitation. A safe and wireless method for transferring energy is inductive coupling [11, 12]. The main drawbacks of IPT systems include the difficult mechanical coupling process between the primary and the secondary side, and also the fact that the air gap is less than 60 mm, and its typical high power density [13]. Another disadvantage of this technology is the fact that the transmitter and receiver need to be aligned [14]. The advantages of inductive-based wireless power transfer are its safety and high efficiency at short distances. According to the features and drawbacks of other technologies, inductive coupling is preferred for the wireless transformation of power [11].

2 Fundamental principles of the inductive coupling approach

Figure 1 illustrates an inductive coupling-based WPT. This is a near-field transmission technique. The transmitting side generates an alternating magnetic field that leads to inducing a voltage across the receiving side. Power transmission efficiency is higher when the transmitter (Tx) and the receiver (Rx) are nearby and aligned.

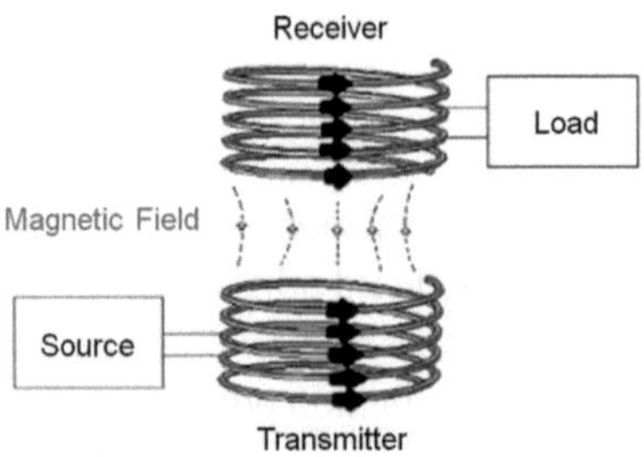

Fig. 1: Near-field transmission technique.

Inductive power transfer depends on a number of parameters, such as the distance between the coils, frequency, current excitation and the geometry of the coils [15]. The basis of inductive coupling includes the use of the fundamental laws of physics. The calculation of the magnetic field, generated by current distribution throughout the coils, can be completed by using Biot–Savart's law [14],

$$B = \frac{\mu_0}{4\pi} \oint \frac{I d_r \times r}{|r|^3} \tag{1}$$

where μ_0 is the permeability of free space, I is the current in the transmitter coil, d_I is the length of the differential element of the wire and r is the full displacement vector from the wire element.

By assuming a homogeneous flux density distribution in the air gap and neglecting fringing flux, the air gap reluctance can be calculated as:

$$R_g = \frac{l_g}{\mu_0 A_g} \tag{2}$$

where l_g, A_g and μ_0 are the air gap length, air gap cross-section and the permeability of free space, respectively. The relation (2) is only accurate when the distance between the coils is very small compared to the dimension of the air gap cross-section [16].

The loss of a ferrite core is a function of three components: temperature T, frequency f and flux density B. This loss is dependent on the ferrite material and core shape. The temperature dependence may be approximated by means of a third-order polynomial, as below approximation applies for the frequency dependence,

$$P(f) \sim \hat{f}((1+a)) \qquad 0 \le a \le 1 \tag{3}$$

Below approximation applies for the flux density dependence,

$$P(B) \sim \hat{B}((2+b)) \qquad 0 \le b \le 1 \tag{4}$$

Core shape and the type of material influence the coefficients a and b. Also, the coefficients of the defined quantity and the relevant parameter have a mutual dependence on another [17].

The induced voltage over the receiver coil can be computed by using Faraday's law [14]. The relation can be written as:

$$V_{\text{Ind}} = -\frac{\partial}{\partial t} \oint B \cdot ds \tag{5}$$

An inductive power transfer system is comparable to that of a transformer [18]. An alternating current in a transmitting coil generates a varying magnetic field that induces a voltage across the terminals of a receiving coil. Power transmission efficiency is higher when the transmitter coil and the receiver coil are close and aligned. Inductive coupling is a popular and widely utilized technology, with numerous applications, ranging from everything to electric toothbrushes, cell charging and medical implants [19].

3 Modeling and simulation of inductive power transfer (IPT) systems

3.1 Resonant inductive coupling as a potential means for WPT to printed spiral coils

This section proposes an inductive coupled wireless power transfer system in order to analyze the relationship between the induced voltage and the distance of resonating inductance in printed circuit spiral coils. The resonant frequency produced by the circuit model of the proposed receiving and transmitting coils is analyzed by simulation and laboratory experiments. The outcome of the two results is compared to verify the validity of the proposed inductive coupling system. Experimental measurements are consistent with simulations over a range of frequencies spanning the resonance. By optimizing the coil design, the quality factor can be improved [6, 20]. In this work, we design the shape of a spiral coil presented in [6] so that it is efficient for transferring high powers. The approximate expressions for the inductance of circular planar inductors have been obtained using accurate expressions for planar spiral inductances, which were partially verified by approximately 60 measurements reported in [21]. By analyzing the mathematical expression regarding determining the circuit capacitors and inductors, the resonant coupling in [22] was obtained. The inductively coupled system was obtained by modifying this procedure for our aims, and it was verified with simulation results. The proposed diameters for the coils are 45.2 mm and 36.4 mm, and the air gap is considered to vary from 5 mm to 15 mm. The experimental results were compared with simulations, and the design procedure was verified. This work provides the fundamental concept of resonant near-field power transmission and the relation of induced electromotive force and the air gap along the transceiver.

3.1.1 Design procedure of spiral coils

The proposed coil shown in Figure 2 was optimized by evolving the shape of the spiral coils. In this section, the design of the spiral coil is described, and coil specifications for the effective permeability are presented [23]. The variables that parameterize the shape of the coil are the wire thickness (t), inner diameter (d_{in}), outer diameter (d_{out}), number of turns (N) and the space between turns (s). Transmitting and receiving coil specifications can be seen in Table 1.

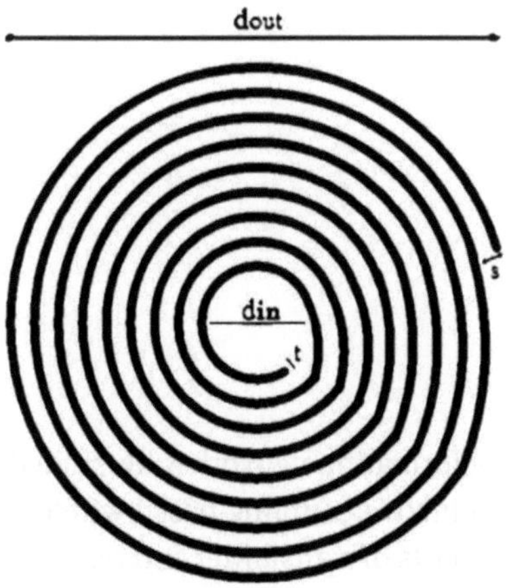

Fig. 2: Printed spiral coil.

Tab. 1: Specifications of transmitter and receiver.

Parameter	Transmitter	Receiver
Inner diameter (d_{in}) (mm)	10	10
Outer diameter (d_{out}) (mm)	45.2	36.4
Number of turns (N)	8	6
Trace width (t) (mm)	0.8	0.8
Spacing (s) (mm)	1.4	1.4

Several equations have been proposed for approximating L [20]. To design the method mentioned, we adopted equation (6) from [21],

$$L = \frac{\mu_0 n^2 d_{avg}}{2}\left[\ln\left(\frac{2.46}{\gamma}\right) + 0.20\gamma^2\right] \tag{6}$$

$$\gamma = \frac{d_{out} - d_{in}}{d_{out} + d_{in}} \tag{7}$$

where n is the number of turns, and d_{out} and d_{in} are the outer and inner diameters of the coil, respectively; $d_{avg} = (d_{out} + d_{in})/2$ and γ is a parameter defined as fill factor [20].

3.1.2 Circuit analysis of a resonant coupled system

The operation of an inductive power transfer system can be compared to an air core transformer. The resonant coupling will boost power transfer efficiency, and then capacitors, which have been connected to both sides, make a resonant inductive coupling system [18]. The circuit can be classified according to the mode of connection of the capacitor to the coil: series-series, series-parallel, parallel-series and parallel-parallel [18, 24]; here the series-parallel mode is proposed. A simplified schematic of wire-

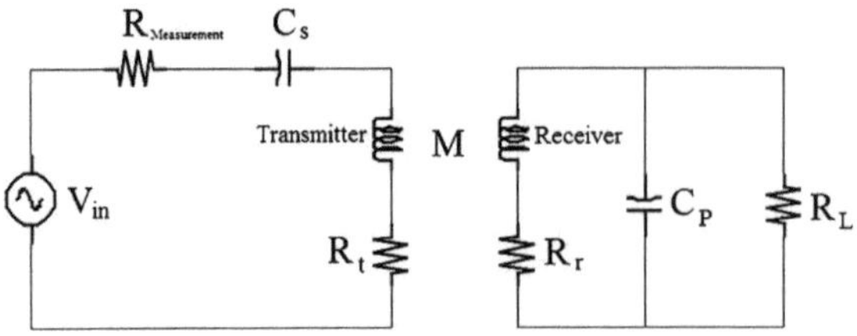

Fig. 3: Equivalent circuit of an inductive coupling system.

less power transfer system using Series-Parallel transformation network is illustrated in Figure 3. This figure is the equivalent circuit of inductive coupling-based wireless power transfer where resistors R_t and R_r are resistive parts of transmitter and receiver. The resistor $R_{measurement}$ is considered as the measurement resistance; C_S, C_P and R_L are the capacitance of the transmitter, the capacitance of the receiver and resistive load, respectively.

In this work, capacitances are chosen such that identical resonant frequency is yielded. As a result, the simple means in (8) are represented so as to achieve resonant coupling along both sides [22, 25],

$$\omega_0 = \frac{1}{\sqrt{L_S C_S}} = \frac{1}{\sqrt{L_P C_P}} \tag{8}$$

To express the circuit equations, the following parameters are defined. Component values are shown in Table 2.

Tab. 2: Design values of proposed coils.

Element	Value
C_S (nf)	1.8
C_P (nf)	3.6
L_t (μH)	1.589
L_r (μH)	0.802
Input voltage amplitude (p–p) (V)	19.6
Resistive load R_L (Ω)	10

3.1.3 Modeling of transceivers for finite element analysis and simulation results

The finite element method is a useful analysis tool that generates a mesh. The different types of elements produced by the mesh generator can be converted to another type. The size of the elements varies from region to region [26]. In this work, the two-dimensional finite element method (2-D FEM) COMSOL software is employed to sim-

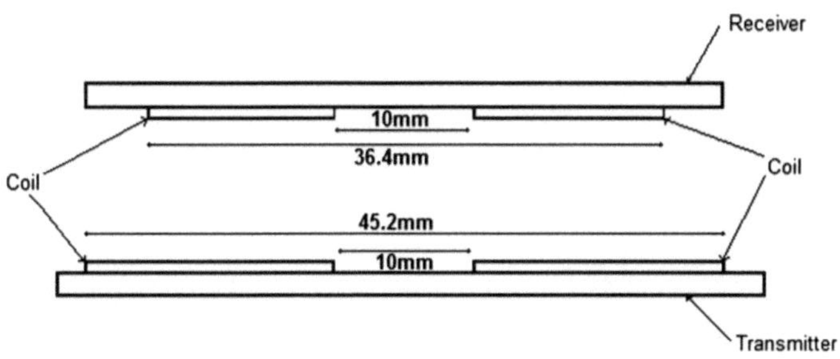

Fig. 4: Transceiver.

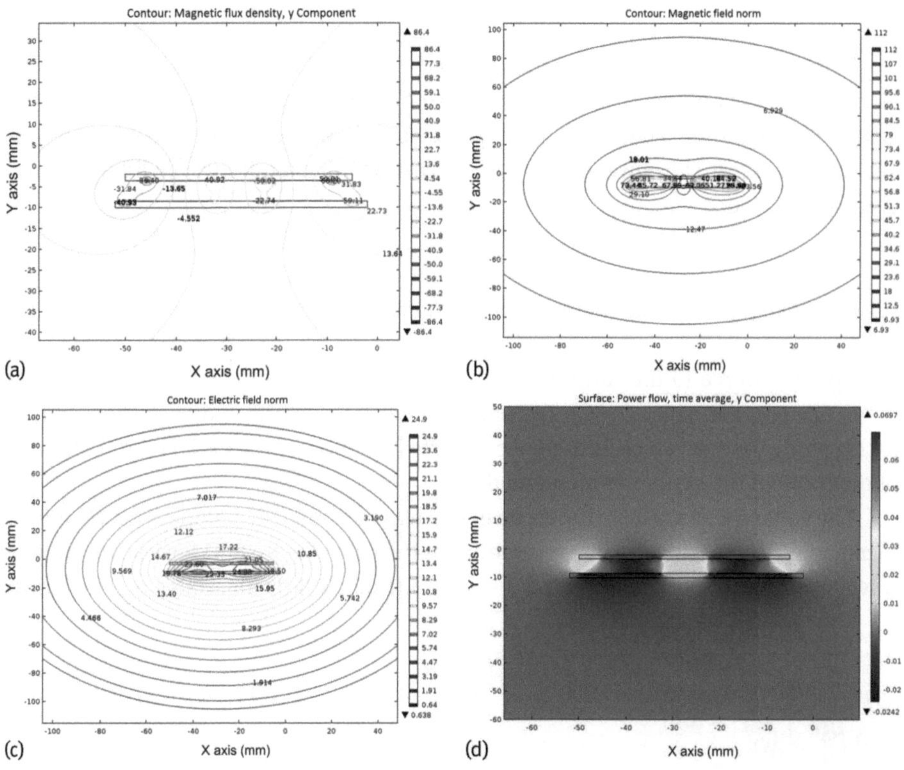

Fig. 5: Simulation results. (a) Magnetic flux density norm of proposed transceiver at 3 MHz (μT); (b) magnetic field norm of proposed transceiver at 3 MHz (A/m); (c) electric field norm of proposed transceiver at 3 MHz (V/m); (d) power density of the proposed transceivers, *y*-component at 3 MHz (W/cm^2).

ulate the system. The geometry of the transceiver and the corresponding windings as in Figure 4 are used to simulate the various configurations possible.

Figure 5 illustrates a set of simulated results of magnetic flux density, magnetic field norm, electric field norm and power flow for the proposed model.

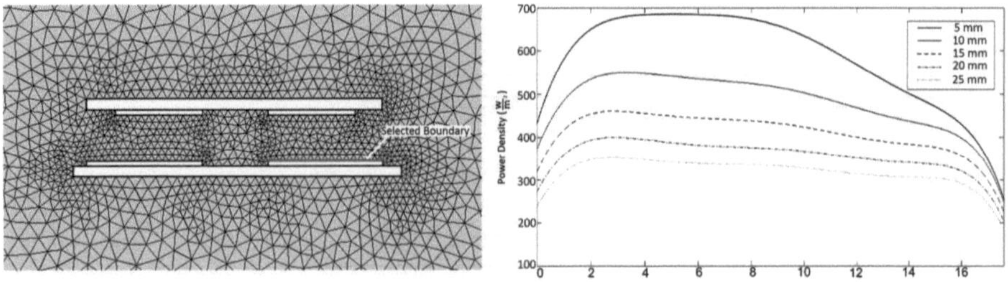

(a) (b)

Fig. 6: Assessment of the effect of the air gap on power density. (a) Selected boundary; (b) power density of the proposed transceiver, *y*-component at 3 MHz (W/m^2).

In this section, a boundary condition has been selected, as Figure 6(a), for assessment of the effect of the air gap on the power density. Figure 6(b) illustrates that the air gap changes from 5 mm to 25 mm. As the air gap reluctance increases, the power density decreases. The power density will have different magnitudes on different parts of the transceiver.

3.1.4 Investigation of experimental measurements

The voltage source to the transceiver is provided by function generator with 3 MHz frequency and 19.6 of V_{p-p} (AC). To record the exact waveform of an electrical signal, in this study we used an oscilloscope connected to software on a PC. Figure 7 shows a photograph of the system implemented.

The voltage waveform on the receiver coil was measured at distances from 5 mm to 15 mm while increasing the distance between the coils. Figure 8(a) represents voltage waveform on the transmitter coil. Figure 8(b) through 8(d) illustrates a set of mea-

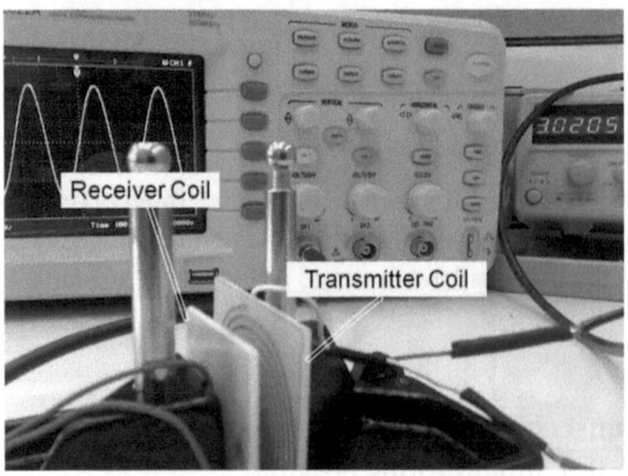

Fig. 7: Experimental setup showing the transmitting and receiving coils.

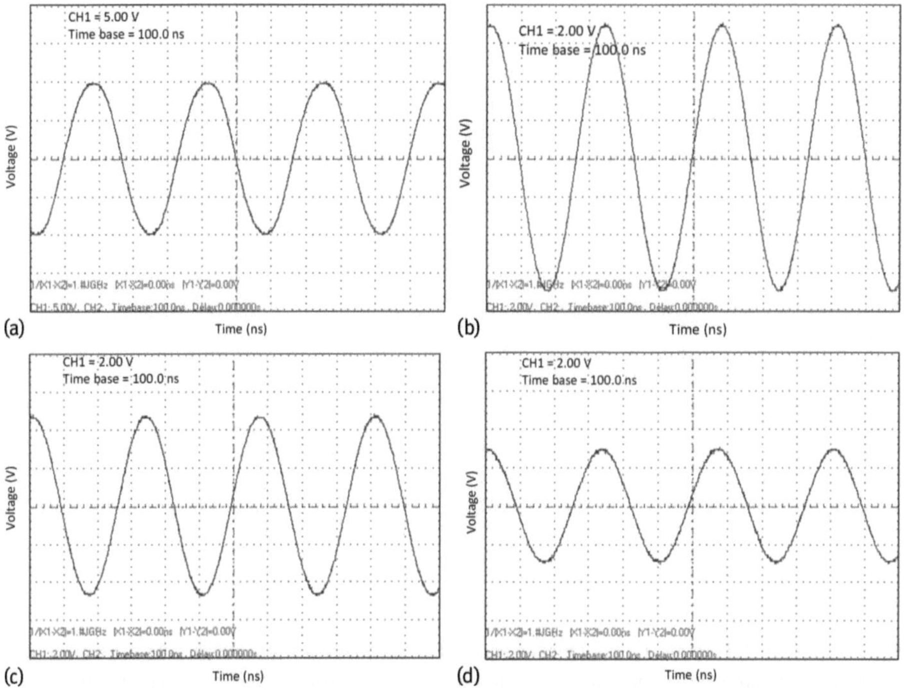

Fig. 8: Voltage waveform. (a) On the transmitter coil; (b) on the receiver coil, at a distance of 5 mm; (c) on the receiver coil, at a distance of 10 mm; (d) on the receiver coil, at a distance of 15 mm.

sured voltage waveforms through the secondary coil changes with distance between the transceiver. Figure 9 represents measured and simulated results for the case that the two coils are separated by different distances.

From the figures, it is obvious that the wireless power transmission is higher when the distance is nearer. This point was demonstrated by drawing the RMS voltage through investigated distances, both from experiments and theory as Figure 9(a).

The simulation model demonstrates the experiments. The voltages were sampled and transferred to MATLAB in order to carry out a curve fitting process. Most of the measured values closely follow the calculated value from MATLAB.

Figure 9(b) shows measured and simulated results for the case that the two coils are separated by 5 mm. It is apparent from the operating frequency displayed in Figure 9(b) that the measured response matches the situational response of the circuit. Finally, the power transmission efficiency through the proposed distance was analyzed. The calculation of efficiency can be completed by assuming that the input current and the input voltage generate active power. The maximum measured efficiency is 38.79% at a distance of 5 mm. The measured output power is 772.8 mW. The measured power transmission efficiency for the case that the two coils are separated by (5–15 mm) is shown in Figure 9(c).

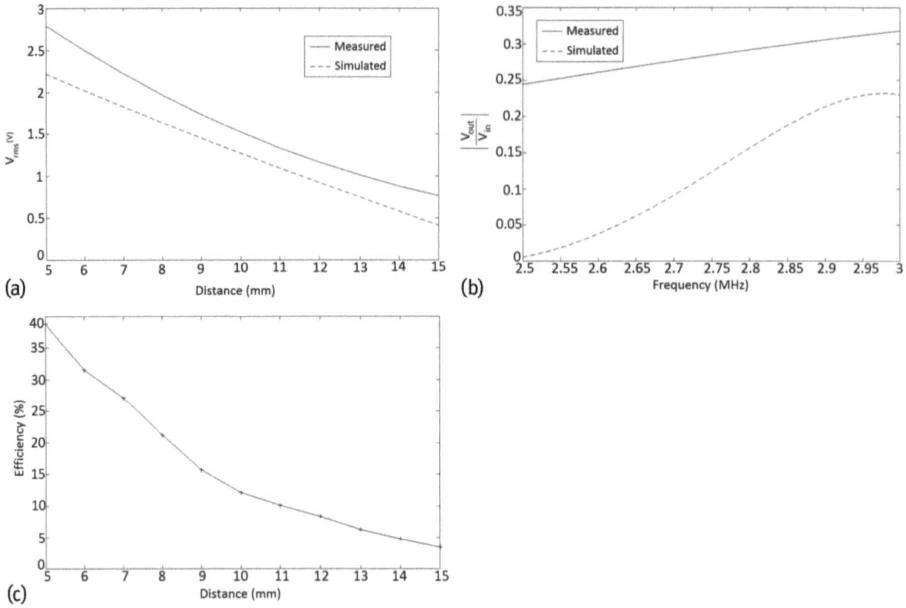

Fig. 9: Experimental results. (a) Root means square (RMS) voltage on secondary coil were measured and simulated at distances ranging from 5 mm to 15 mm.; (b) comparison of transfer function $|V_{out}/V_{in}|$ calculated experimentally, as well as through simulation, near the resonant peak (d = 5 mm); (c) power transfer efficiency measured at distances ranging from 5 mm to 15 mm while increasing the distance between the coils.

3.2 Investigation of the receiving pot core effect on the magnetic flux density in inductive coupling-based wireless power transfer

An inductive-based wireless power transfer system for low-power applications is studied in this section. The transferring power system was modeled, simulated and analyzed via the finite element method. An inductive power transmission (IPT) system can consist of a transmission line, mutually coupled coils, tuning capacitors and cores [27]. This work studies the magnetic flux density of inductive coupling-based WPT systems with discussions on the effect of the air gap and the shape of the receiving core. Based on the simulation performed the use of the pot core as a receiving core is proposed. It is concluded that this type of core can reduce the dependency of the magnetic flux density on the air gap in the receiving side. The operation of an IPT system can be compared to an air core transformer. To compensate the leakage, capacitance was applied to both sides of the transformer. This increases the power transmission efficiency [18]. The circuit can be classified according to the mode of connection of the capacitor to the coil: series–series, series–parallel, parallel–series and parallel–parallel [18, 24]. In this work, the parallel-parallel mode was proposed. Inductive power transfer depends on many parameters, such as the distance between coils, frequency, current excitation, the geometry of the coils [15] and the shape of the cores for different applicants. A great variety of applications in the field of power electronics has led to the use of different core shapes and materials. Due to the geometry of the cores, a magnetic ferrite core, such as a coil, can change the magnetic flux [17]. The shape of the circular core was designed as presented in [12]. The outer diameter of the cores (210 mm) and the air gap determined for proposed system (10–210 mm) is of the order of the diameters of the cores. The shape of the pot core was designed in this research. This work presents the whole process of designing cores and building an inductive power transfer system with COMSOL Multiphysics 5.1 software in detail. To determine the flux magnetic density, the magnetic field and the electric field between sets of coils, we used a finite element method partially based on the magnetic and electric field tool contained in the software. We expect this work to form a basis for understanding the effect of the air gap and the behavior of different cores on the flux magnetic density without using wires through inductive coupling.

3.2.1 Modeling of core for FEM analysis

The finite element method (FEM) was employed in a number of physical problems where the governing differential equations are obtainable. The fundamental principle in the physical interpretation of FEM is the subdivision of the mathematical model. The manner comprises supposing the continuous function for the solution. Then, this method obtains the parameters of the functions in a manner that reduces the error in the solution procedure [28]. FEM studies electromagnetic results for misaligned and aligned transceivers. The two-dimensional FEM (2-D FEM) COMSOL software is employed to simulate the various possible configurations. In this work, the shape of the circular core was designed as presented in [12]. The geometry of the transceiver and the corresponding windings as in Figure 10 were used to simulate the various possible configurations [29]. Figure 11 shows the equivalent circuit of inductive coupling-based wireless power transfer, where resistors R_t and R_r are resistive parts of the transmitter and the receiver; C_p, C_s and R_L are the capacitance of the transmitter and the capacitance of the receiver and resistive load, respectively.

The variables that parameterize the shape of the coils and cores are the inner diameter of the coil, the outer diameter of the coil, the diameter of the core, the outer width of pot core, an outer wall, the coil thickness, the ferrite thickness and the air gap. Transmitter and receiver specifications can be seen in Table 3.

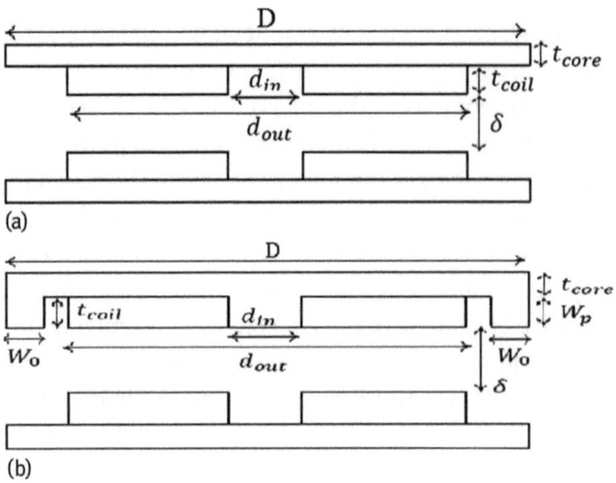

Fig. 10: (a) Receiving circular core; (b) receiving pot core.

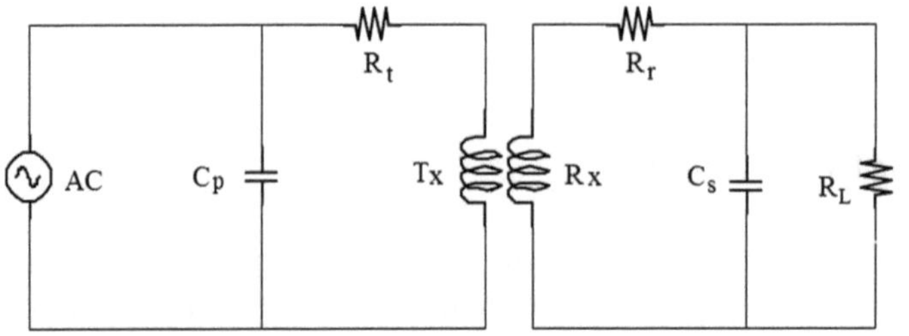

Fig. 11: Equivalent circuit of inductive coupling-based WPT (parallel–parallel mode).

Tab. 3: Transmitter and receiver specifications.

Variable	Value (mm)
Inner diameter of coil (d_{in})	30
Outer diameter of coil (d_{out})	160
Diameter of core (D)	210
Outer width of pot core (W_O)	15
Outer wall (W_p)	4.9
Coil thickness (t_{coil})	4.9
Core thickness (t_{core})	3.75
Air gap (δ)	(10–210)

3.2.2 Simulation results and discussion

Twelve turns of the coil for primary winding and secondary winding are considered, the air gap is 10 mm; the capacitances were 100 nF and the load resistance 20 Ω was considered. Figures 12 and 13 show a set of simulated results of the magnetic flux density, the magnetic field norm and the electric field norm for the different shapes of the receiver cores.

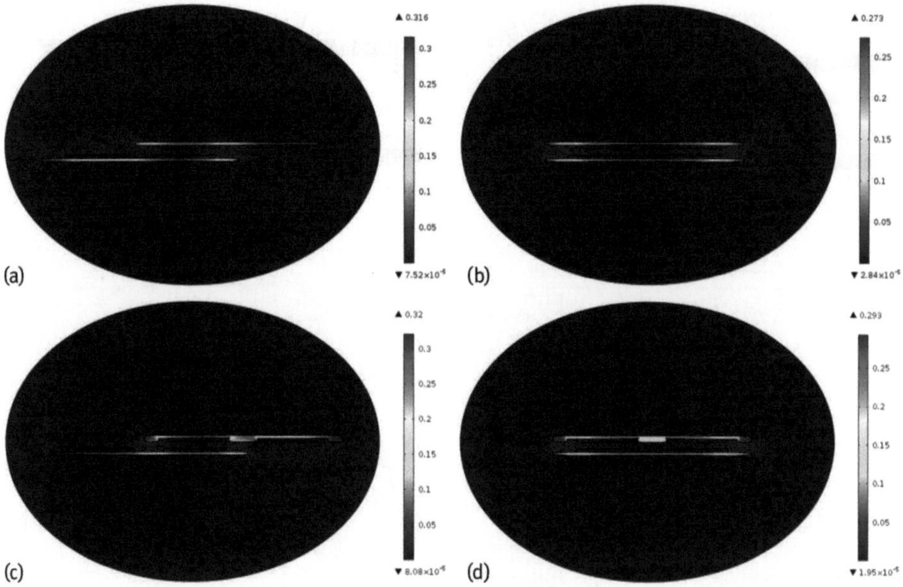

Fig. 12: Magnetic flux density norm at 100 kHz (T). (a) Circular core receiver – misaligned; (b) circular core receiver – aligned; (c) pot core receiver – misaligned; (d) pot core receiver – aligned.

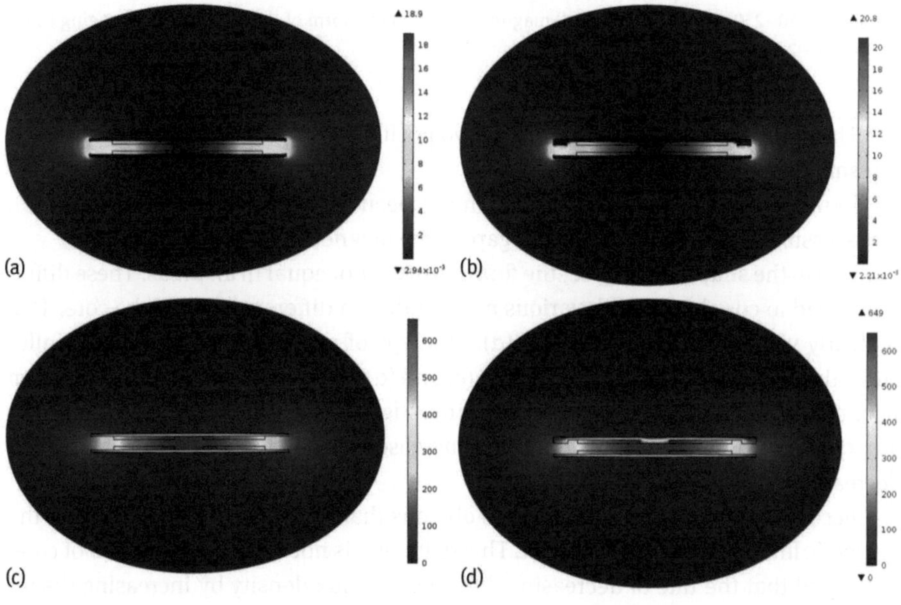

Fig. 13: Electromagnetic results. (a) Magnetic field norm of the circular core receiver at 100 kHz (A/mm); (b) magnetic field norm of the pot core receiver at 100 kHz (A/mm); (c) electric field norm of the circular core receiver at 100 kHz (V/m); (d) electric field norm of the pot core receiver at 100 kHz (Vm).

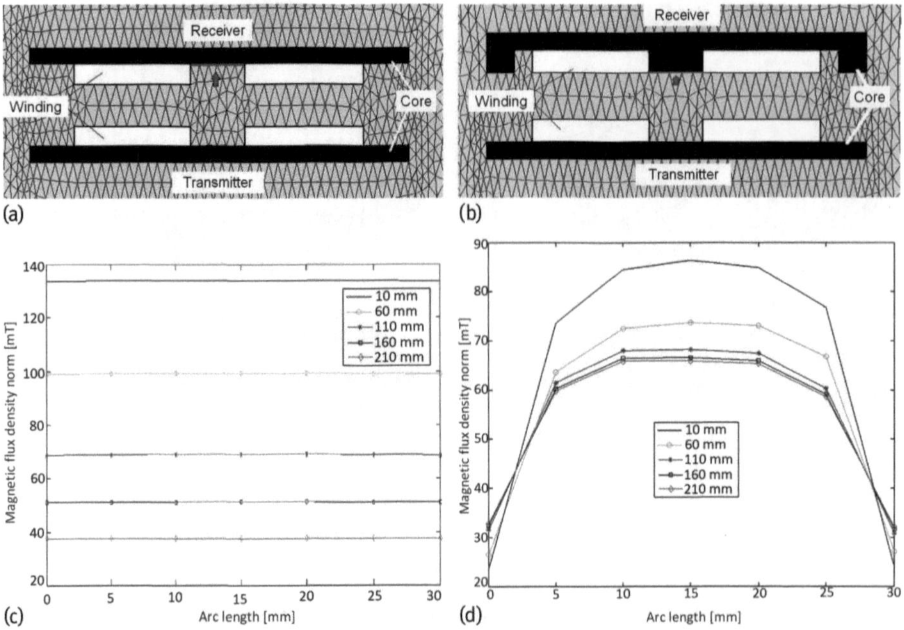

Fig. 14: Assessment of the effect of the air gap on the magnetic flux density. (a) Receiver with circular core; (b) receiver with pot core; (c) magnetic flux density norm of the circular core by changing the air gap (10–210 mm) at 100 kHz; (d) magnetic flux density norm of the pot core by changing the air gap (10–210 mm) at 100 kHz.

The magnetic flux density norm for two positions of the receiver core against the transmitter core was obtained.

In this section, a boundary condition has been selected, as Figure 14(a) and (b), for assessment of the effect of the air gap on the magnetic flux density.

Due to the shape of the core, the flux density is not equal in all areas. These differences lead to core losses with various magnitudes on different parts of the core. This is clearly visible in Figure 13(c) and (d). The type of material used in the core influences the hysteresis losses [30]. Figure 14(c) and (d) illustrate that δ has changed from 10 mm to 210 mm for two types of cores to investigate the effect of air gap on magnetic flux density. The air gap reluctance will increase, so the magnetic flux density will decrease.

According to the Figures 12–14, it is obvious that by increasing the distance, the magnetic flux density has decreased. This reduction is not equal for two types of core. It is found that the rate of decreasing the magnetic flux density by increasing the air gap in the selected boundary, is about 48% for the circular core receiver, while for the pot core receiver it is about 10%. One advantage of the pot core shape is the wide flux area; this feature of the pot core leads to low magnetic leakage [17].

3.3 Electromagnetic analysis of different geometries of transmitting coils for wireless power transmission applications

An inductive system can include parts like coils, cores and coupling capacitances [27]. The operation of such a system can be compared to an air core transformer [18]. The transmitting coil, which is excited by means of an alternating current, generates an electromagnetic field that is dependent on the dimensions of the coil, the drive current and the frequency [31]. There is an inductive coupling between the transmitting and receiving coils [32]. Inductive-based wireless power transmission is dependent on different parameters, such as the air gap between the transmitter and receiver, the frequency and the current excitation [15]. The power quality is dependent on the geometry of coil [33]. Many of the coils are designed based on the classic theories, and this method does not work for the complex shape of coils [34]. The objective of this section is to present an obvious understanding of the various geometric forms coils takes when being used as a transmitters. This objective is achieved by the analysis and comparison of various coils via electromagnetic results.

This work can provide a useful perspective for designing innovative coils. The different geometries of coils are accessible and reasonable. The diameter of a wire is standard, as used in [35]. The design process of coils and its dimensions are presented in detail. 2D-tools of COMSOL Multiphysics 5.1 software were used to simulate the problems [36]. The magnetic field, electric field, magnetic flux density and current density are presented and compared. The behavior of various geometries of coils versus changing frequency is illustrated. The experimental results are added to verify the validity of simulation results. Inductive power transfer has recently become a common method for transferring power. As the modern technologies need to become more efficient and updated, this technology is developing. The power transfer efficiency has the potential to improve. There are different ways to achieve a desirable efficiency. In this research, the suitable geometry of a coil for transferring power as a transmitting coil is examined. Three type of geometries are designed. Frequency analysis at the frequency range (10–50 kHz) is done to investigate the behavior of various geometries. The magnetic field, electric field, magnetic flux density and current density for various geometries are presented and compared. Magnetic flux density is measured via an experimental setup and is compared to a simulated one to verify the validity of the simulation results.

3.3.1 Modeling of a different geometry of coils for finite element analysis

FEM can be used to solve different physical problems. This method makes differential equations solvable. This method approaches the problem by reducing errors [28]. FEM is used in different works for various purposes, such as modeling and parameter iden-

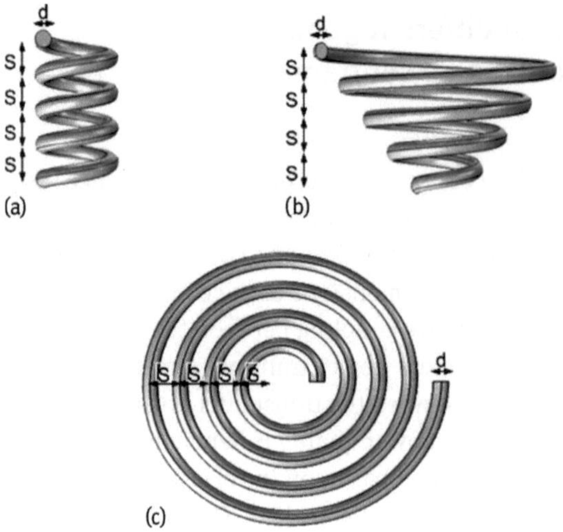

Fig. 15: Geometry of coils. (a) Helical; (b) conical; (c) circular.

Tab. 4: Design values of the proposed coils.

Parameter	Value
Diameter (*d*)	2 (mm)
Space (*S*)	5 (mm)
Turns	4
Voltage source (p–p)	20 (V)

tification [37–43]. An acceptable solution from Maxwell's equations for most practical cases is neither possible nor accurate. Thus, FEM is often used to calculate physics quantities [44]. The 2D-tools of COMSOL Multiphysics 5.1 software was used to solve problems via FEM. In this work, the diameter of a wire is considered as used in [35]. New geometries of coils are designed as presented in [45, 46]. These models are shown in Figure 15. The design values of the proposed coils are presented in Table 4.

3.3.2 Assessment of simulation and experimental results

For all type of coils, four turns are considered. The coils are connected to a signal generator as shown in Figure 16.

To analyze the behavior of the coils, a transmitting coil connected to a source is assumed. Figures 17 through 20 illustrate the magnetic field, the electric field, the magnetic flux density and the current density at frequency 10 kHz, respectively. In most models, the effect of high frequency is ignored, although this effect is an important source of losses [47].

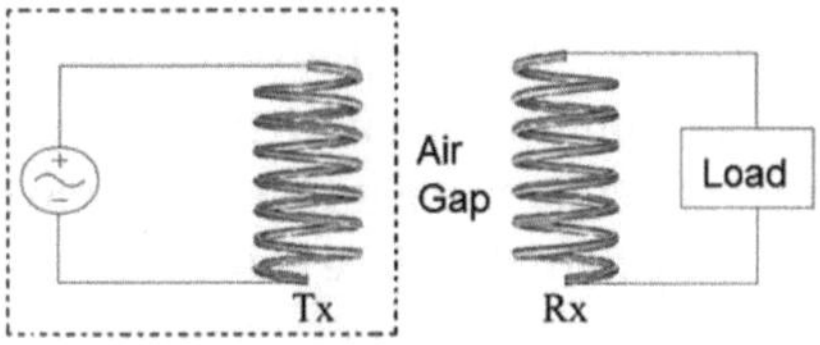

Fig. 16: Inductive coupling system.

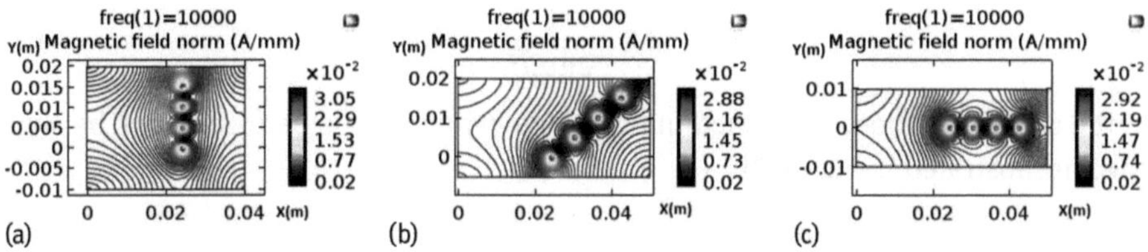

Fig. 17: Magnetic field norm at 10 kHz (A/mm). (a) Helical; (b) conical; (c) circular.

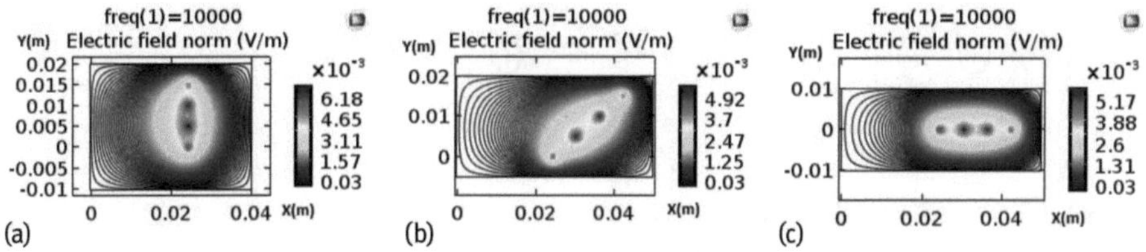

Fig. 18: Electric field norm at 10 kHz (V/m). (a) Helical; (b) conical; (c) circular.

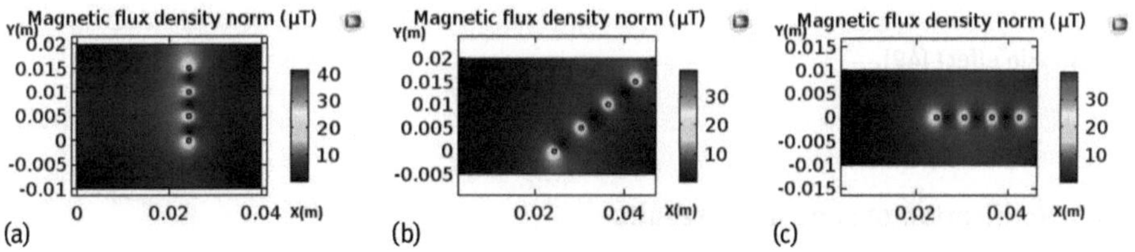

Fig. 19: Magnetic flux density norm at 10 kHz (μT). (a) Helical; (b) conical; (c) circular.

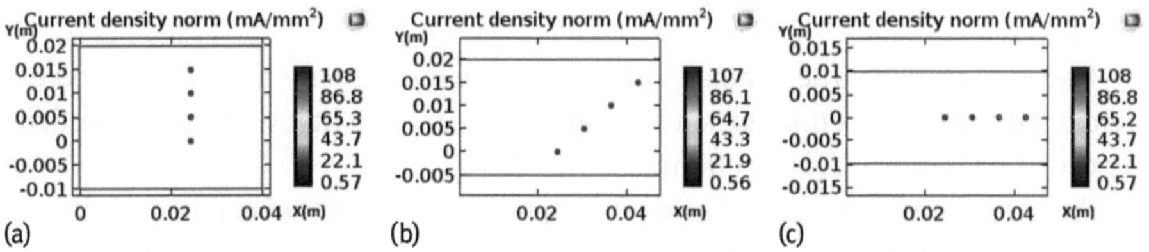

Fig. 20: Current density norm at 10 kHz (mA/mm^2). (a) Helical; (b) conical; (c) circular.

The behavior of the coil changes at high frequencies. Skin and proximity effects lead to the reduction of coil inductance. Parasitic capacitors are not negligible at high frequencies [48]. The skin effect leads to an internal magnetic field in a conductive wire. This internal field pushes electric current to the external surface of the conductor. There is an expression to calculate skin depth (δ) [49]:

$$\delta = \frac{1}{\sqrt{\pi \sigma \mu f}} \quad (9)$$

In the above equation, (σ) is medium conductivity and (μ) is permeability. The skin effect is illustrated in Figure 21.

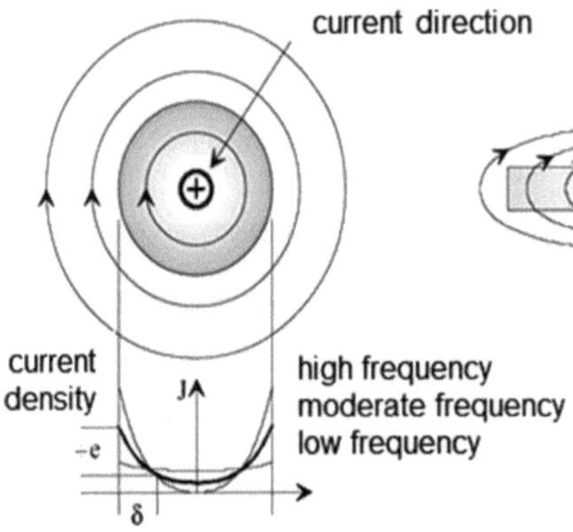

Fig. 21: Skin effect [49].

3.3.3 Verification of simulation results

To verify the validity of simulation results, the magnetic flux density is measured practically and compared to simulated values. The experimental setup is shown in Figure 22.

Figure 23 presents the comparison of magnetic flux density norm calculated experimentally, as well as through simulation. This figure illustrates magnetic flux density versus changing frequency from 5 kHz to 15 kHz.

The error occurs when comparing the experimental results of magnetic flux density with simulations ones. It increases as the frequency increases, and this is due to the measuring instrument. Figure 24 shows a comparison of the measured current and voltage versus changing frequency from 10 kHz to 50 kHz for a different geometry of coils.

To calculate the current density practically, it is assumed that the electric current is distributed uniformly within the cross-section. The comparison of the simulation via calculated values is presented in Table 5.

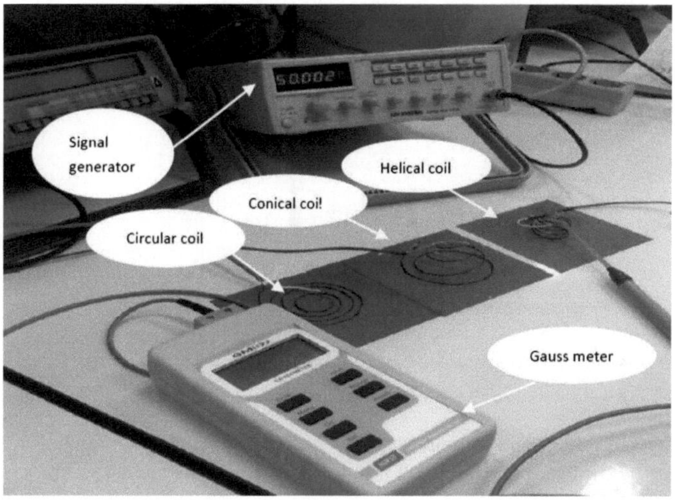

Fig. 22: Experimental setup.

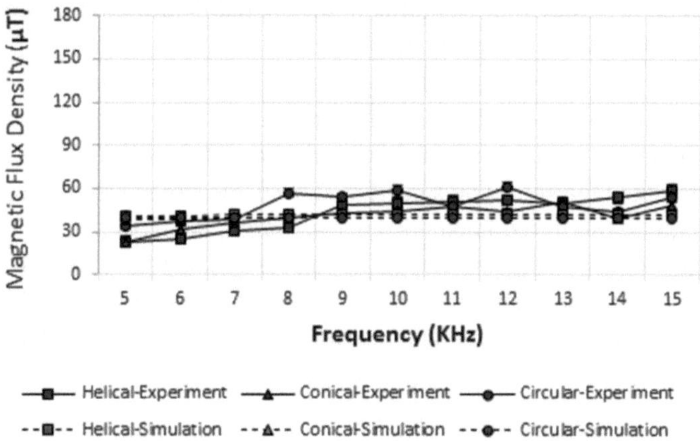

Fig. 23: Comparison of magnetic flux density calculated experimentally, as well as through simulation.

Tab. 5: Comparison of simulation via calculated values.

	Current density (J)	
Type	Maximum point-simulation ($\frac{mA}{mm^2}$) at 10 kHz	Uniform-measured ($\frac{mA}{mm^2}$) at 10 kHz
Helical	108	103.95
Conical	107	105.61
Circular	108	105.73

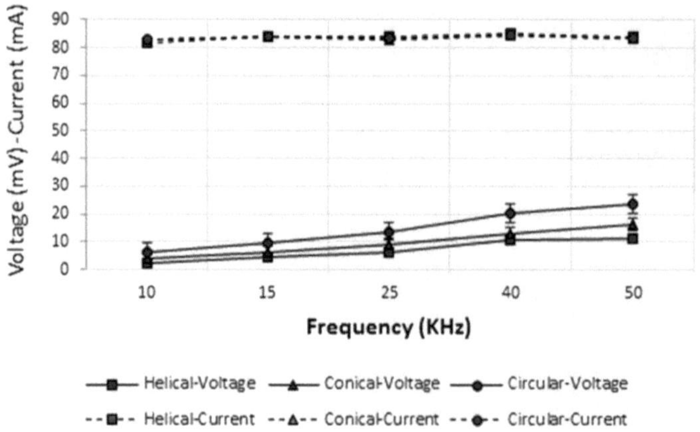

Fig. 24: Comparison of measured voltage and current.

About 3% error occurs at frequency 10 kHz when the experimental results of current density are compared with the simulations. The design of inductive power transfer systems is based on an accurate understanding of the spatial distribution of the magnetic field produced by a certain coil geometry [44].

Magnetic and electric fields have been compared to evaluate the stored magnetic and electric energies produced by the coils. Assuming an equal volume around each coil, the one with higher magnetic and electric fields has the larger magnetic and electric energies [50]. According to the values presented in Table 6, the stored electric energy is negligible compare to the stored magnetic energy. Storing the most magnetic energy is an important factor in all wireless transfer systems based on inductive technique because storing more magnetic energy leads to higher power transmission efficiency. Table 6 shows that the helical coil can be recognized as an efficient geometry among the proposed coils, based on storing magnetic energy.

Tab. 6: Comparison of maximum electromagnetic characteristics at 10 kHz.

Type	Magnetic field (A/mm)	Magnetic energy density (J/m³)	Electric field (V/m)	Electric energy density (J/m³)
Helical	$3.05 \cdot 10^{-2}$	$3.24 \cdot 10^{-4}$	$6.18 \cdot 10^{-3}$	$1.05 \cdot 10^{-16}$
Circular	$2.92 \cdot 10^{-2}$	$2.92 \cdot 10^{-4}$	$5.17 \cdot 10^{-3}$	$7.36 \cdot 10^{-17}$
Conical	$2.88 \cdot 10^{-2}$	$2.87 \cdot 10^{-4}$	$4.92 \cdot 10^{-3}$	$6.67 \cdot 10^{-17}$

4 Conclusion

This chapter studied an inductive-based wireless power transfer system for low-power applications at short distances. The transferring power system has been modeled, simulated and analyzed via FEM. This chapter presented a resonant inductive coupling system with printed spiral coils along with its design procedure. The comparison of the experimental output voltage with the simulation ones shows an error of 20.14% at 5 mm. This work has provided the fundamental concept of resonant near-field power transmission, as well as the relationship between the induced voltage and the air gap along the transceiver. Using a full-bridge rectifier at the receiver output, small devices such as mobile phone can be charged easily. Moreover, based on the modulation and simulation performed the use of the pot core as the receiving core is proposed. It is concluded that this type of core can improve the magnetic flux density in the receiving side. In this chapter, various geometries of coils for inductive power transfer applications are analyzed. A set of simulation results, the magnetic field, the electric field, the magnetic flux density and the current density are presented. The COMSOL Multiphysics software was used to simulate results. The simulations were verified with empirical results. This work has presented an efficient perspective to coil designers.

Bibliography

[1] K. Inoue, T. Nagashima, and X. Wei. Design of High-efficiency Inductive-Coupled Wireless Power Transfer System with Class-DE Transmitter and Class-E Rectifier. In *Industrial Electronics Society. 39th Annual Conference of the IEEE*, pages 613–618, 2013.

[2] S. Li and C. C. Mi. Wireless Power Transfer for Electric Vehicle Applications. *IEEE Journal of Emerging and Selected Topics in Power Electronics*, 3(1):4–17, 2015.

[3] U. K. Madawala and D. J. Thrimawithana. Current sourced bi-directional inductive power transfer system. *IET Power Electron*, 4(4):471–480, 2011.

[4] E. Abel and S. Third. Contactless power transfer-An exercise in topology. *IEEE Trans. Magn*, 20(5):1813–1815, 1984.

[5] D. Ahn and S. Hong. A Study on Magnetic Field Repeater in Wireless Power Transfer. *IEEE Transactions on Industrial Electronics*, 60(1):360–371, 2013.

[6] A. Rajagopalan, A. K. RamRakhyani, D. K. Schurig, and G. Lazzi. Improving Power Transfer Efficiency of a Short-Range Telemetry System Using Compact Metamaterials. *IEEE Transactions On Microwave Theory and Techniques*, 1–9, 2014.

[7] B. Lee and M. Kiani. A Triple-Loop Inductive Power Transmission System for Biomedical Applications. *IEEE Transactions on Biomedical Circuits and Systems*, 1–11, 2015.

[8] T. Imura and Y. Hori. Maximizing Air Gap and Efficiency of Magnetic Resonant Coupling for Wireless Power Transfer Using Equivalent Circuit and Neumann Formula. *IEEE Transactions on Industrial Electronics*, 58(10):4746–4752, 2011.

[9] M. Pinuela, D. C. Yates, S. Lucyszyn, and P. D. Mitcheson. Maximising DC to Load Efficiency for Inductive Power Transfer. *IEEE Transactions on Power Electronics*, 1–11, 2012.

[10] B. G. A. Covic and J. T. Boys. Inductive Power Transfer. *Proceedings of the IEEE*, 101(6), 2013.
[11] L. Olvitz, D. Vinko, and T. Švedek. Wireless Power Transfer for Mobile Phone Charging Device. *PMIPRO*, 141–145, 2012.
[12] J. Mühlethaler, J. W. Kolar, and R. Bosshard. Optimized Magnetic Design for Inductive Power Transfer Coils. In *Proceedings of the 28th Applied Power Electronics Conference and Exposition (APEC)*, 2013.
[13] N. D. Madzharov and A. T. Tonchev. Inductive High Power Transfer Technologies for Electric Vehicles. *Journal Of Electrical Engineering*, 65(2):125–128, 2014.
[14] X. Mou and H. Sun. Wireless Power Transfer: Survey and Roadmap. In *2015 IEEE 81st Vehicular Technology Conference (VTC Spring)*, pages 1–5, 2015.
[15] B. Kallel, T. Keutel, and O. Kanoun. Miso configuration efficiency in inductive power transmission for supplying wireless sensors. In *11th International Multi-Conference on (SSD)*, pages 1–5, 2014.
[16] J. Mühlethaler, J. W. Kolar, and A. Ecklebe. A Novel Approach for 3D Air Gap Reluctance Calculations. In *8th International Conference on Power Electronics – ECCE Asia*, pages 446–452, 2011.
[17] Ferrites and accessories. EPCOS data book, 2013.
[18] V. Prasanth. Wireless Power Transfer for E-Mobility. Master's thesis, faculty of electrical engineering, mathematics and computer science electrical power processing, delft university of technology, Delft, Netherlands, 2012.
[19] L. Xie, Y. Shi, Y. T. Hou, and W. Lou. Wireless power transfer and applications to sensor networks. *IEEE Wireless Communications*, 20(4):140–145, 2013.
[20] U. Jow and M. Ghovanloo. Design and Optimization of Printed Spiral Coils for Efficient Transcutaneous Inductive Power Transmission. *IEEE Transactions on Biomedical Circuits and Systems*, 1(3):193–202, 2008.
[21] S. S. Mohan, M. Hershenson, S. P. Boyd, and T. H. Lee. Simple accurate expressions for planar spiral inductances. *IEEE J. Solid-State Circuits*, 34(10):1419–1424, 1999.
[22] C. B. Provo. *Magnetic Resonate coupling as a means for Wireless emf transfer*, pages 1–8. 2013.
[23] M. Haerinia and E. Afjei. Resonant inductive coupling as a potential means for wireless power transfer to printed spiral coil. *Journal of Electrical Engineering, Romania*, 16(2), 2016.
[24] C. Duan, C. Jiang, A. Taylor, and K. Bai. Design of a zero-voltage-switching large-air-gap wireless charger with low electric Stress for electric vehicles. *Power Electronics, IET*, 6:1742–1750, 2013.
[25] B. L. Cannon, J. F. Hoburg, D. D. Stancil, and S. C. Goldstein. Magnetic Resonant Coupling As a PotentialMeans for Wireless Power Transfer to Multiple Small Receivers. *IEEE Trans. on Power Electronics*, 24(7):1819–1825, 2009.
[26] K. Ho-Le. Finite element mesh generation methods: A review and classification. *Computer-Aided Design*, 20(1):27–38, 1988.
[27] G. A. J. Elliott, J. T. Boys, and A. W. Green. Magnetically coupled systems for power transfer to electric vehicles. In *Proceedings of 1995 International Conference on Power Electronics and Drive Systems*, volume 2, pages 797–801, 1995.
[28] U. S. Dixit. *Finite Element Method: An Introduction*. Department of Mechanical Engineering, Indian Institute of Technology Guwahati, India, 2007.
[29] M. Haerinia and E. Afjei. Investigation of Receiving Pot Core Effect on Magnetic Flux Density in Inductive Coupling-Based Wireless Power Transfer. In *IEEE conference, International Symposium on Power Electronics, Electrical Drives, Automation, and Motion (SPEEDAM 2016), Italy*, 2016.
[30] R. Berglund. Frequency Dependence of Transformer Losses. Master's thesis, Chalmers University of Technology, Gothenburg, Sweden, 2009.

[31] P. Apoorva, K. S. Deeksha, N. Pavithra, M. N. Vijayalakshmi, B. Somashekar, and D. Livingston. Design of a Wireless Power Transfer System using Inductive Coupling and MATLAB programming. *International Journal on Recent and Innovation Trends in Computing and Communication*, 3(6):3817–3825, 2015.
[32] S. H. Hwang, C. G. Kang, Y. H. Son, and B. J. Jang. Software-Based Wireless Power Transfer Platform for Various Power Control Experiments. *Energies*, 8(8):7677–7689, 2015.
[33] M. Kiani. *Wireless Power and Data Transmission to High-performance Implantable Medical Devices*. PhD thesis, Georgia Institute of Technology, USA, 2014.
[34] R. Chang, L. Quan, X. Zhu, Z. Zong, and H. Zhou. Design of a Wireless Power Transfer System for EV Application Based on Finite Element Analysis and MATLAB Simulation. *ITEC Asia-Pacific*, 1–4, 2014.
[35] J. Kim and Y. J. Park. Approximate Closed-Form Formula for Calculating Ohmic Resistance in Coils of Parallel Round Wires With Unequal Pitches. *IEEE Trans. on Industrial Electronics*, 62(6):3482–3489, 2015.
[36] Version 5.1 of COMSOL Multiphysics Software, *User Manual*, COMSOL Ltd., 2015.
[37] E. Afjei, A. Siadatan, and H. Torkaman. Analytical Design and FEM Verification of a Novel Three-Phase Seven Layers Switched Reluctance Motor. *Progress In Electromagnetics Research*, 140:131–146, 2013.
[38] H. M. Cheshmehbeigi, E. Afjei, and B. Nasiri. Electromagnetic Design Based on Hybrid Analytical and 3-D Finite Element Method for Novel Two Layers BLDS Machine. *Progress In Electromagnetics Research*, 136:141–155, 2013.
[39] H. Torkaman and E. Afjei. Comparison of Three Novel Types of Two-Phase Switched Reluctance Motors Using Finite Element Method. *Progress In Electromagnetics Research*, 125:151–164, 2012.
[40] H. Torkaman and E. Afjei. Radial Force Characteristic Assessment in a Novel Two-Phase Dual Layer SRG Using FEM. *Progress In Electromagnetics Research*, 125:185–202, 2012.
[41] E. Afjei and H. Torkaman. Comparison of Two Types of Dual Layer Generator in Field Assisted Mode Utilizing 3D-FEM and Experimental Verification. *Progress In Electromagnetics Research B*, 23:293–309, 2010.
[42] H. Torkaman and E. Afjei. FEM Analysis of Angular Misalignment Fault in SRM Magnetostatic Characteristics. *Progress In Electromagnetics Research, PIER*, 104:31–48, 2010.
[43] H. Moradi, E. Afjei, and F. Faghihi. FEM Analysis for a Novel Configuration of Brushless DC Motor without Permanent Magnet. *Progress In Electromagnetics Research, PIER*, 98:407–423, 2009.
[44] B. A. Esteban. A Comparative Study of Power Supply Architectures In Wireless Electric Vehicle ChargingSystems. Master's thesis, University of Windsor, Windsor, Ontario, Canada, 2014.
[45] M. Haerinia, A. Mosallanejad, and E. Afjei. Electromagnetic Analysis of Different Geometry of Transmitting Coils for Wireless Power Transmission Applications. *Progress In Electromagnetics Research M, USA*, 50:161–168, 2016.
[46] M. Haerinia. Assessment of New Method for Inductive-Based Wireless Power Transmission System. Master's thesis, faculty of electrical engineering, Shahid Beheshti University, Tehran, Iran, 2016.
[47] N. Hasan. Optimization and control of lumped transmitting coil-based in motion wireless power transfer systems. Master's thesis, Utah State University, Logan, Utah, 2015.
[48] G. Grandi, M. K. Kazimierczuk, A. Massarini, and U. Reggiani. Stray capacitances of single-layer air-core inductors for high-frequency applications. In *Industry Applications Conference, 1996. Thirty-First IAS Annual Meeting, IAS '96., Conference Record of the 1996 IEEE*, volume 3, pages 1384–1388, 1996.

[49] K. V. Schuylenbergh and R. Puers. *Inductive Powering: Basic Theory and Application to Biomedical Systems*. Springer Science, Leuven, Belgium, 2009.

[50] C. Jimmy Li. A Planarized, Capacitor-loaded and Optimized Loop Structure for Wireless Power Transfer. Master's thesis, University of Texas at Austin, Austin, USA, 2013.

Bilel Kallel, Ghada Bouattour, Olfa Kanoun, and Hafedh Trabelsi

Wireless power transmission via a multi-coil inductive system

Abstract: Wireless power transmission via inductive links is widely used to power wireless sensors, having power consumption in the range of micro to milliwatts. It presents many advantages, such as independency of environmental effects (e.g., humidity, temperature), easy accessibility (i.e., power can be sent to electronic devices having micro and/or nano-geometry), and high flexibility (i.e., the system can operate even at small lateral and/or axial misalignments). The idea behind the use of multi-coil inductive systems, in which the sending and/or the receiving sides have multiple coils connected in series or in parallel, is to increase the efficiency of the inductive system in the case of lateral and/or misalignment, to provide energy to movable receivers and to increase their flexibility. In this chapter, we present, first, a general idea about the different types and classification of multi-coil inductive systems, followed by some examples from the literature. In the second section, we propose different analytical expressions of the equivalent parameters, such as the equivalent inductance of coupled coils, equivalent mutual inductance, induced voltage and induced current. These parameters are needed and are important for the optimization of all the inductive systems. In the last part, we propose the optimization of both transmitted power to the load and transmission efficiency of an example of a multi-coil inductive system by the study of its coil configuration and excitation, its compensation topologies, and the receiver's detection in the case of movable systems. To validate our propositions, simulations and experimental results are explored and thoroughly discussed.

Keywords: Wireless power transmission, inductive link, multi-coil system, mutual inductance, receiver detection

1 Introduction

The history of wireless power transmission began when Nicola Tesla started his first experiment at the end of the nineteenth century. In the beginning, his idea did not generate big interest because of its low efficiency and relatively high cost. Afterward, and because of the increased need to send power to movable, micro and underwater systems, wireless technology became more and more interesting. Researchers proved

Bilel Kallel, Ghada Bouattour, Olfa Kanoun, National School of Engineers of Sfax, Tunisia
Bilel Kallel, Ghada Bouattour, Hafedh Trabelsi, Technische Universität Chemnitz, Germany

https://doi.org/10.1515/9783110445053-013

Tab. 1: Overview on wireless power transmission techniques [1].

Type of WPT		Transmitted energy	Typical distance
Electromagnetic waves and radio frequency	LF	high quantity	< 50 cm
	HF	30–40 W	up to 2 m
	UHF	1–3 W	3.5–10 m
	Microwave	1–75 W	up to 30 m
Laser		high quantity	long distance
Capacitive link		up to 1–3 W	< 1 mm
Inductive link		1 mW–few KW	< 2 m

that power can be transferred wirelessly through different technologies: via laser, via electromagnetic induction (inductive), via electrostatic induction (capacitive) and via microwaves (RF).

Wireless power transmission via inductive links is useful not only for systems where energy transfer is to take place in hazardous, humid and wet areas, but also for movable and small sized systems. Generally, these systems should be maintenance free. In order to achieve a good power transfer efficiency and high transmitted power, the sending and receiving sides should be perfectly aligned and close to each other. Unfortunately, these conditions cannot be considered in all inductive systems. So, a misalignment between the sender and the receiver is an unavoidable phenomenon, especially for systems with movable parts. One solution that has been proposed is the use of multi-coil inductive systems.

2 Effect of vertical and lateral distances on Inductive Power Transfer (IPT) systems

Generally, inductive systems are dependent on the distance separating the sending and receiving side d and the lateral distance Δ (Figure 1) [1, 2]. In the case of lateral misalignment, not all the magnetics field issued by the sending coil pass through the receiving one, thereby causing high magnetic flux leakage. This phenomenon becomes worse in the case of a large air gap between the two sides. In the aforementioned cases, we have a weak coupling between the coils and low mutual inductance that cause low-power transmission efficiency.

An example of a conventional two-coil inductive system is shown in Figure 1. The diameter of the sending and receiving coils equals 60 mm. We remark that with an air gap corresponding to 17% of the receiving coil's diameter and in the case of lateral misalignment corresponding to 83.33% of the sending coil's diameter, the coupling coefficient decreases from 0.18 to 0.06 at 0 mm to 20 mm lateral misalignment, respectively.

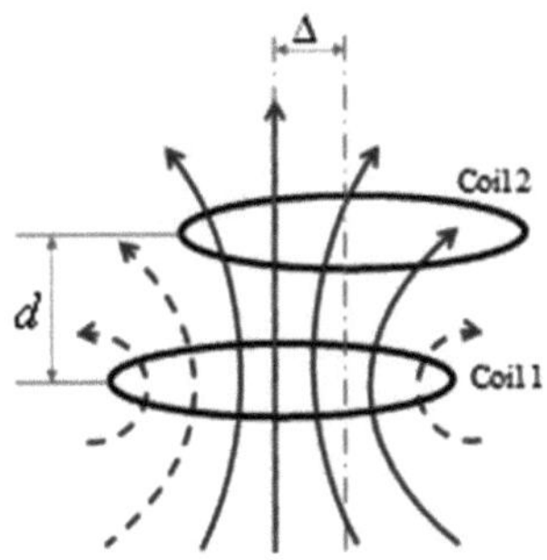

Fig. 1: Lateral misalignment *Δ* in the case of a two-coil inductive system [2].

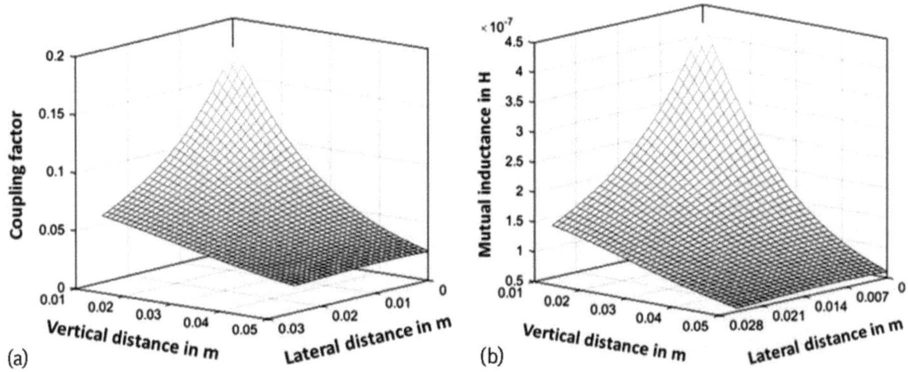

Fig. 2: (a) Coupling factor and (b) mutual inductance for different lateral and vertical distances.

With a larger air gap, corresponding to more than 160% of the sending coil's diameter, the coupling factor decreases also by 8% of its initial value in the same lateral misalignment range. The mutual inductance is also affected by both lateral and vertical distances. It shows a sharp decrease of more than 85% and 59% of its initial value with a large air gap, and in the case of lateral misalignment, respectively.

The transmitted power is also affected by both vertical (air gap) [1] and lateral [2] (misalignment) distances. It decreases by more than 64% of its initial value in the case of a large air gap and by more than 80% of its initial value in the case of lateral misalignment. Multi-coil systems, in which we use additional coils compared to conventional two-coil systems, present an effective solution to the lateral misalignment problem in the case of a large air gap separating the sending and receiving sides.

3 Classification of the multi-coil inductive systems

Multi-coil inductive systems can be categorized into three different types (Figure 4): three-coil systems and four-coil systems, and the rest are just called multi-coil systems, themselves, are classified into three sub-categories: multiple senders and one

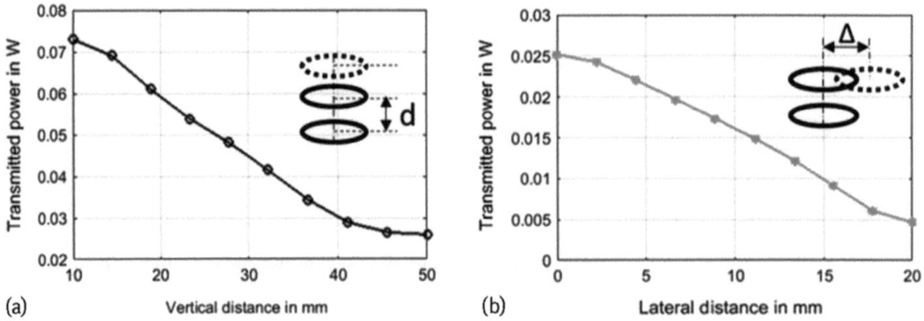

Fig. 3: Transmitted power with respect to (a) vertical [1] and (b) lateral [2] distances.

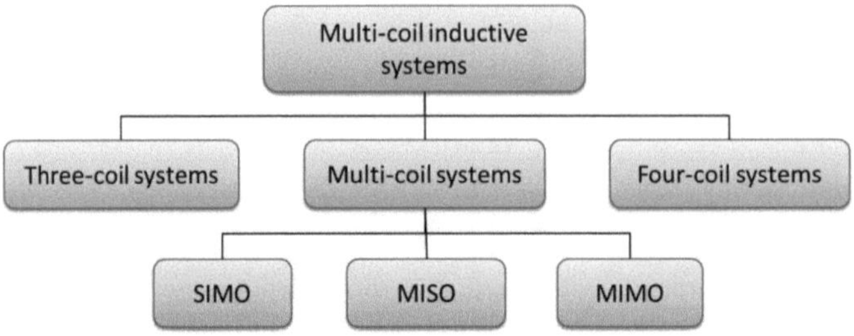

Fig. 4: Classification of multi-coil inductive systems.

receiver known as multi input single output (MISO), one sender and multiple receivers known as single input multi output (SIMO) and multiple senders and multiple receivers known as multi input multi output (MIMO).

Three-coil and four-coil systems are destined only for systems having a large vertical distance (large air gap) or systems with small lateral misalignment, but they are not suitable for applications with a movable receiver. However, multi-coil systems are used to solve the problem of movable devices to enlarge the misalignment tolerance and to increase, in several cases, the power transmission efficiency.

4 Examples of the use of multi-coil inductive systems

In this section, we will give a short overview on the most interesting examples from the state of the art dealing with multi-coil inductive systems.

4.1 Three-coil systems

The three-coil system is one kind of multi-coil system, in which we add a third coil to the conventional two-coil system. This added coil resonates at the same resonance fre-

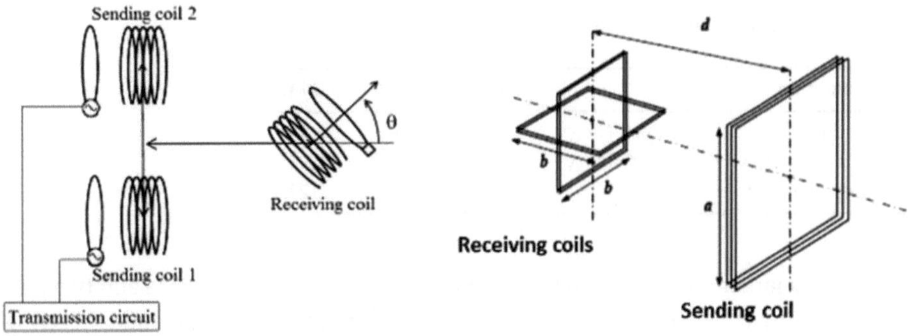

Fig. 5: Examples of three-coil systems [4, 5].

quency of the system. Different examples have been investigated in the literature [3–5]. By adopting a system as described in [3], the power transmission efficiency is improved twice. It was also improved in [4] by introducing the out-of-phase excitation technique, in which the sending coils are excited in opposite directions. In [5], by adopting two receiving coils placed orthogonally, the system becomes less sensitive to lateral misalignment. Three-coil systems show an improvement of power transmission efficiency with both large air gap and lateral misalignment. However, they are not suitable for systems dealing with a movable receiver. They not show only a strong dependency on the position of the receiver and its orientation but also a high controlling complexity, high manufacturing difficulty and controlling complexity. Examples of three-coil systems are shown in Figure 5.

4.2 Four-coil systems

Four-coil systems [6–8] also present a perfect solution to enlarge the misalignment tolerance and to increase the vertical distance between the sending and receiving sides. Examples of four-coil systems are shown in Figure 6. They can be in different configuration. The sending and receiving coil with the same size and two parallel intermediate

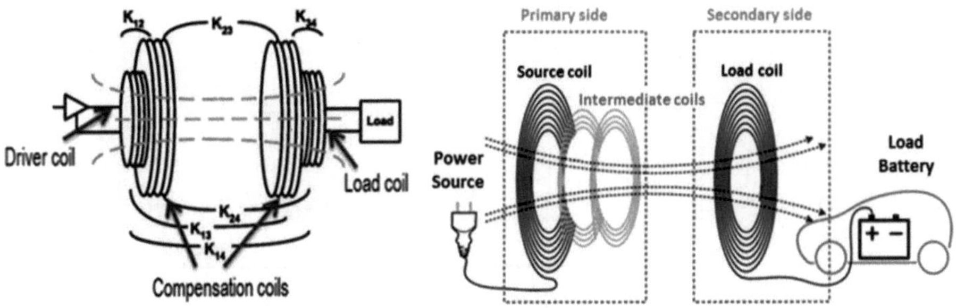

Fig. 6: Examples of four-coil systems [6, 7].

Tab. 2: Efficiency comparison between two-coil and four-coil systems.

Load condition	1.1 kW	2.2 kW	3.3 kW
Conventional two-coil system efficiency	89.8	90.31	90.45
Asymmetric four-coil system efficiency	95.63	96.21	96.56
% of improvement	6%	6.13%	6.32%

coils with a larger size [6], two parallel sending coils with the same size and two parallel receiving coils with a bigger or a larger size [8], and the sending and receiving coil have the same size and two parallel intermediate coils have a smaller size [7]. This kind of inductive system is used to improve the coupling coefficient at large vertical distances by adding a magnetic enhanced resonator and using intermediate coils with a high quality factor (see Table 2).

Like three-coil systems, four-coil systems cannot be considered for systems with a movable receiver, as they show a high dependency on the position of the receiver. They present a high cost solution, because the manufacturing process of high quality factor coils is expensive.

4.3 Multi-coil systems

Different solutions have been proposed for inductive power transmission systems with large air gap, lateral misalignment and movable receivers. However, there is no recent investigation dealing with all these problems at the same time. The majority of multi-coil systems [9–14] is suitable for systems with a movable receiver. They even provide free positioning [9, 10, 12, 15–17] or position detection of the receiver [16, 17]. Some investigations deal with the large air gap using air-core coils, and others do not because they use printed coils. All these proprieties are candidates for increasing both the control's complexity and the manufacturing cost of the systems. Examples of multi-coil systems are shown in Figure 7.

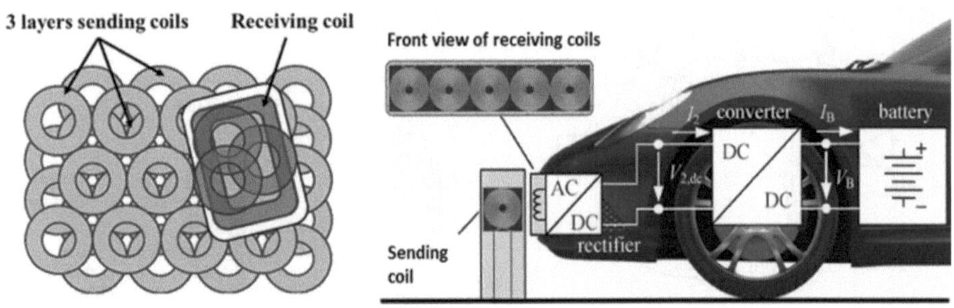

Fig. 7: Examples of multi-coil systems [9, 11].

5 Modeling of multi-coil inductive systems

5.1 Basic laws

5.1.1 Faraday's law

Faraday's law [18] states that a changing magnetic flux through a surface A induces a voltage (emf) in any boundary path of that surface, and a changing magnetic field $\vec{B}$ induces a circulating electric field $\vec{E}$,

$$\text{emf} = \int_C \vec{E}\,\mathrm{d}\vec{l} = -\frac{\mathrm{d}}{\mathrm{d}t}\int_A \vec{B}\,\mathrm{d}\vec{A} \tag{1}$$

where C is a conductor loop, $\mathrm{d}\vec{l}$ is an incremental segment of path C and $\mathrm{d}\vec{A}$ is the differential of the surface area. In other words, if a conducting material is present along that boundary, the induced electric field provides an (emf) that drives a current through the material. The differential form of Faraday's law is given by:

$$\nabla \times \vec{E} = -\frac{\partial \vec{B}}{\partial t} \tag{2}$$

It states that a circulating electric field is produced by a magnetic field that changes with time.

5.1.2 The Ampere–Maxwell law

The Ampere–Maxwell law [18] states that a magnetic field is produced along a path if any current is enclosed by the path, or if the electric flux through any surface bounded by the path changes over time,

$$\oint \vec{B}\,\mathrm{d}\vec{l} = \mu_0 \left(I + \epsilon_0 \frac{\mathrm{d}}{\mathrm{d}t} \int_A \frac{\mathrm{d}\vec{E}}{\mathrm{d}t}\,\mathrm{d}\vec{A} \right) \tag{3}$$

where μ_0 is the vacuum permeability, I is the enclosed electric current and ϵ_0 is the electric permittivity of the vacuum. The differential form of the Ampere–Maxwell law is:

$$\nabla \times \vec{B} = \mu_0 \left(\vec{J} + \epsilon_0 \frac{\mathrm{d}\vec{E}}{\mathrm{d}t} \right) \tag{4}$$

where $\vec{J}$ is the total electric current density. Equation (4) states that by an electric current and an electric field that changes with time, we produce a circulating magnetic field.

5.1.3 The Biot–Savart law

The Biot–Savart law [18] allows the calculation of $\vec{B}$ due to wire-shaped conductors of arbitrary geometry. Referring to Figure 8, the contribution at a specific point P from a small element of electric current is given by:

$$\mathrm{d}\vec{B} = \mu_0 \frac{I\,\mathrm{d}\vec{l} \times \hat{r}}{4\pi r^2} \tag{5}$$

where I is the current, $\mathrm{d}\vec{l}$ is a vector with the length of the current element and pointing in the direction of the current, $\hat{r}$ is a unit vector pointing from the current element to the point P at which the field is calculated and r is the distance between the current element and P.

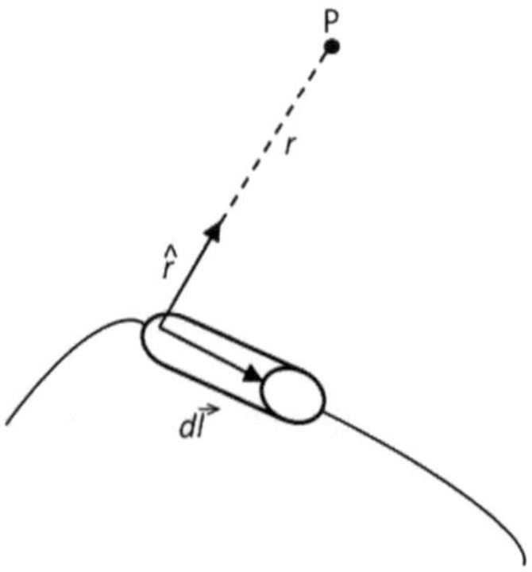

Fig. 8: Geometry for the Biot–Savart law [18].

Considering a circular loop of radius r and with a circulating current I (see Figure 9), from equation (5), we obtain the magnetic field across x-axis as follows:

$$\mathrm{d}B_x = \frac{\mu_0 I r\,\mathrm{d}l}{4\pi\sqrt{(r^2 + x^2)^3}} \tag{6}$$

where x is the distance along the x-axis

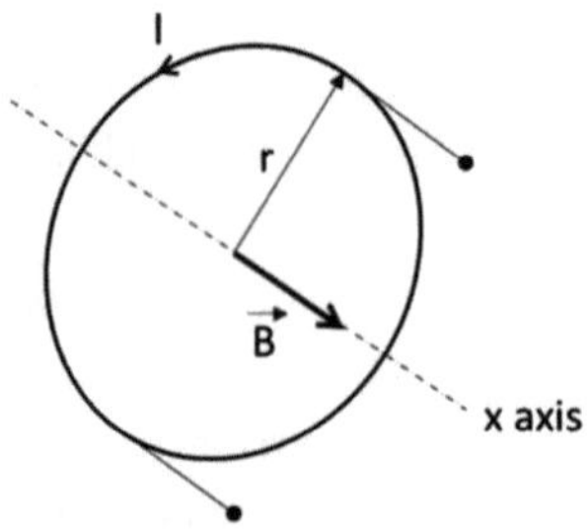

Fig. 9: Circular loop with radius r and current I.

5.2 Analytical modeling

In order to extract the critical parameters of inductive systems and to optimize the transmitted power and the transmission efficiency, an accurate analytical model is needed. For this, in Figure 10, we consider an equivalent circuit of a multi-coil inductive system consisting of four sending coils and one receiving one [19]; L_1, L_2, L_3 and L_4 represent the inductance of sending coils, and L_r represents the inductance of the receiving coil; R_{eq} is the equivalent resistance of the sending coils, which can be connected in parallel or in series, and R_r is the receiving coil resistance. The capacitor C_1 and C_r are added in parallel or in series at the primary and secondary sides, respectively, to work in resonance. The load is modeled as a resistor R_L; V_1 and R_s are the source voltage and its resistance; M_{eq} represents the equivalent mutual inductance between the sending and receiving coils.

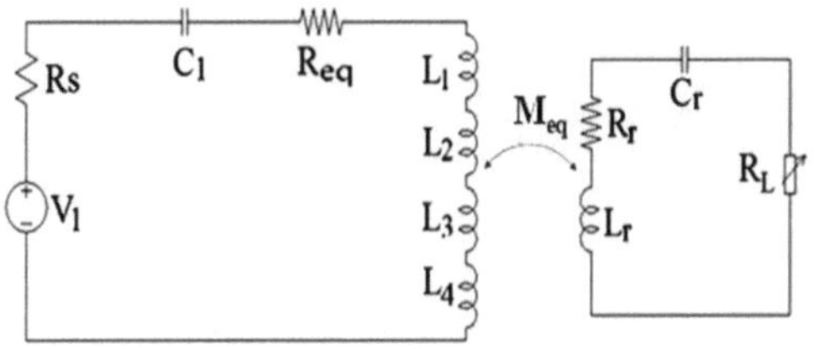

Fig. 10: Equivalent circuit of a multi-coil inductive system with a resistive load.

5.3 Analytical expression for equivalent inductance L_{eq}

The sending coils can be powered in the same or in the opposite direction. They can be in parallel or in series. The equivalent inductance L_{eq} of the sending coils, which are connected in series and in the same direction, is given by:

$$I_{eq} = \sum_i L_i + \sum_{i \neq j} M_{ij} \tag{7}$$

where L_i is the coil self-inductance of coil i and M_{ij} is the mutual inductance between coil i and coil j. In the case of coils that are connected in series but in opposite direction, the expression of equivalent inductance is given by the following:

$$L_{eq} = \sum_i L_i - \sum_{i \neq j} M_{ij} \tag{8}$$

By powering the sending coils in parallel in the same direction, the expression of the equivalent inductance is:

$$L_{eq} = \frac{\sum_{i \neq j} \frac{1}{2}(L_i L_j - M_{ij}^2)}{\sum_i L_i - \sum_{i \neq j} M_{ij}} \tag{9}$$

In the case of coils that are connected in parallel but in opposite direction, the equivalent inductance expression becomes as follows:

$$L_{\text{eq}} = \frac{\sum_{i\neq j} \frac{1}{2}(L_i L_j - M_{ij}^2)}{\sum_i L_i + \sum_{i\neq j} M_{ij}} \tag{10}$$

5.4 Analytic expression for equivalent mutual inductance M_{eq}

The equivalent mutual inductance M_{eq} can be defined as the sum of all mutual inductances between the sending and the receiving coils $M_{i\text{r}}$. Its expression is given as follows:

$$M_{\text{eq}} = \sum_{i=1}^{n} M_{i\text{r}} \tag{11}$$

The equivalent mutual inductance is equal to:

$$M_{\text{eq}} = \mu_0 N_{\text{r}} r_{\text{r}} \sum_{i=1}^{n} \frac{N_i r_i}{r_i(r_{\text{r}} - \Delta_i)} G(h_i) \tag{12}$$

where

$$G(h_i) = \left(\frac{2}{\sqrt{\frac{4r_i(r_{\text{r}}+\Delta_i)}{(r_i+r_{\text{r}}+\Delta_i)^2+d^2}}} - \sqrt{\frac{4r_i(r_{\text{r}}+\Delta_i)}{(r_i+r_{\text{r}}+\Delta_i)^2+d^2}} \right) K(h_i) - \frac{2}{\sqrt{\frac{4r_i(r_{\text{r}}+\Delta_i)}{(r_i+r_{\text{r}}+\Delta_i)^2+d^2}}} E(h_i) \tag{13}$$

N number of turns
r radius of the coil
d coil to coil vertical distance (the air gap)
Δ lateral distance between coils (lateral misalignment)

The subscripts r and i refer to the receiving coil r and the sending coil number i, respectively. Referring to equation (6), we can conclude the analytical expression of the coupling coefficient k between the sending and receiving sides is:

$$k = \frac{\mu_0 N_{\text{r}} r_{\text{r}} \sum_{i=1}^{n} \frac{N_i r_i}{r_i(r_{\text{r}}-\Delta_i)} G(h_i)}{\sqrt{L_{\text{eq}} L_{\text{r}}}} \tag{14}$$

where L_{eq} is the equivalent inductance defined in Section 5.1. For example, in the case of four sending coils connected in parallel and in the same direction, the expression of the coupling factor equals:

$$k = \frac{\mu_0 N_{\text{r}} r_{\text{r}} \sum_{i=1}^{n} \frac{N_i r_i}{r_i(r_{\text{r}}-\Delta_i)} G(h_i)}{\sqrt{\left(\frac{\sum_{i\neq j} \frac{1}{2}(L_i L_j - (\mu_0 N_{\text{r}} r_{\text{r}} \sum_{i=1}^{n} \frac{N_i r_i}{r_i(r_{\text{r}}-\Delta_i)} G(h_i))^2)}{\sum_i L_i - \sum_{i\neq j} \mu_0 N_{\text{r}} r_{\text{r}} \sum_{i=1}^{n} \frac{N_i r_i}{r_i(r_{\text{r}}-\Delta_i)} G(h_i)} \right)}} \tag{15}$$

5.5 Analytical expression for induced voltage and current

The expression of the induced voltage and the induced current are derived from Faraday's law. The induced voltage V_{ind} in the receiving coil is given by equation (10):

$$V_{\text{ind}} = -L_{\text{r}}\frac{\mathrm{d}I_{\text{r}}}{\mathrm{d}t} + \sum_{i} M_{i\text{r}}\frac{\mathrm{d}I_{i}}{\mathrm{d}t} \tag{16}$$

$M_{i\text{r}}$ mutual inductance between the receiving coil r and the sending coil number i
n number of sending coils
I_i sending coil current
L_{r} self-inductance of the receiving coil

The expression of the induced current I_{ind} in the receiving coil is given as follows:

$$I_{\text{ind}} = \frac{j\omega \sum_i M_{i\text{r}} I_i}{Z_2} \tag{17}$$

where Z_2 is the impedance of the receiving side that depends on the topology of the inductive system.

6 Optimization of multi-coil inductive systems

The received power at the load and the system efficiency of multi-coil inductive systems can be optimized in different ways. In this section, we propose the optimization of a multi-coil inductive system by the study of the excitation and arrangement of its coils, the choice of the appropriate topology and also by the implementation of a detection circuit for movable receivers in which only the sending coils that are in the proximity of the receiver will be activated and the others will be kept switched off.

6.1 Received power optimization

One of the solutions proposed to increase the received power at the load was investigated in [1, 2, 20]. It is based on the orientation of the magnetic field lines, issued by the active sending coils, directly to the receiving coil. This is performed by the use of the two nearest neighbor coils. These two coils are excited with less current and in the opposite direction, which is known by the out-of-phase excitation (see Figure 11). Experimental (dashed lines) and simulation (solid lines) results are illustrated in Figure 12. Here, the received power at the load is determined as a function of different levels of lateral misalignment (varying from 0 mm to 20 mm) and at a 5 cm coil-to-coil distance (at this large air gap, we expect high magnetic field losses).

Both experimental and simulation results show that the received power at the load is increased by 18% by using a multi-coil system with oriented magnetic field

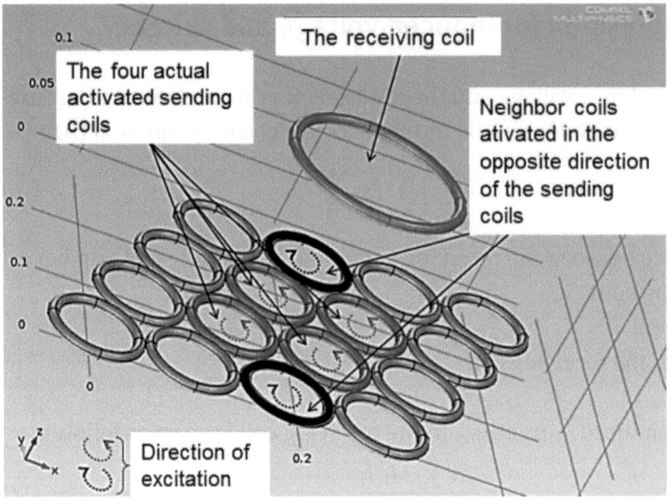

Fig. 11: Multi-coil inductive system.

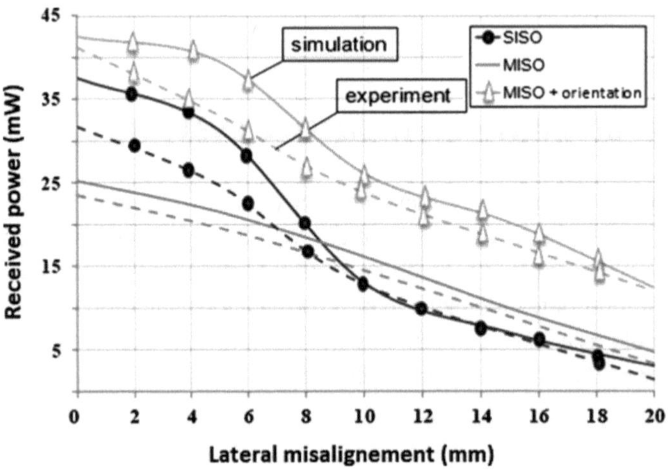

Fig. 12: Simulation (solid lines) and experimental (dashed lines) results of the received power at the load at 50 mm coil-to-coil vertical distance [20].

(MISO + orientation) instead of the conventional two-coil system single input single output (SISO). It can be seen that, even at bigger levels of misalignment, corresponding to 50% of the sending coil diameter, the system is capable of transmitting sufficient power to meet the power consumption requirement of most sensor systems (usually this varies from some microwatts to few milliwatts). In this case, the transmission efficiency equals 15%. According to these results, we can conclude that the multi-coil system with oriented magnetic field is, first, a tolerable system because of its smaller sensitivity to the lateral misalignment between the sending and receiving coils and, secondly, an efficient system because it gives sufficient power to the load at large air gaps.

6.2 Topology optimization

Similarly to the conventional two-coil system, multi-coil inductive systems are also optimized by the study of their topologies. So, in this section, we will show the impact of the system topology on the performance of multi-coil IPT systems. Figure 13 depicts the four topologies studied: series-series SS, series-parallel SP, parallel-series PS and parallel-parallel PP. The choice of the appropriate capacitance is dependent on the inductance (L_{eq} and L_r) and resistance (R_{eq} and R_{L_r}) of the sending and receiving coils, their mutual inductance (M_{eq}), the angular frequency (ω), the distance separating the two sides (d), the lateral misalignment (Δ) and the load (R_L). The idea behind the use of these topologies is to work at resonance to nullify the reflected impedance. It can be seen that systems with parallel–parallel 'PP' compensation topology, followed by the SP topology, offer the highest power at the load in both perfect and lateral misalign-

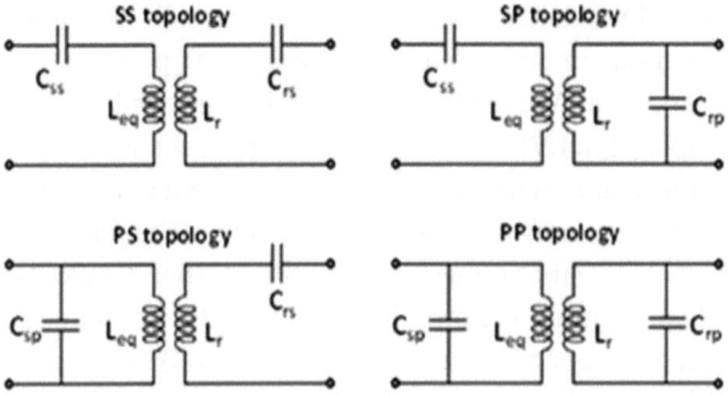

Fig. 13: SS, SP, PS and PP topologies of the multi-coil inductive system.

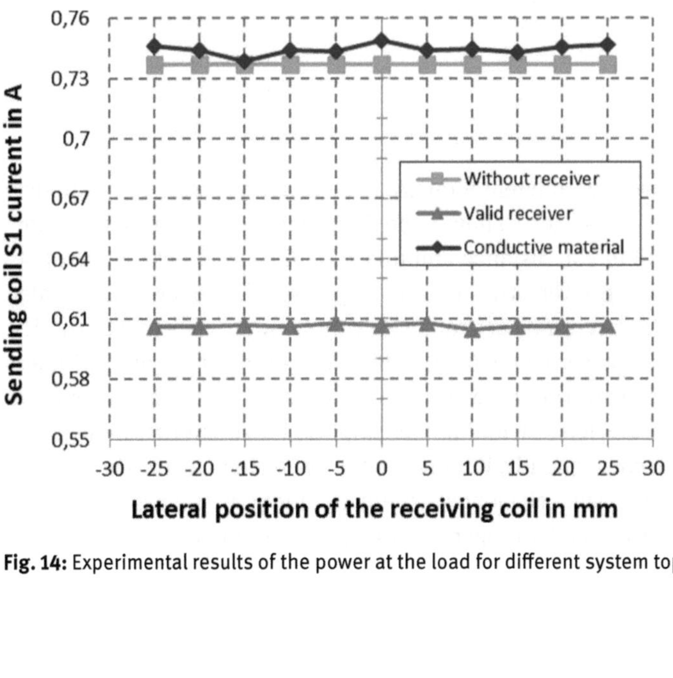

Fig. 14: Experimental results of the power at the load for different system topologies.

ment cases. The same investigation was demonstrated for the case of the conventional two-coil system in [21, 22] and [23], which demonstrate that the highest performance is given by SP and PP topologies, respectively.

6.3 Detection of the receiver

A method proposed to reduce the energy consumption and increase the system efficiency of multi-coil inductive systems is to implement a detection circuit for the receiver, in which we power only the sending coils that are under the receiving coil and we switch off the other ones. Only a few detection methods have been proposed in literature, e.g., in [24–26]. In [24], the authors proposed the use of additional sensor coils in each sending coil. However, this method shows some drawbacks due to the necessity of using ferrite materials and the low detection height. In [26], a sensing principle is based on the measurement of the impedance change of the sending coils. However, this solution requires extra estimation circuits and other electronic components that increase the system complexity and cost. A detection method using the sending coils themselves as detectors was proposed in [27]. It is based on the measurement of the peak of the AC current of the sending coils. It first converts the AC current of the sending coils to a DC voltage and then compares it to a reference voltage. The reference voltage is chosen within the gap between the voltage thresholds corresponding to the case of a valid and an invalid load. Figure 14 shows that the current amplitude decreases only in the case of the presence of a valid receiver (receiving coil) in the proximity. Based on these results, a detection threshold can be chosen between 0.61 A and 0.73 A. For a threshold current of 0.67 A, a detection voltage threshold of 2.7 V can be chosen.

7 Conclusion

Inductive systems are used for different industrial, medical and smart home applications. Conventional two-coil systems show some limitations for systems with a movable receiver, a large coil to coil distance and unavoidable lateral misalignment between the sender and the receiver. With multi-coil systems, the field is concentrated and better directed to the receiver by activating only the sending coils that are underneath the receiver and by powering the neighboring coils of the active ones with less current and in the opposite direction, respectively. The use of multi-coil systems enhances most system parameters, including the coupling factor, mutual inductance, transmitted power and, consequently, it improves the system efficiency. The received power at the load and the system efficiency of multi-coil inductive systems can be optimized by the right choice of coil arrangement and how to excite them, by the choice of the appropriate topology (PP, PS, SP or SS) and also by the implementation of a detection circuit for movable receivers. The idea behind the detection of the receiver comes as an attempt to save more energy, during which only the sending coils in the proximity of the receiver will be activated and all the others will be kept switched off.

Bibliography

[1] B. Kallel, T. Keutel, and O. Kanoun. MISO Configuration Efficiency in Inductive Power Transmission for Supplying Wireless Sensors. In *IEEE 11th International Multi-Conference on Systems, Signals and Devices (SSD'14), Barcelona, Spain, March 2014*, 2014.

[2] B. Kallel, O. Kanoun, T. Keutel, and C. Viehweger. Improvement of the efficiency of MISO configuration in inductive power transmission in case of coils misalignment. In *2014 IEEE International Instrumentation and Measurement Technology Conference. I2MTC, Montevideo, Uruguay, May 2014*, 2014.

[3] R. E. Hamam, A. Karalis, and J. Joannopoulos. Efficient weakly-radiative wireless energy transfer: An EIT-like approach. *Annals of Physics*, 324(8):1783–1795, 2009.

[4] N. Oodachi, K. Ogawa, H. Kudo, et al. Efficiency improvement of wireless power transfer via magnetic resonance using transmission coil array. In *IEEE International Symposium on Antennas and Propagation, Spokane, Washington, USA, July 2011*, pages 1707–1710, 2011.

[5] J. P. W. Chow, N. Chen, H. S. H. Chung, et al. Misalignment tolerable coil structure for biomedical applications with wireless power transfer. In *IEEE International Conference of the Engineering in Medicine and Biology Society, Osaka, July 2013*, pages 775–778, 2013.

[6] S. M. A. K. RamRakhyani and M. Chiao. Design and optimization of resonance-based efficient wireless power delivery systems for biomedical implants. *IEEE Transaction on Biomedical Circuits and Systems*, 5(1):48–63, 2011.

[7] S. C. Moon and G.-W. Moon. Wireless Power Transfer System with an asymmetric four-coil resonator for electric vehicle battery chargers. *IEEE Transactions on Power Electronics*, 31(10):6844–6854, 2016.

[8] X. Li, H. Zhang, F. Peng, et al. Wireless magnetic resonance energy transfer system for micro implantable medical sensors. *Sensors*, 12:10292–10308, 2012.

[9] S. Y. R. Hui. Inductive battery charger system with primary transformer windings formed in a multi-layer structure. U.S. Patent 7 164255, Jan. 16 2007.

[10] S. Y. R. Hui, W. C. Ho, X. Liu, and W. C. Chan. Localized charging, load identification and bi-directional communication methods for a planar inductive battery charging pad. U.S. Patent 7915858, Mar. 29 2011.

[11] C. Joffe, A. Roßkopf, S. Ehrlich, C. Dobmeier, and M. März. Design and Optimization of a Multi-Coil System for Inductive Charging with Small Air Gap. In *IEEE Applied Power Electronics Conference and Exposition, Tampa, Florida, March 2016*, pages 1741–1747, 2016.

[12] Z. Zhang and K. T. Chau. Homogeneous Wireless Power Transfer for Move-and-Charge. *IEEE Transactions on Power Electronics*, 30(11):6213–6220, 2015.

[13] B. Kallel, K. Sasmal, O. Kanoun, and H. Trabelsi. Analytical modeling of multi-coil system for inductive powering of movable wireless low-power devices. In *IEEE 12th International Multi-Conference on Systems, Signals and Devices (SSD'15), Mahdia, Tunisia, March 2015*, 2015.

[14] K. Sasmal. Energy management for an inductive coil system with detection of the receiving coil. In *Master thesis, Chemnitz University of Technology, October 2015*, 2015.

[15] H. R. Ahn, M. S. Kim, and Z. J. Kim. Inductor array for minimising transfer efficiency decrease of wireless power transmission components at misalignment. *Electronics Letters*, 50(5):393–394, 2014.
[16] C. L. W. Sonntag, J. L. Duarte, and A. J. M. Pemen. Load Position Detection and Validation on Variable-Phase Contactless Energy Transfer Desktops. In *IEEE Energy Conversion Congress and Exposition, San Jose, USA*, pages 1818–1825, 2009.
[17] C. L. W. Sonntag, E. A. Lomonova, and J. L. Duarte. Variable-Phase Contactless Energy Transfer Desktop Part I: Design. In *International Conference on Electrical Machines and Systems, Hankou Wuhan, China*, pages 4460–4465, 2008.
[18] D. Fleisch. *A Student's Guide to Maxwell's Equations*. Cambridge, 2008.
[19] B. Kallel, K. Sasmal, O. Kanoun, and H. Trabelsi. Analytical modeling of a multi-coil system for inductive powering of movable low-power wireless devices. In *12th International Multi-Conference on Systems, Signals & Devices. SSD'15, Mahdia, Tunisia, March 2015*, 2015.
[20] B. Kallel, O. Kanoun, and H. Trabelsi. Large air gap misalignment tolerable multi-coil inductive power transfer for wireless sensors. *IET Power Electronics*, 9(8):1768–1774, 2016.
[21] N. Jamal, S. Saat, Y. Yusmarnita, et al. Investigations on capacitor compensation topologies effects of different inductive coupling links configurations. *International Journal of Power Electronics and Drive Systems*, 6(2):274–281, 2015.
[22] M. E. Halpern and D. C. Ng. Optimal tuning of inductive wireless power Links: limits of performance. *IEEE Transactions on Circuits and Systems I, Reg. Papers*, 62(3):725–732, 2015.
[23] T. Campi, S. Cruciani, F. Palandrani, et al. Wireless power transfer charging system for AIMDs and pacemakers. *IEEE Transaction on Microwave Theory and Techniques*, 64(2):633–642, 2016.
[24] Z. N. Low, J. J. Casanova, P. H. Maier, J. A. Taylor, R. A. Chinga, et al. Method of load/fault detection for loosely coupled planar wireless power transfer system with power delivery tracking. *IEEE Trans. Ind. Electron.*, 57:1478–1486, 2010.
[25] E. Waffenschmidt and T. Staring. Limitation of inductive power transfer for consumer applications. In *European Conference on Power Electronics and Applications,Barcelona, Spain, September 2009*, pages 1–10, 2009.
[26] C. L. W. Sonntag, J. L. Duarte, and A. J. M. Pemen. Load position detection and validation on variable-phase contactless energy transfer desktops. *IEEE Energy Conversion Congress and Exposition*, 15–27, 2009.
[27] B. Kallel, O. Kanoun, H. Trabelsi, and M. Roes. Sensing Movable Receiving Coils by Detection of AC Current Changes on the Primary Side of a Multi-Coil System. *Procedia engineering*, 168:991–994, 2017.

Ghada Bouattour, Bilel Kallel, Olfa Kanoun, and Houda Derbel

Energy management for inductive power transmission

Abstract: Energy management circuits are indispensable for wireless power transmission via inductive links to realize high efficiency in spite of the electronic losses and coil position perturbations. This chapter examines the most popular energy management circuits applied to the sending and the receiving side of inductive power transmission systems. First of all, we examine coil shape and orientation. Then, the energy management design of the sending side and receiving side are illustrated, and the system topology is presented in detail, including resonance circuits and the necessary transformations. Finally, different approaches for control systems in the inductive power transmission are summarized.

Keywords: Wireless power transmission, inductive power transmission, DC to AC inverter, rectifier, control

1 Introduction

In the last decades, researchers gave more importance to wireless systems that present a powerful solution to applications where the use of a wired connection is dangerous or not possible or critical using different methods: capacitive links, inductive links and via laser and microwaves. The choice of power transmission method is dependent on the transmission distance, power range and other environmental factors. Inductive power transmission presents several advantages in middle-range applications, such as safety, flexibility of rotation on the same plane, low maintenance, etc. Therefore, it can be used in a harsh environment, but it presents some limitations in terms of power consumption, efficiency, emitted radiation and also high costs caused by the complexity of the necessary electronic circuits [1]. Many applications with different power levels use inductive links for charging systems that can be started from the lowest power in the order of microwatts for implementable devices [2], micro-electronics applications in milliwatts [3], in watts for charging mobiles and electronics [4] and, finally, for high-power applications around kilo-watts, e.g., for charging electric vehicles [5]. Despite the difference of power ranges and the frequency chosen, all these applications, generally, have a similar system design with a difference concerning the energy management circuits to reach the highest output power and efficiency. Figure 1

Ghada Bouattour, Bilel Kallel, Olfa Kanoun, Chair for Measurement and Sensor Technology, Technische Universität Chemnitz, Chemnitz, Germany
Ghada Bouattour, Bilel Kallel, Houda Derbel, National Engineering School of Sfax, Sfax, Tunisia

https://doi.org/10.1515/9783110445053-014

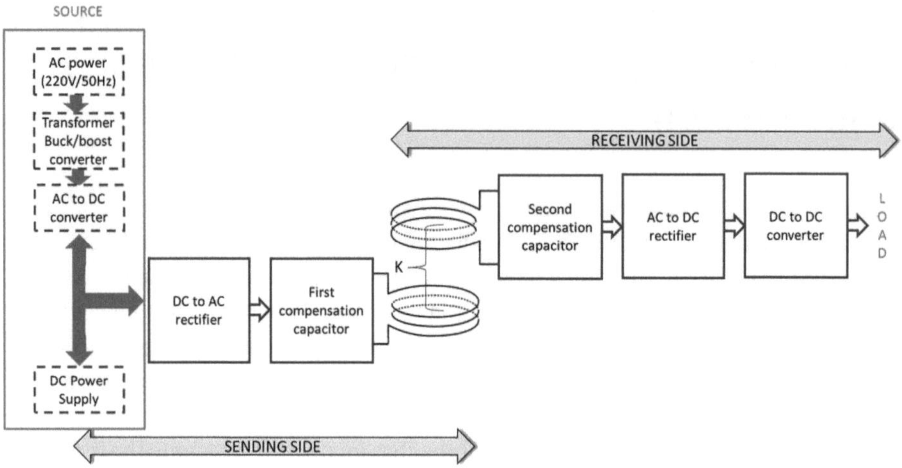

Fig. 1: General structure of the inductive power transmission system.

shows a typical inductive power transmission system. An AC signal with defined amplitude and frequency is generated by different AC/DC and DC/AC converters [6]. The coils are coupled by the magnetic field and have a coupling factor k. In the secondary side and in order to obtain smooth DC power to the load, AC/DC and DC/DC converters are used. To achieve a resonant system, compensation capacitors in both primary and secondary sides are included.

There are many factors that cause energy and magnetic field losses; so that the coupling factor between the coils will be very weak. This affects the efficiency of inductive power transfer systems. Among these factors we find the following [7]:

- the geometry and the parameters of the coil including shape, number of turns, winding thickness, core material and dimension;
- the quality factor of the coil, which has a direct relation to the working frequency;
- both vertical and lateral distances between the coils;
- the use of additional material; ferrite core in the proximity [7].

In this chapter, energy management for both primary and secondary circuits is presented, which is important for improving the power transmission efficiency of an IPT system. The first part is dedicated to the energy management of the sending side (primary circuit) where different inverter circuits are evaluated. Different techniques to achieve a resonant system with a good performance, by adding compensation capacitors on both the primary and the secondary side, are discussed. In the second part, we present different rectifier circuits followed by DC/DC converters to provide smooth DC power for the load. Finally, an overview of charge storage elements and control algorithms is presented.

2 Coil shape and orientation

2.1 Coil shape

The efficiency of an inductive power transmission system depends on different parameters of the coil system, such self-inductance, quality factor, alignment and orientation of the coils, size and coil material. For this, the coil characteristics, including geometry, thickness of windings, number of turns, dimensions, etc., present the key parameters of the inductive system. Selection of the coil shape depends, in general, on the system requirements. In the literature, various coil geometries are presented, which include:

- single circular pad (SCP)
- circular rectangular pad (CRP)
- multiple circular pad (MCP)
- double-D (DDP)
- double-D quadrature pad (DDQP)
- bipolar pad (BPP)

All these coil geometries are presented and compared in Table 1 according to the features: basic structure, operating principles, advantages and drawbacks [8, 9].

Tab. 1: Comparison between different coil architectures [8, 9].

	CRP	SCP	MCP	DDP	DDQP	BPP
Size	**	**	***	*	*	*
Weight	*	*	***	*	**	**
Cost	*	*	***	**	***	**
Transmission distance	*	*	**	**	***	***
Transferrable power	**	**	*	***	***	***
Misalignment immunity	*	*	***	**	**	**

Legend: *: low or small **: medium ***: high or large

2.2 Coil orientation

In practice, the ideal coil position is difficult to reached because of coil movements and manufacturing errors. Thus, different misalignment cases may occur during the power transmission process, as shown in Figure 2:

- Angular misalignment by a planar variation with an angle θ presented in Figure 2(b), which conserves the alignment of coil centers.
- Lateral misalignment by a horizontal variation d between sending and receiving coils (Figure 2(c))

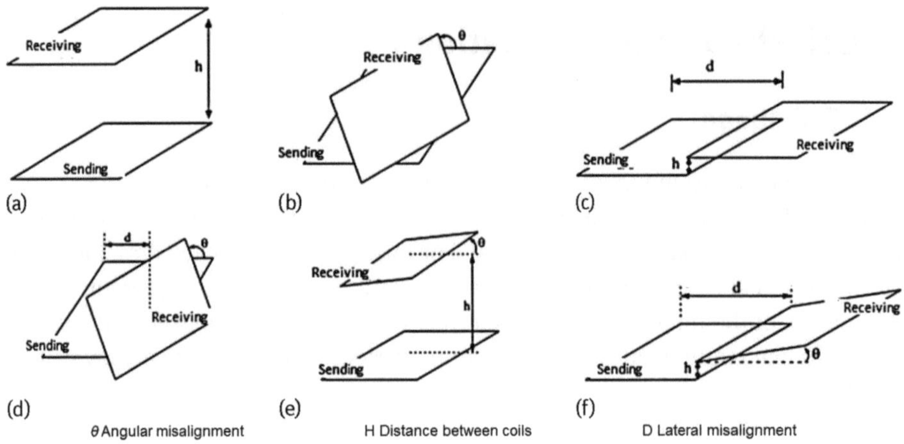

Fig. 2: Examples of misalignment cases [10].

Even combinations between two or three kinds of basic misalignments can occur. For example, a planar and horizontal variation (Figure 2(d)), a combination of angular misalignment and vertical variation (Figure 2(e)) and a combination of angular misalignment, horizontal and vertical variation (Figure 2(f)).

3 Energy management at the sending side

As signal generator in inductive power transmission systems, different DC power supplies can be used, such as battery or rectified grid power. For an adjusted voltage amplitude of the system, an elevator, step-down transformer or boost-buck converters can be associated to the power supply [6, 11]. In this part, a description of the energy management circuit used on the sending side of inductive power transmission systems is presented. First of all, a DC to AC inverter is needed to convert the DC input to an adequate AC output with a suitable frequency. Then, the main system topologies are presented. They realize resonance between sending and receiving coils by associating compensation capacitors and/or inductors to the coil system to form an LC resonant circuit.

3.1 DC/AC inverters

The inverter circuits built into inductive power transmission systems are presented. They realize resonance between sending and receiving coils by the selection of the circuit frequency and the system control. To this end, many studies focus on the reduction of losses due to the conversion of current from DC to AC. The inverter circuits

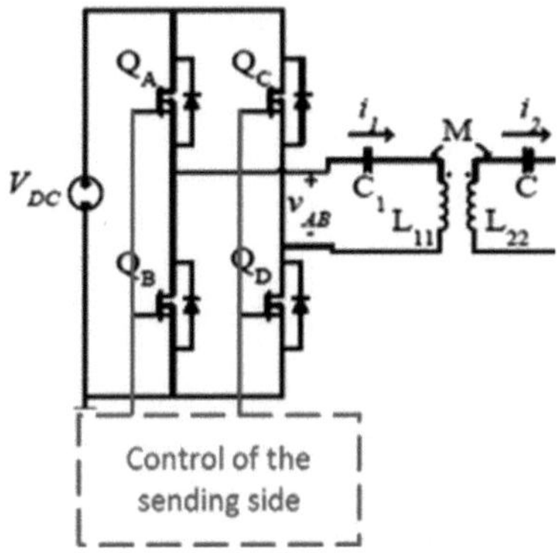

Fig. 3: Full-wave inverter [12].

that are used most frequently are the bridge inverter, the half-wave inverter and the class E inverter. Those circuits are different concerning the system architecture, control strategy, frequency range, efficiency, cost and complexity of implementation. The control of the inverter circuit uses mostly a square wave signal, which turns the switch on and off. The switch generates a pulse-width modulation (square wave, a modified sine wave, etc.) The full-wave inverter presented in Figure 3 is composed of four Mosfet switches connected as a bridge and controlled by a micro-controller, which generates a square wave or a modified sine wave pulse width modulated signal (PWM). The bridge can generate a high efficiency for low-range and medium-range frequencies; the four switches make it a low current stress on the switch, but this solution is relatively complex in its structure and control requirements. The bridge inverter as presented in Figure 4 is controlled by a synchronization of "on" and "off" of a Mosfet switch:

- In the first half period, the Mosfet switches "S_1" and "S_4" on but the other switches are off.
- In the second half period, the Mosfet switches "S_1," and "S_4" are off, the others are on.

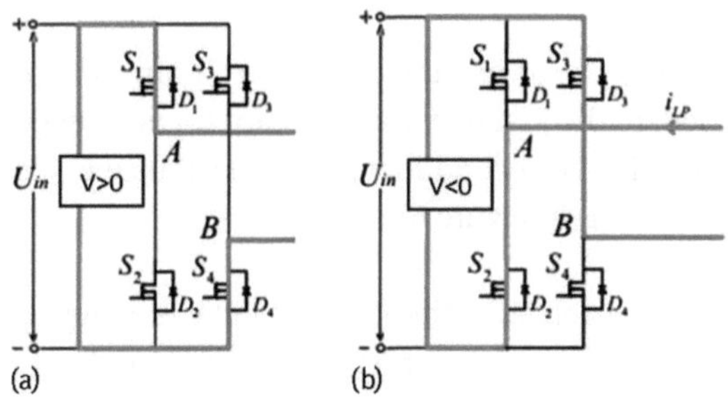

Fig. 4: Output voltage generation from the full-wave inverter. (a) Positive period; (b) negative period [13].

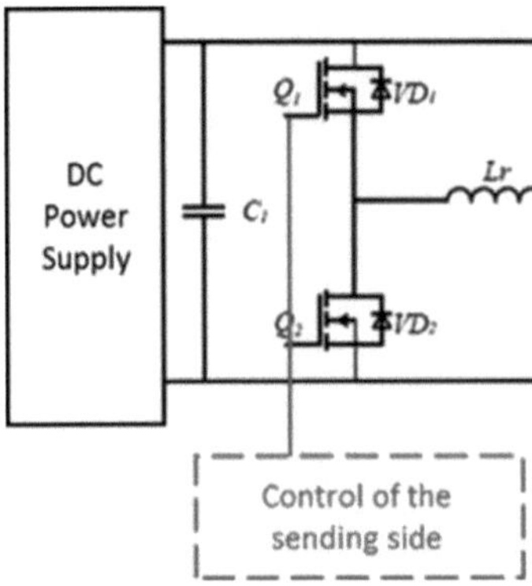

Fig. 5: Half-wave inverter [14].

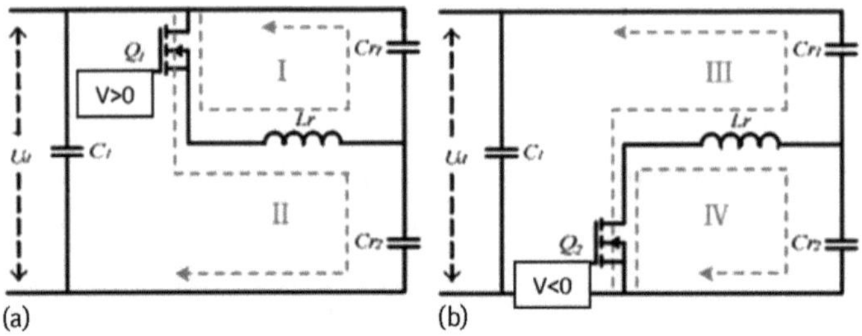

Fig. 6: Output voltage generation of the half-wave inverter (a) Q_1 on, (b) Q_2 on [14].

The half-wave inverter presented in Figure 5 is composed of a couple of Mosfet switches associated to a couple of capacitors in a bridge form. This architecture is balanced concerning efficiency, complexity of implementation and system control and costs. Nevertheless, it presents more voltage stress in the Mosfet switch compared to the bridge rectifier generally used for low and medium ranges of frequency applications. The square wave signals of the half-wave Mosfet inverter are synchronized as shown in Figure 6

- In the positive half period, the switch Q_1 turns on and the switch Q_2 turns off. In this case, two loops are formed. In the first one, a resonant circuit discharges the capacitor C_{r1}, and in the second one, a DC power supply charges the capacitor C_{r2} (see (I) in Figure 6).
- In the negative half period, the switch Q_2 turns on, and the switch Q_1 turns off. In this case, two other loops are formed. In the first one, the capacitor C_{r1} is charged from the DC power supply. However, in the second loop, the capacitance C_{r2} is discharged by the resonant circuit (see (II) in Figure 6).

For high frequency applications, which can reach MHz, the class E inverter is preferred (Figure 7). It is composed of only one Mosfet switch in order to reduce noise due to the high frequencies on the switches This leads to a drawback concerning the voltage stress across the Mosfet switch [15].

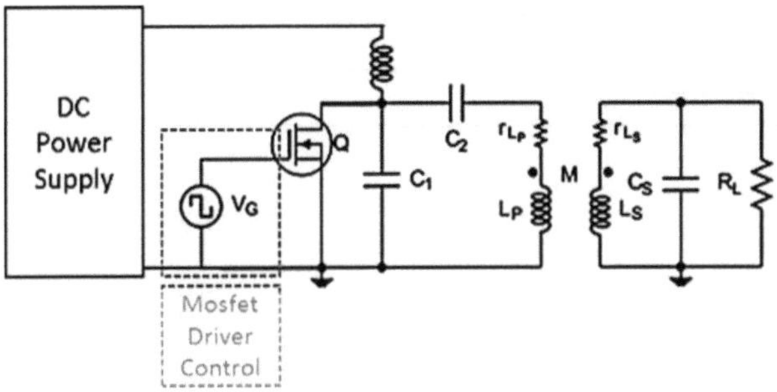

Fig. 7: Class E inverter applied to the inductive power transmission system [15].

To reach a high efficiency for inverters circuits, diodes are connected in parallel to the Mosfet switches to form a free-wheel loop in the dead time, and also a capacitance is added to filter the square wave inverter output.

3.2 Resonance topologies

The basic inductive power transmission system has properties of reactive power that reduce the transmission efficiency. To reach a high load power an LC resonance is realized. The basic method uses an additional compensation capacitor at the sending and receiving sides of the inductive power transmission system. In the literature, many system topologies have been presented for different positions, such as in series or in parallel and selection of the compensation capacitors values. The most common topologies are presented in Figure 8. They use only one capacitance at the sending and/or receiving sides. The first method of compensation aims to reduce the reactive power at the sending side by fixing the second compensation capacitor C_S and realiz-

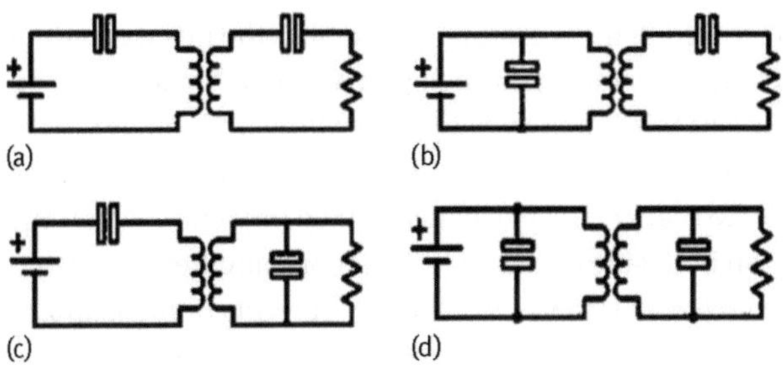

Fig. 8: Configurations of compensation capacitors. (a) Serial–serial; (b) serial–parallel; (c) parallel–serial; (d) parallel–parallel.

Tab. 2: Primary equivalent capacitance for different system topologies [16].

Topology	Primary compensation capacitance (C_p)
SS	$\frac{1}{\omega^2 L_p}$
SP	$\frac{L_p}{\left(R_p + \frac{\omega^2 M^2}{R_l + R_s}\right)^2 + \omega^2 L_p^2}$
SP	$\frac{1}{\omega^2 \left[L_p - \frac{\omega^2 M^2 L_s}{R_s^2 + \omega^2 (L_s + C_s R_s R_L)^2}\right]}$
PP	$\frac{\left[L_p - \frac{\omega^2 M^2 L_s}{R_s^2 + \omega^2 (L_s + C_s R_s R_L)^2}\right]}{\left[R_p + \frac{\omega^2 M^2 (R_s + R_L + \omega^2 C_s^2 R_L^2 R_s)}{R_s^2 + \omega^2 (L_s + C_s R_L R_s)^2}\right]^2 + \omega^2 \left[L_p - \frac{\omega^2 M^2 L_s}{R_s^2 + \omega^2 (L_s + C_s R_s R_L)^2}\right]^2}$

ing the resonance with the receiving coil, as presented by Equation(1) [16]. The primary compensation capacitance C_p value is presented in Table 2 [16], with

$$C_s = \frac{1}{L_s \omega} \tag{1}$$

The second method of calculation the compensation capacitors aims to eliminate the reactive power on both sides [17, 18]. The equations of the compensation capacitors and also the equivalent resistance expressions on the primary and the secondary sides are presented in Tables 3 and 4.

In this case, the equations of the SP and PP topologies need to fulfill the condition given in equation (2),

$$R_L > 2\omega L_r \tag{2}$$

Other topologies have been presented in the literature, which filter the inverter output with additional capacitances and inductances in the system to reach resonance on the both sides [19]. The structure of this topology is by associating an inductor in series with the input voltage connected to a parallel and series capacitances to form the LCC topology (inductance with two capacitances), which can be implemented on the sending side, receiving side or on both sides. This system design combines the advantages of series and parallel topologies in terms of voltage and current compensations. The equivalent circuit of the inductive power transmission system corresponding to this system topology is presented in Figure 8, where L_{f1}, L_{f2} are the primary and the secondary compensation inductors, C_1, C_{f1} are the primary compensation capacitors, C_2', C_{f2}' are the secondary compensation capacitors, L_{s1}, L_{s2}', L_m are the equivalent inductance model of the coil system and L_1, L_2 are the self-inductances of the coil system.

Tab. 3: Primary and secondary equivalent capacitances for different system topologies.

Topology	Primary capacitance C_s	Secondary capacitance C_r
SS	$\frac{1}{\omega^2 L_s}$	$\frac{1}{\omega^2 L_r}$
SP	$\frac{1}{\omega^2 L_s}$	$\frac{R_L^2 + \sqrt{Re(Zp)}}{2L_r(\omega R_L)^2}$, $Zp = R_L^4 - (2\omega R_L R_s)^2$
PS	$\frac{L_s}{(L_s\omega)^2 + R_0^2}$ with $R_0 = R_r + R_{L_s}$, $R_r = \frac{(M\omega)^2}{R_2}$	$\frac{1}{\omega^2 L_r}$
PP	$\frac{L_s}{(L_s\omega)^2 + R_0^2}$ with $R_0 = R_r + R_{L_s}$, $R_r = \frac{(M\omega)^2}{R_2}$	$\frac{R_L^2 + \sqrt{Re(Zp)}}{2L_r(\omega R_L)^2}$, $Zp = R_L^4 - (2\omega R_L R_s)^2$

Tab. 4: Primary and secondary equivalent resistances for different system topologies.

Topology	Primary resistance R_1	Secondary resistance R_2
SS	$R_s + R_{L_s}$	$R_{L_r} + R_L$
SP	$R_s + R_{L_s}$	$R_{L_r} + \frac{R_L}{1 + (R_{LC_{rp}}\omega)^2}$
PS	$R_s + \frac{R_0}{A^2 + B^2}$ with $A = 1 - L_s C_{ss}\omega^2$, $B = R_0 C_{ss}\omega$, $R_0 = R_r + R_{L_s}$	$R_{L_r} + R_L$
PP	$R_s + \frac{R_0}{A^2 + B^2}$ with $A = 1 - L_s C_{ss}\omega^2$, $B = R_0 C_{ss}\omega$, $R_0 = R_r + R_{L_s}$	$R_{L_r} + \frac{R_L}{1 + (R_{LC_{rp}}\omega)^2}$

The equations of the LCC topology are presented by the following equations [19]:

$$L_{f1} C_{f1} = \frac{1}{\omega_0^2} \tag{3}$$

$$L_{f2} C_{f2} = \frac{1}{\omega_0^2} \tag{4}$$

$$L_1 - L_{f1} = \frac{1}{C_1 \omega_0^2} \tag{5}$$

$$L_2 - L_{f2} = \frac{1}{C_2 \omega_0^2} \tag{6}$$

Different system topologies have been presented in this part. The selection of the optimum system topologies depend on the system and application requirements.

4 Energy management on the receiving side

In the first part of this section, we will present different rectifier circuits used for inductive systems. The choice of the appropriate circuit is decisive for system efficiency. The load or the storage element can, in general, be a battery or a supercapacitor. In the second part, the DC/DC converter providing smooth DC power to the load or the storage element is studied. We focus here on maximum power point tracking (MPPT) and voltage regulator systems.

4.1 Rectification stage

Generally, to charge systems, a DC output is required. Many types of AC to DC converters have been studied, such as bridge rectifiers, diode voltage multipliers, active bridge rectifiers and active voltage multipliers.

Passive rectifiers

There are two common passive rectifiers:
- The passive bridge rectifier (Figure 9) composed by four Schottky diodes connected as a bridge and a smoothing capacitor in parallel to the load in order to eliminate the ripple of rectifications. To generate the DC output using the bridge rectifier, the diode works with an AC input, as shown in Figure 10.

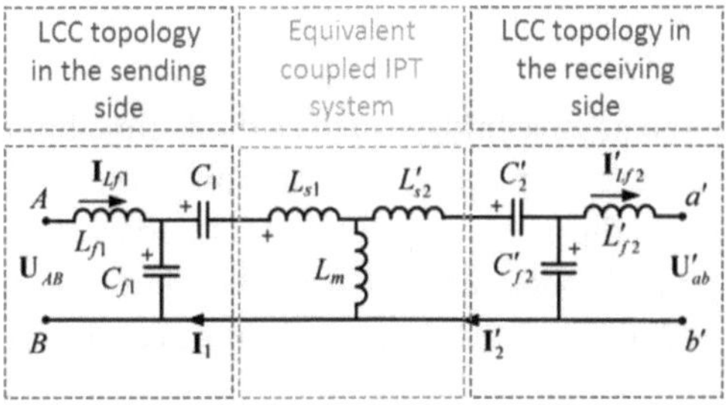

Fig. 9: Equivalent circuit of the IPT system with LCC topologies on both sides [19].

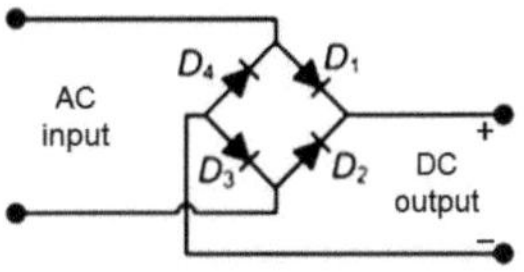

Fig. 10: Passive bridge rectifier circuit [20].

This circuit works as follows:

- During the positive half cycle, the current passes through the diodes D_1 and D_3.
- During the negative half cycle, the current passes through the diodes D_4 and D_2.

The bridge rectifier is often used in inductive power transmission systems due to the simplicity of implementation, the efficiency caused by the low drop-out voltage and the low cost, despite its high leakage current compared to other rectifier circuits [21].

- The voltage multiplier is a circuit that amplifies the output voltage of the system. Generally, it is applied for low voltage applications. The voltage multiplier circuit is composed by cascades of voltage doubler stages. In the voltage doubler circuit in Figure 11, the input peak voltage passes through C_{IN} and D_{N} to give a sine wave output with a doubled voltage value. Then, it passes directly to D_{p} and C_{p} to transform the AC input to a doubled DC output.

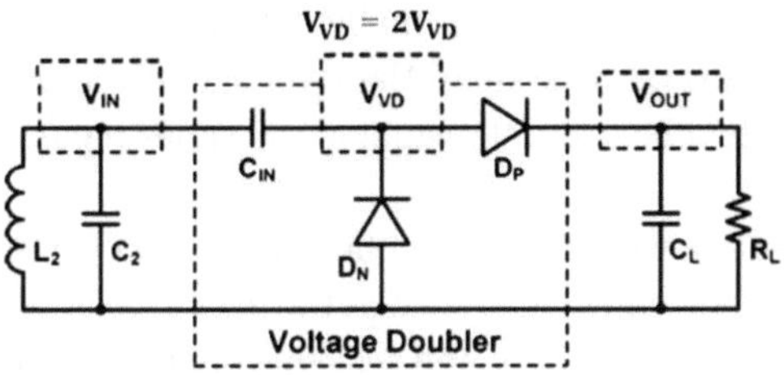

Fig. 11: Passive voltage multiplier circuit on the receiving side of the inductive power transmission system [21].

Active rectifier

In order act against the disadvantages of the passive bridge, such as diode threshold voltage, power losses and high forward voltage drop, in [22], the authors present an active rectifier as shown in Figure 12, in which the passive diodes are replaced by Mosfet switches in the bridge and the voltage multiplier circuit. Mosfet switches offer a high power conversion efficiency and a reduction of the drop-out voltage caused by the diode effect [21, 22].

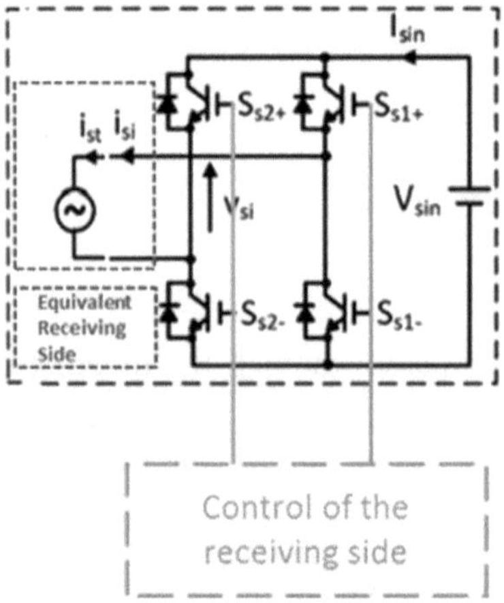

Fig. 12: Full-wave active bridge rectifier [22].

4.2 DC to DC converter

To improve system performance, with a high efficiency and stable output voltage even under movements of the receiving coil, load value variation or other parameter variations, a control system is required, such as passive voltage regulator and an active DC to DC (boost) converter.

- Passive voltage regulators: The passive voltage regulator circuit keeps a constant output voltage without to possibility to control the system behavior. It is placed just before the load on the receiving side of the system. Many voltage regulators have been proposed in [23], such as the linear shunt regulator and series regulators.
- Active DC-DC converters: The power achieved from the rectifier stage can be regulated and stabilized through active DC-DC converters, such as the buck converter, boost converter, buck-boost converter, Cuk converter and zeta converter [24]. The most used active DC-DC converter in inductive power transmission systems is

the boost converter, which can achieve MPPT. This converter is used in order to achieve a constant voltage or current output that can be controlled under load and coupling factor variation:

- The coupling factor changes caused by lateral, angular and longitudinal movement between coils.
- The load varies when the number of receivers changes [25, 26].

The boost converter is generally applied for inductive charging applications (Figure 13). It is composed of a coil, a capacitor, a Mosfet switch and a diode. It is connected between the load and the rectifier output.

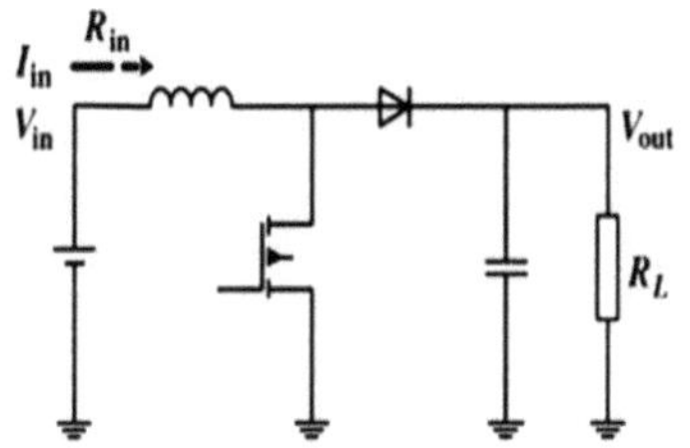

Fig. 13: Boost converter circuit [24].

4.3 Charge storage

Various storage elements can be used, such as batteries, double-layer capacitors or other storage elements [27].

- A simple battery equivalent circuit model is illustrated in Figure 14. It is composed of a constant source E_{eq} connected to a capacitance with a series resistance and a parallel resistance. The typical battery charging profile is presented in Figure 15. It contains a constant current (cc) mode and a constant voltage (cv) mode. When the battery is totally discharged, charging with a constant current I_b is used until the voltage reaches its nominal value V_b. Then, constant voltage charging starts, and the battery current decreases exponentially until the end of the charging process [28].

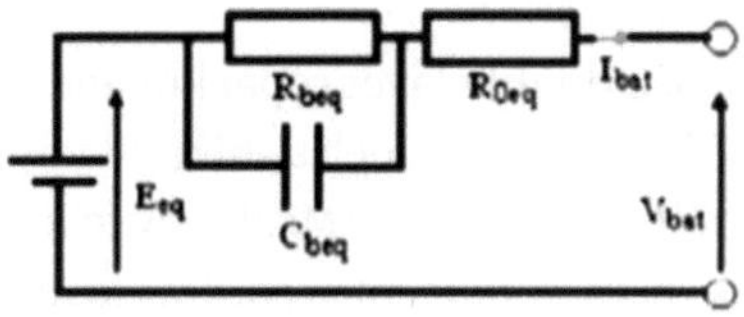

Fig. 14: Equivalent model of a battery [29].

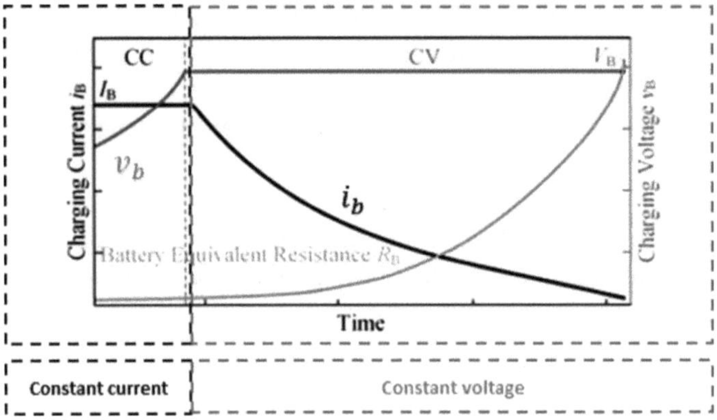

Fig. 15: Typical charging process of a lithium-ion battery [28].

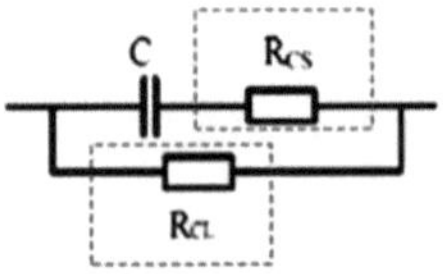

Fig. 16: Equivalent model of a supercapacitor [30].

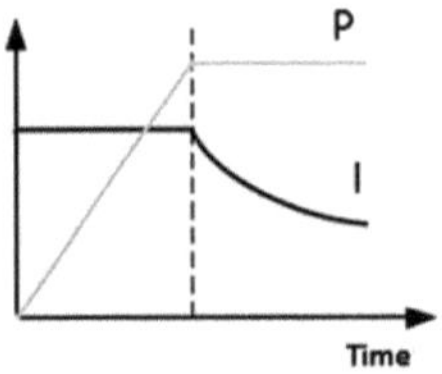

Fig. 17: Supercapacitor charging process.[30].

- Double-layer capacitance/supercapacitor: The equivalent model of the supercapacitor presented in Figure 16 has the same characteristics as a classic capacitor, which is composed of an equivalent capacitor C connected in parallel and in series to the equivalent resistance R_{cs} and R_{cl} [30]. The charging process of the supercapacitor (Figure 17) begins with a constant current (cc) equal to I_{sc} charging mode, until reaching the related voltage. Then it is implemented to a constant power (cp) equal to the P_{sc} charging mode. The supercapacitor is generally applied in applications that require faster charging compared to the battery.
- Hybrid storage elements contain a combination of a double-layer capacitor and a battery. This combines the advantages of both of the elements, such as high energy and high power density [31]. The charging process is controlled by Mosfet switches connected to the rectifier stage, which selects the storage element according to the load demand [31].

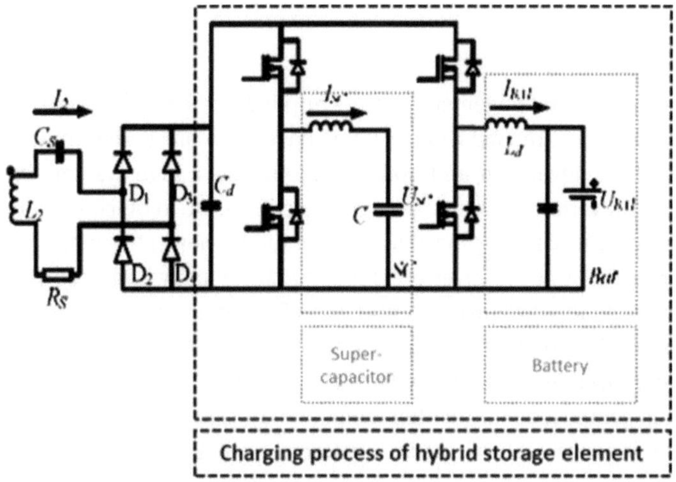

Fig. 18: Receiving side inductive power transmission system with a hybrid storage element [31].

5 Control system on inductive power transmission systems

Generally, the controller is connected to the Mosfet switch of the boost converter in the inductive power transmission system in order to achieve a power transfer to the load with controlled frequencies and currents [32]:

- Control by a fixed frequency: This is a simple control of the secondary side of the inductive power transfer system. The frequency is selected after the construction of the coil system and is applied for a fixed coupling factor system where the coil is fixed in an optimal position. This method presents drawbacks of increasing voltage/ampere of the power supply and difficulties of soft switching [32].
- Control by detection of a zero phase angle between the primary current and the voltage: This method is applied to the sending side of the inductive power transmission system, but it presents many drawbacks with respect to system bifurcation instability and also unsafe low coupling conditions [32, 33].
- Control by the use of the perturb and observe method applied on the MPPT system, which compares the output power to a previous value and a result system frequency perturbation to reach the maximum power. This method requires a longer response time then it is not practical for high-speed movable systems, but it can be used in applications with contains two power peaks [32].

6 Conclusion

In this chapter, common energy management circuits applied to the sending and receiving sides of inductive power transmission systems have been presented. In general, many stages need to be developed and optimized to realize a high system efficiency, including DC to AC inverters, system topologies, rectifiers, DC to DC converters and the energy storage element. The sending side of the inductive power transmission system is composed first of a DC to AC inverter, which can have many architectures, such as full-wave inverters, half-wave inverters and class E inverters. Then, the resonance system is used by adding compensation capacitors and/or inductors connected to the coil system to form variable topologies with different forms.

The receiving part is composed of a rectifier stage and a DC to DC converter. For the rectifier stage, many structures of circuits are available, such as: passive bridges, voltage multipliers and active circuits. In addition, a DC to DC converter can be used to regulate the output voltage and generate the adequate output voltage and power to charge the system load.

Bibliography

[1] N. Chawla and S. Tosunoglu. State of the Art in Inductive Charging for Electronic Appliances and its Future in Transportation. In *Proc. Conf. on Recent Advances in Robotics*, 2012.

[2] A. Ben Amar, A. B. Kouki, and H. Cao. Power Approaches for Implantable Medical Devices. *Sensors (Basel, Switzerland)*, 15(11):28889–28914, 2015.

[3] A. Nasiri, S. A. Zabalawi, and D. C. Jeutter. A Linear Permanent Magnet Generator for Powering Implanted Electronic Devices. *IEEE Transactions on Power Electronics*, 26(1):192–199, 2011.

[4] S. Y. Chang, S. L. S. Kumar, and Y. C. Hu. Performance of Cognitive Wireless Charger for Near-Field Wireless Charging. In *2017 IEEE 37th International Conference on Distributed Computing Systems (ICDCS), Atlanta, GA*, pages 2555–2556, 2017.

[5] D. H. Tran, V. B. Vu, and W. Choi. Design of a High-Efficiency Wireless Power Transfer System with Intermediate Coils for the On-Board Chargers of Electric Vehicles. *IEEE Transactions on Power Electronics*, 33(1):175–187, 2018.

[6] J. L. Villa, J. Sallan, A. Llombart, and J. F. Sanz. Design of a high frequency inductively coupled power transfer system for electric vehicle battery charge. *Appl. Energy*, 86(3):355–363, 2009.

[7] D. Kurschner and C. Rathge. Integrated contactless power transmission systems with high positioning flexibility. In *2008 13th International Power Electronics and Motion Control Conference, Poznan*, pages 1696–1703, 2008.

[8] C. Liu, C. Jiang, and C. Qiu. Overview of coil designs for wireless charging of electric vehicle. In *2017 IEEE PELS Workshop on Emerging Technologies: Wireless Power Transfer (WoW), Chongqing*, pages 1–6, 2017.

[9] G. Ombach, D. Kurschner, S. Mathar, and W. Chlebosz. Optimum magnetic solution for interoperable system for stationary wireless EV charging. In *2015 Tenth International Conference on Ecological Vehicles and Renewable Energies (EVER), Monte Carlo*, pages 1–8, 2015.

[10] E. R. Joy, A. Dalal, and P. Kumar. Accurate Computation of Mutual Inductance of Two Air Core Square Coils with Lateral and Angular Misalignments in a Flat Planar Surface. *IEEE Transactions on Magnetics*, 50(1):1–9, 2014.
[11] X. Ge, Y. Sun, C. Tang, Z. Wang, and Z. Xu. Loss analysis and efficiency optimization of buck converter in wireless charging system for EVs. In *2017 IEEE PELS Workshop on Emerging Technologies: Wireless Power Transfer (WoW), Chongqing*, pages 329–331, 2017.
[12] Z. U. Zahid, Z. M. Dalala, C. Zheng, R. Chen, W. E. Faraci, J. S. J. Lai, G. Lisi, and D. Anderson. Modeling and Control of Series–Series Compensated Inductive Power Transfer System. *IEEE Journal of Emerging and Selected Topics in Power Electronics*, 3(1):111–123, 2015.
[13] P. Tan, H. He, and X. Gao. Phase compensation, ZVS operation of wireless power transfer system based on SOGI-PLL. In *2016 IEEE Applied Power Electronics Conference and Exposition (APEC), Long Beach, CA*, pages 3185–3188, 2016.
[14] L. Shao, Q. Li, C. Tan, K. Yao, and J. Song. A Study of Magnetic Resonance Wireless Power Transfer System Based on Half Bridge Inverter. In *2016 IEEE Vehicle Power and Propulsion Conference (VPPC), Hangzhou*, pages 1–5, 2016.
[15] S. Aldhaher, P. C. K. Luk, and J. F. Whidborne. Wireless power transfer using Class E inverter with saturable DC-feed inductor. In *2013 IEEE Energy Conversion Congress and Exposition, Denver, CO*, pages 1902–1909, 2013.
[16] J. Sallan, J. L. Villa, A. Llombart, and J. F. Sanz. Optimal Design of ICPT Systems Applied to Electric Vehicle Battery Charge. *IEEE Transactions on Industrial Electronics*, 56(6):2140–2149, 2009.
[17] G. Bouattour, B. Kallel, O. Kanoun, and N. Derbel. Primary Side Circuit Design of a Multi-Coil Inductive System for Powering Wireless Sensors. *Procedia Engineering*, 168:920–923, 2017.
[18] G. Bouattour, B. Kallel, K. Sasmal, O. Kanoun, and N. Derbel. Comparative study of resonant circuit for power transmission via inductive link. In *015 IEEE 12th International Multi-Conference on Systems, Signals and Devices (SSD15), Mahdia*, pages 1–62, 2015.
[19] S. Li, W. Li, J. Deng, T. D. Nguyen, and C. C. Mi. A Double-Sided LCC Compensation Network and Its Tuning Method for Wireless Power Transfer. *IEEE Transactions on Vehicular Technology*, 64(6):2261–2273, 2015.
[20] J. Albesa and M. Gasulla. Occupancy and Belt Detection in Removable Vehicle Seats Via Inductive Power Transmission. *IEEE Transactions on Vehicular Technology*, 64(8):3392–3401, 2015.
[21] H. M. Lee and M. Ghovanloo. A High Frequency Active Voltage Doubler in Standard CMOS Using Offset-Controlled Comparators for Inductive Power Transmission. *IEEE Transactions on Biomedical Circuits and Systems*, 7(3):213–224, 2013.
[22] M. J. Neath, A. K. Swain, U. K. Madawala, D. J. Thrimawithana, and D. M. Vilathgamuwa. Controller Synthesis of a Bidirectional Inductive Power Interface for electric vehicles. In *2012 IEEE Third International Conference on Sustainable Energy Technologies (ICSET), Kathmandu*, pages 60–65, 2012.
[23] J. Albesa, L. Reindl, and M. Gasulla. Inductive power transmission for autonomous sensors: Voltage regulation effects on efficiency. In *10th International Multi-Conferences on Systems, Signals and Devices 2013 (SSD13), Hammamet*, pages 1–6, 2013.
[24] M. Fu, C. Ma, and X. Zhu. A Cascaded Boost–Buck Converter for High-Efficiency Wireless Power Transfer Systems. *IEEE Transactions on Industrial Informatics*, 10(3):1972–1980, 2014.
[25] B. Lee, M. Kiani, and M. Ghovanloo. A Triple-Loop Inductive Power Transmission System for Biomedical Applications. *IEEE Transactions on Biomedical Circuits and Systems*, 10(1):138–148, 2016.
[26] W. Zhang and C. C. Mi. Compensation Topologies of High-Power Wireless Power Transfer Systems. *IEEE Transactions on Vehicular Technology*, 65(6):4768–4778, 2016.

[27] Y.-S. Seo, M. Q. Nguyen, Z. Hughes, S. Rao, and J. C. Chiao. Wireless power transfer by inductive coupling for implantable batteryless stimulators. In *2012 IEEE/MTT-S International Microwave Symposium Digest, Montreal, QC, Canada*, pages 1–3, 2012.

[28] R. Mai, Y. Chen, Y. Li, Y. Zhang, G. Cao, and Z. He. Inductive Power Transfer for Massive Electric Bicycles Charging Based on Hybrid Topology Switching With a Single Inverter. *IEEE Transactions on Power Electronics*, 32(8):5897–5906, 2017.

[29] M. B. Camara, H. Gualous, F. Gustin, A. Berthon, and B. Dakyo. DC/DC Converter Design for Supercapacitor and Battery Power Management in Hybrid Vehicle Applications—Polynomial Control Strategy. *IEEE Transactions on Industrial Electronics*, 57(2):587–597, 2010.

[30] Y. Geng, B. Li, Z. Yang, F. Lin, and H. Sun. A high efficiency charging strategy for a supercapacitor using a wireless power transfer system based on inductor/capacitor/capacitor (LCC) compensation topology. *Energies*, 10:135, 2017.

[31] Y. Geng, Z. Yang, F. Lin, and Y. Wang. Maximum power and efficiency transmission using parallel energy storage load for wireless power transfer systems. In *2017 IEEE PELS Workshop on Emerging Technologies: Wireless Power Transfer (WoW), Chongqing*, pages 40–47, 2017.

[32] E. Gati, G. Kampitsis, and S. Manias. Variable Frequency Controller for Inductive Power Transfer in Dynamic Conditions. *IEEE Transactions on Power Electronics*, 32(2):1684–1696, 2017.

[33] C.-S. Wang, G. A. Covic, and O. H. Stielau. General stability criterions for zero phase angle controlled loosely coupled inductive power transfer systems. In *Industrial Electronics Society, 2001. IECON '01. The 27th Annual Conference of the IEEE, Denver, CO*, volume 2, pages 1049–1054, 2001.

Part IV: **Energy saving and management strategies**

Rym Chéour, Mohamed Wassim Jmal, Olfa Kanoun, and Mohamed Abid

Towards energy-efficient power management for wireless sensors networks

Abstract: The wireless sensor network (WSN) sets a very real benchmark for the Internet of Things (IoT) by utilizing the latest technology to optimize communication and safety systems in hazardous environments. Indeed, WSN is the key technology for IoT. Thanks to the advantages offered by these technologies, in particular, collecting data in real time at low cost, WSNs are gaining popularity in a multitude of fields of application. However, the energy challenge remains persistent, as WSNs suffer from energy scarcity. In this paper, we present an energy-saving design for WSN monitoring systems known as HEEPS (hybrid energy-efficient power manager scheduling). DVFS (dynamic voltage and frequency scaling) and DPM (dynamic power management) are jointly leveraged to reduce the total energy consumption, as well as a scheduling policy that exploits effectively the resources at the global level. The experiments were conducted with real network traces and hardware models.

Keywords: Wireless sensor networks, modeling, simulation, energy saving, power consumption, DVFS, DPM, EDF, scheduling, QoS, Internet of Things

1 Introduction

Nowadays, there are more and more sensors have reached a priority status in our everyday lives and, in particular, connected sensors. These connected devices form the IoT and should reach 50 billion by 2020 [1]. They have invaded several fields of application, including home automation, smart cities, smart grids and many others. As a result, they contribute to analyzing the information collected in real time of both the surrounding context and environment. Besides, they make rapid responses to abnormal status and guarantee system security [2]. However, relying on WSNs that are configured to access the Internet raises new challenges, which need to be addressed before taking advantage of the many benefits of such integration, such as improving energy efficiency, contributing to environmental monitoring and enhancing social services [3].

Rym Chéour, Mohamed Wassim Jmal, Mohamed Abid, National School of Engineers of Sfax, Computer and Embedded System Lab, University of Sfax, Tunisia
Olfa Kanoun, Chair for Measurement and Sensor Technology, Technische Universität Chemnitz, 09126 Chemnitz, Germany

https://doi.org/10.1515/9783110445053-015

The first challenge of a WSN is to save as much as possible the energy consumed by sensor nodes and to ensure optimum performance for users [4]. A sensor node would drain its battery faster in the absence of efficient energy optimization techniques [5]. Indeed, the energy consumption has a major impact on all levels: application software, motes (software and hardware) routing and communication channels. This fact has motivated researchers to design protocols and mechanisms to reduce its use in all layers of the protocol stack [6]. In the physical layer, the energy of the nodes can be minimized by reducing the data size, an efficient throughput and through an optimal energy model. In the media access control (MAC) layer, an energy-efficient duty cycle design and packet scheduling undoubtedly reduces power consumption. On the other hand, relying on routing protocols would also reduce the energy consumption in the network layer [7]. However, sensor nodes are increasingly equipped with advanced capabilities, as well as additional services, such as the use of a global positioning system (GPS) [8], multimedia or ultrasound sensors, etc. These features allow them to invade complex and energy-intensive applications, such as cryptography [9]. In this case, and due to heavy processing power and energy supply constraints, it becomes difficult to apply the algorithms of data security in some applications because the process of encryption and decryption consume a lot of time and energy [10]. These enhanced features of the WSN have their setbacks. Indeed, the applications of WSNs harden their energy requirements because the ever-increasing performance requires a significant amount of energy that becomes rare due to the reduced capacity of the battery, which does not keep up with the growth pace of the applications. As a result, the large computing capabilities can quickly exhaust the capacity of the battery, which could make the node inoperable and disrupt the communication within the network.

This paper proposes an heterogeneous resource management mechanism to manage task scheduling in a processor. Besides, it explores intertask DVFS and time-out two DPM energy saving techniques for CPU by considering the processor time for each task without violating performance requirements. In our case, we will use the "Global Earliest Deadline First" (GEDF) strategy. In the evaluation phase, we will rely on the STORM simulator.

The use of simulation rather than hardware prototyping is marked by many advantages like significant cost savings, compressed validation time, and improved component analysis [11]. Moreover, the choice of a typical simulator is not trivial and is influenced by which aspect and application of the WSN is selected to simulate among the physical layer issues, the mobility, the IP layer, etc.

The remainder of the paper is organized as follows. In Section 2, we will give a brief summary of the related work. It includes an overview and a comparison of the existing techniques to save WSN energy and to improve the efficacy of networks. In the next section, will describe the HEEPS power manager. Section 4 will give some energy-saving performance results. Finally, a conclusion and future work are given.

2 Energy saving techniques in wireless sensor networks

WSNs have become an equipment key in many industrial applications [12]. Indeed, they have proved their efficiency in monitoring, tracking, or controlling phenomena where wiring is difficult or cost prohibitive. The future applications of WSN will gradually invade our day to day lives, going from smart home to smart textiles [13]. The use of advanced strategies decreases the energy consumption of the WSN, but it significantly increases the complexity related to the methodology impact, such as the design, the architecture, the verification and the implementation [5].

The slack time (the amount of time that a task can be delayed without increasing the total) of the timing penalties and delays can be managed by the scheduler. Table 1 illustrates the tradeoff associated with various power management techniques such as clock and power gating, and the performance metrics such as timing penalties, power benefit and the impact of those technologies. For instance, even when a low duty cycle increases the network's lifetime, it has a negative impact on the reactivity of the nodes, which can increase the communication delay. So, energy conservation becomes more complex when the application has other constraints, which can be as critical as the lifetime of the nodes.

Tab. 1: Trade-offs among power minimization techniques.

Power minimization techniques	Power benefit	Timing penalties	Methodology impact			
			Architecture	Design	Verification	Implementation
Clock gating	Medium	Little	Low	Low	None	Low
Activity based Clock Gating	Medium	Little	Low	Low	Low	Low
Power gating	High	Some	High	High	High	High
Multi-Vdd	High	Some	High	Medium	Low	Medium
Multi-threshold voltages optimization	Medium	Little	Low	Low	None	Low
DVFS	High	Some	High	High	High	High
DPM	High	Some	High	High	High	High

As a result, the comparison established in Table 1 showing different techniques allowed us to choose DPM and DVFS for their high impact on energy savings. Our contribution is the combination of several strategies usually used individually, which apply time-out2 DPM and the intertask DVFS that are appropriate to WSN and on a global EDF (earliest deadline first) scheduler.

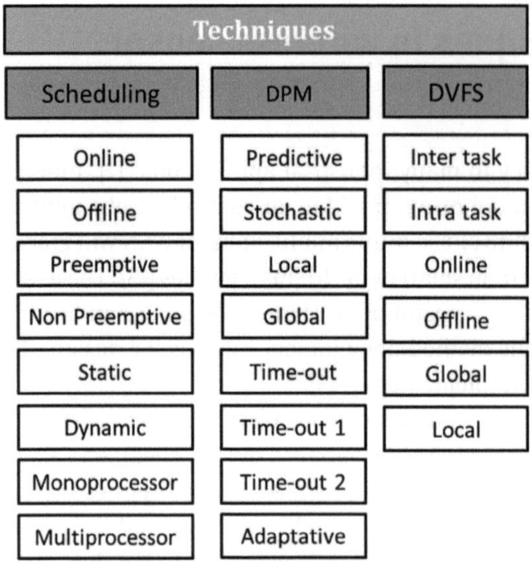

Fig. 1: Power management taxonomy.

The energy-efficiency and the scheduling correctness of real-time multiprocessor systems are, therefore, tightly connected issues, which should be tackled in conjunction for the best results [14]. So, to achieve better energy savings, we based our work on scheduling. In Figure 1, we have established a taxonomy of the different types of energy management techniques, like DPM, DVFS, scheduling and energy harvesting, and the purpose of their use.

To provide performance enhancement as well as energy efficiency in WSNs we made a comparison of the most relevant contributions in the area of energy optimization. Recent studies have moved towards the employment of energy harvesting techniques as a solution to improve energy efficiency [15]. The energy-scavenge, in particular that coming from intermittent and variable sources (e.g., implantable batteries and capacitive methods), poses additional challenges like the availability. Moreover, most methods are still under investigation and provide low energy savings. They may also cause hazards and skin infections when deployed in healthcare monitoring [16].

In [17], the authors present an implementation of DPM and DVFS techniques. However, these techniques alone are not sufficient to ensure sustainable operation [11]. The contributions of Le et al. [18] deal with the use of energy harvesting coupled with an energy manager. They propose a global power management (PM) that incorporates supercapacitors and stores energy via a light source. With a periodic energy source, this PM reduces the variation due to waking up nodes. However, it provides low energy savings.

In [19], the authors focus on application changes over time, on the one hand, and environmental conditions, on the other hand, to provide a new optimal policy based on MDP (Markov decision process). This technique is based on adjustable parameters other than voltage and frequency, namely frequency and transmission power. [20] proposes to reduce the power consumption of the processing unit by "undervolting" by supplying the electrical circuits by a voltage below the specified voltage levels. Nevertheless, it is limited by the environment in which it is deployed and by a threshold value of the voltage beyond which a malfunction appears. [21] describes a stochastic modeling or an MDP scheme for DPM, which dynamically manages wireless sensor node operations in order to minimize power consumption. It was implemented at the operating system (OS) level. This paper shows that the energy efficiency is independent of any specific power management hardware.

The solution proposed in the work [22] takes advantage of an ideal battery model to save energy. However, this model overestimates lifetime due to its constant voltage and its inability to model the non-linear properties of real batteries very different from those found in sensor nodes.

The next section gives an overview of proposed solution HEEPS and describes how the design goals can be achieved.

3 HEEPS: Hybrid energy-efficient power manager scheduling

Energy efficiency is a crucial issue to be pursued at both the node and the network level. Over time, a single energy reduction strategy cannot always change the profile of power consumption of a processor (often because of the costs of transition) as demonstrated previously. To achieve these goals, we propose an online power manager, HEEPS, which will be described in detail in this section. However, our PM switches from one policy to another, and eventually through an interplay between the DPM, DVFS and GEDF results in better energy savings. It must be able to decide which algorithm to use. The power model implemented relies on power states as a methodology for power consumption modeling. The scheduling algorithm carries out the tasks, so that the duration of idle periods is optimized to activate low-consumption states through time-out2 DPM. As a result, during periods of activity, the voltage and frequency couple is modified while taking into account the deadlines of the tasks. The energy accounting starts from the allocation of the first task on the first node to the end of the last task. It is deployed as a mid layer at the application level, as shown in Figure 2. This hybrid and multi-objective infrastructure is portable to other platforms.

When combining DPM and DVFS, the compromise between the two techniques should be considered. Indeed, it should be noted that in the context of WSNs, studies that focused on this concept are rare. When DVFS is used, the frequency is decreased in order to reduce the power consumption while performing the tasks, while the execution time increases, and the idle time decreases [23]. Decreasing the frequency may

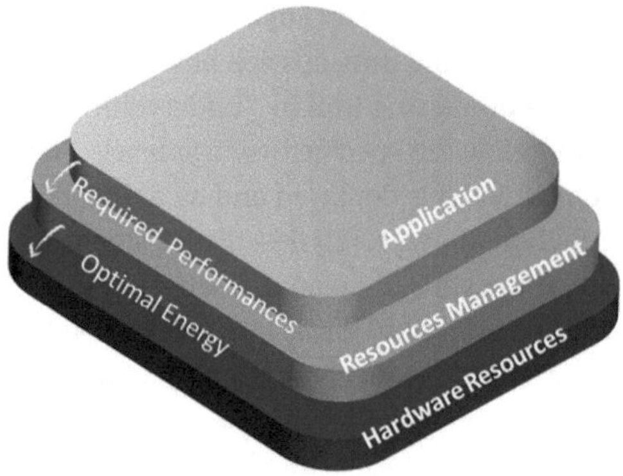

Fig. 2: General architecture of HEEPS.

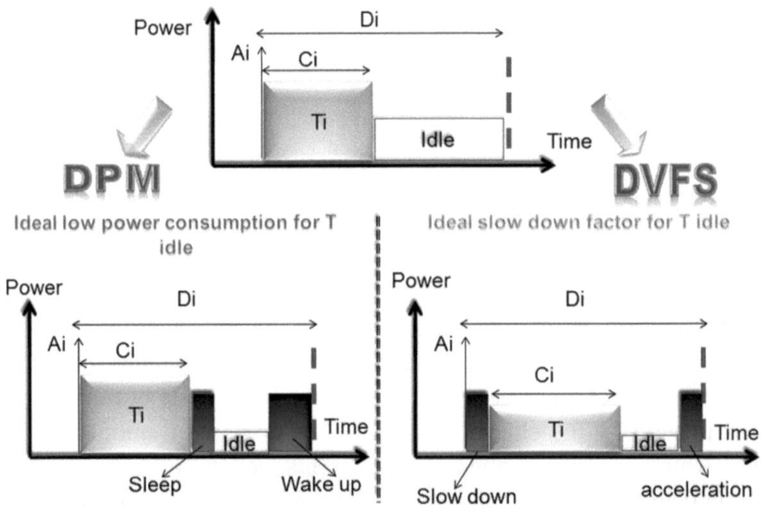

Fig. 3: Optimal combination of DPM and DVFS.

reduce the duration of a period of inactivity to such an extent that its length is less than the time beyond which we switch to the low-power mode of the DPM technique, as illustrated in Figure 3. Therefore, there is an interaction between DVFS and DPM, which should be examined carefully to minimize energy consumption using both these techniques together.

The architecture of the power manager model is made up of four basic components. Figure 4 illustrates the different phases to elaborate the HEEPS power model. The first phase is to assign tasks to the nodes through the extensible markup language (XML) file. The inputs of the model are the data related to the parameters of time, number of tasks, type of processor, etc. The tasks are scheduled with the GEDF policy in the next phase. We consider n independent, periodic real-time tasks $T_1 \ldots T_n$ allocated to

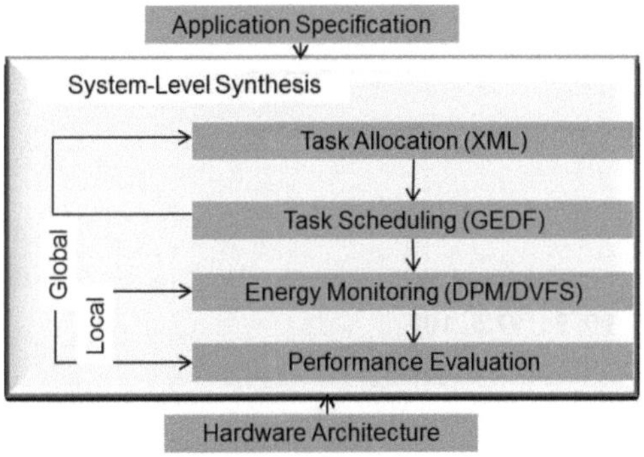

Fig. 4: Model of the HEEPS Technique.

m identical processors $P_1 \ldots P_m$ with possible migration of tasks between the processors in a centralized way and with an overall view of the state of the system.

Applied at the global level, the GEDF scheduler uses a model of Liu and Layland, which is preemptive with deadlines on request (the period is equal to the deadline). For more optimality, we invest in a new condition to bypass the Dhall effect [24]. The Dhall effect shows the limitation of global EDF scheduling, stipulating that some low utilization task sets can be unscheduable regardless of how many processors are used. So, in order to efficiently reduce the energy without violating temporal constraints, our scheduler is distinguished by the combination of the Goossens, Funk and Baruah (GFB) conditions and those of Srinivasan and Belkadi.

At the local level (node or processor), the energy management component tries to control the resource usage by selecting between DPM and DVFS. DPM becomes inactive when the CPU is executing a task. DVFS becomes active during the periods of inactivity of the processor. Ideally, a power manager component with more than two power down states and its switching to deep sleep state can reduce more power consumption, which is known as the time-out 2 technique [11]. Finally, the power evaluation component aims to give feedback about the performance criteria, such as the energy gain or with respect to time constraints during the design phase of the application to improve the overall energy efficiency. The local power management (energy monitoring and performance evaluation) and the global power management(task allocation and task scheduling) can achieve an efficient power consumption in a wireless sensor node.

The evaluation will be done and validated by the STORM simulator. The simulation technique aims to describe the behavior of the node [25]. It also impacts the development of the WSN, as it simulates several tens of thousands of nodes where only the simulation is effective. In addition, it provides a basis for constructing sensor network models in other formalisms. Selecting a simulator for experimentation is not an easy task. The investigation of more than 25 simulation environments that we conducted in [26], led us to the STORM simulator. Its multiple benefits meet our need to model temporal execution of tasks where each CPU emulates the functioning of a real

Tab. 2: Simulation parameters.

Parameters	Values
Duration of simulation	1000 ms
Precision	10^{-9} s
Number of tasks (n)	2, 5, 10, 20
Number of processors (m)	2, 3, 5, 10
$\frac{U_{total}}{m}$	70, 75, 80, 85, 90, 95, 97.5, 100
Model of execution time	WCET and AET
Behavior in case of exceeding of deadlines	Task abortion
Scheduler	GEDF
Distribution of periods	[2, 100] ms
Time penalties	None for DVFS but taken into account for DPM

node. In our case, a processor is the equivalent of one node. The objective is that all the tasks check their deadlines during the simulation. We assume that the overhead of a preemption is zero.

The frequency and voltage are linearly related. The frequency change directly influences the execution time. So, scaling the frequency by x, the execution time becomes $1/x$.

4 Performance evaluation

The various simulations and measurements make it possible to highlight several metrics, which will help us to evaluate the power consumption generated by the processor. The simulation measurements were conducted with MICA2 to determine the improvement in the WSN lifetime. MICA2 is better suited to our experiments for enabling low power, WSN and for allowing us to manage a much larger number of variables with even more flexibility. The simulation parameters are chosen in order to satisfy the condition of scheduability of both, which, as mentioned above, is the combination of both the GFB and the Srinivasan and Belkadi conditions.

During the simulation, we changed the size of the network (number of nodes) and studied its impact on the entire system. As a result, we noted that the larger the number of processors increases the processing time, and the makespan (i.e., the date of completion of the last scheduled task) are reduced. The simulation also shows how adjusting more frequencies can lead to more energy at the global level. Each transition has the potential to miss an interesting event. As we have shown, the transition between the different power modes generates additional costs, although this is rarely considered in the related works. In the same way, each power mode corresponds to a particular consumption and to a related delay; therefore, not considering them induces an error, as we need an accurate model and not an approximation.

In this context, we rely on the AET (actual execution time) and not on the WCET (worst case execution time), which offers more realistic conditions. In terms of gain, the use of WCET provides a gain equal to 11.82%, however, when we relied on AET, we saved 19.5%.

Figure 5 illustrates the experiment results obtained with the DPM strategy. The time is represented in units of 50 seconds (x-axis). Each task is represented by a different color. We noticed that it is not desirable to keep nodes inactive for too long, because this can impact the network quality-of-service. When we applied the DPM policy, major improvement was seen, such as the elimination of both dynamic and static power dissipation. Besides, the delay's transition was set up to avoid the potential impact of missing the execution of any interesting task. Moreover, the transition between the different power configurations showed extra energy and latency costs.

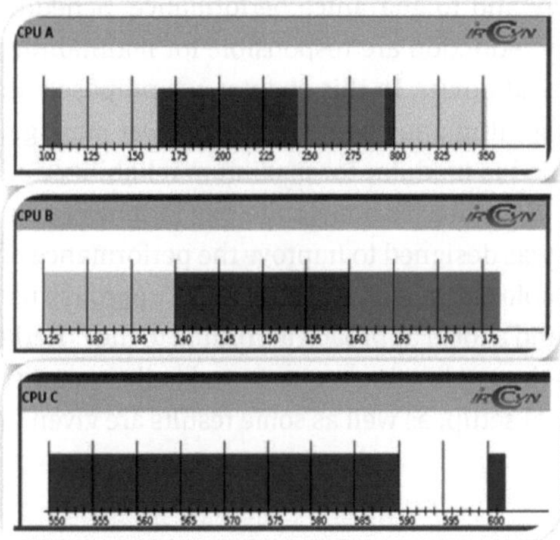

Fig. 5: EDF-DPM experiment results.

However, we noticed that the costs of transition between modes and also the time spent by the CPU in each mode had a significant impact on the total consumption of energy of a sensor node. A gain of energy was observed every time we changed the frequency and mainly when the sleep mode is not applicable. Also, we noticed that applying the scheduling when taking into account the current energy level and both the power consumption and the priorities of the tasks (done through the global EDF scheduler) enabled graceful degradation. The results are scaled between 0.0 and 1.0, respecting the values that are not optimized. Thus, power consumption is very likely larger at 16 MHz than at 8 MHz, as shown in Figure 6.

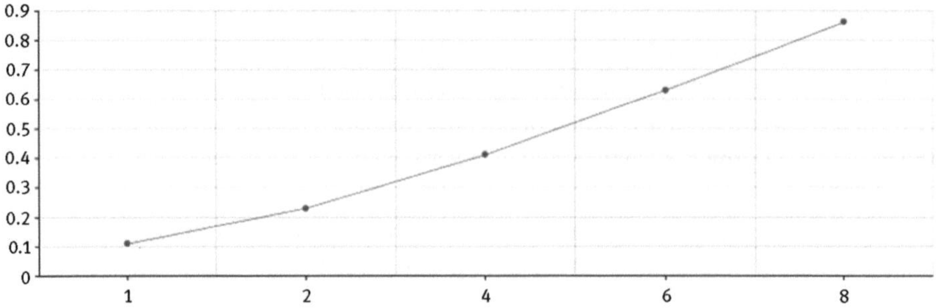

Fig. 6: DVFS energy gain.

5 Conclusion

The overall capture of the behavior of the sensor node by the model makes it possible to develop an appropriate energy strategy and to guarantee performance. Scheduling policies together with those of energy reduction are responsible for maintaining a compromise between time constraints and power. In this chapter, we proposed an implementation of the EDF scheduling algorithm, augmented with a power management policy for WSN. The HEEPS developed is modular for maximum reliability and scalability. The power model relies on power states as a methodology for power consumption modeling. The power manager was designed to improve the performance of the WSN in terms of energy thanks to a double gait: a global and dynamic approach using the analysis of the behavior of nodes and a local approach according to the criteria selected customizing their approaches to the application requests and to the quantity of energy that is available. An experimental setup, as well as some results are given to illustrate the gain obtained.

Bibliography

[1] R. Kitchin. The real-time city? Big data and smart urbanism. *GeoJournal*, 79(1):1–14, 2014.

[2] P. L. Yinbiao, Shu and F. Jianbin. Internet of Things – Wireless Sensor Networks. *White Paper. International Electrotechnical Commission (IEC)*, 2014.

[3] D. Christin, A. Reinhardt, P. S. Mogre, R. Steinmetz, et al. Wireless sensor networks and the internet of things: selected challenges. In *Proceedings of the 8th GI/ITG KuVS Fachgespräch Drahtlose Sensornetze*, pages 31–34, 2009.

[4] M. Conti and S. Giordano. Mobile ad hoc networking: milestones, challenges, and new research directions. *Communications Magazine, IEEE*, 52(1):85–96, 2014.

[5] R. Cheour, F. Derbel, O. Kanoun, and M. Abid. Wireless sensor networks with power management for low energy consumption. In *IEEE International Multi-Conference on Systems, Signals and Devices (SSD'13), (Tunisia), March 2013*, 2013.

[6] R. Guru. Energy efficiency mechanisms in wireless sensor networks: A survey. *International Journal of Computer Applications*, 139(14), 2016.
[7] W. Jerbi, A. Guermazi, and H. Trabelsi. A novel clustering algorithm for coverage a large scale in wsn. arXiv preprint arXiv:1605.03079, 2016.
[8] F. K. Shaikh and S. Zeadally. Energy harvesting in wireless sensor networks: A comprehensive review. *Renewable and Sustainable Energy Reviews*, 55:1041–1054, 2016.
[9] H. U. Yildiz, K. Bicakci, B. Tavli, H. Gultekin, and D. Incebacak. Maximizing wireless sensor network lifetime by communication/ computation energy optimization of non-repudiation security service: Node level versus network level strategies. *Ad Hoc Networks*, 37:301–323, 2016.
[10] S. Bartariya and A. Rastogi. Security in wireless sensor networks: Attacks and solutions. *environment*, 5(3), 2016.
[11] R. Chéour, M. Jmal, and M. Abid. Modeling and simulation of an autonomous power manager for sensor networks. *International Journal of Modeling and Optimization*, 6(4):256–260, 2016.
[12] F. Karray, A. Garcia-Ortiz, M. W. Jmal, A. M. Obeid, and M. Abid. Earnpipe: A testbed for smart water pipeline monitoring using wireless sensor network. *Procedia Computer Science*, 96:285–294, 2016.
[13] O. Ojuroye, R. Torah, S. Beeby, and A. Wilde. Smart textiles for smart home control and enriching future wireless sensor network data. In *Sensors for Everyday Life*, pages 159–183. Springer, 2017.
[14] D. Niyato, X. Lu, P. Wang, D. I. Kim, and Z. Han. Distributed wireless energy scheduling for wireless powered sensor networks. In *Communications (ICC), 2016 IEEE International Conference on*, pages 1–6. IEEE, 2016.
[15] J. Amaro, R. Cortesao, J. Landeck, and F. J. Ferreira. Harvested power wireless sensor network solution for disaggregated current estimation in large buildings. *IEEE Trans. Instrum. Meas.*, 64(7):1847–1857, 2015.
[16] M. A. Hannan, S. Mutashar, S. A. Samad, and A. Hussain. Energy harvesting for the implantable biomedical devices: issues and challenges. *Biomed. Eng. Online*, 13(1):79, 2014.
[17] V. T. Hoang, N. Julien, and P. Berruet. Increasing the autonomy of wireless sensor node by effective use of both dpm and dvfs methods. In *Faible Tension Faible Consommation (FTFC), 2013 IEEE*, pages 1–4. IEEE, 2013.
[18] T. N. Le. *Global power management system for self-powered autonomous wireless sensor node.* PhD thesis, Université Rennes 1, 2014.
[19] A. Munir and A. Gordon-Ross. *Optimization approaches in wireless sensor networks*, pages 313–338. Sustainable Wireless Sensor Networks, 2010.
[20] U. Kulau, F. Busching, and L. Wolf. Undervolting in wsns: A feasibility analysis. In *IEEE World Forum on Internet of Things (WF-IoT)*, pages 553–558. IEEE, 2014.
[21] A. Pughat and V. Sharma. A review on stochastic approach for dynamic power management in wireless sensor networks. *Human Centric Computing and Information Sciences*, 5(1):1–14, 2015.
[22] W. Dron, S. Duquennoy, T. Voigt, K. Hachicha, and P. Garda. An Emulation-Based Method for Lifetime Estimation of Wireless Sensor Networks. In *2014 IEEE International Conference on Distributed Computing in Sensor Systems, Marina Del Rey, CA*, pages 241–248, 2014.
[23] V. Devadas and H. Aydin. On the interplay of voltage/frequency scaling and device power management for frame-based real-time embedded applications. *Computers, IEEE Transactions on*, 61(1):31–44, 2012.
[24] S. K. Dhall and C. L. Liu. On a real-time scheduling problem. *Operations Research*, 26(1):127–140, 1978.

[25] E. Lattanzi, V. Freschi, M. Dromedari, L. S. Lorello, R. Peruzzini, and A. Bogliolo. A fast and accurate energy source emulator for wireless sensor networks. *EURASIP Journal on Embedded Systems*, 1:18, 2016, 2017.

[26] R. Chéour, M. W. Jmal, O. Kanoun, and M. Abid. *Evaluation of simulators tools and power-aware scheduling model for wireless sensor networks*. IET Computers & Digital Techniques, 2017.

Steffi Knorn and Daniel E. Quevedo

Optimal energy allocation in energy harvesting and sharing wireless sensor networks

Abstract: Wireless sensors offer great advantages when monitoring the environment or industrial applications due to their flexible and fast implementation. In order to power those sensors, the use of energy harvesting to extract power from their immediate surroundings is a viable option. Even wireless energy transfer between neighboring sensors is becoming popular due to huge advances in the field. Here, we introduce a system engineering approach to model such systems. With our framework, optimization techniques can be used in order to derive policies on when and how to use the available energy in order to achieve the best possible performance. A network of energy harvesting and energy sharing sensors is investigated by means of theoretical results and illustrative simulations.

Keywords: Wireless sensor network (WSN), energy harvesting, energy sharing, power control, energy allocation, optimization

1 Introduction

Energy harvesting devices are able to extract energy from their immediate environment; examples include solar cells, wind mills, vibration harvesters, Peltier elements, that harvest energy from temperature gradients and radio frequency energy harvesting units. Such technologies have developed significantly over recent years. Not only have the efficiencies of energy harvesters increased and the size of the devices decreased, but energy harvesting devices have also become more affordable. Hence, using energy harvesters has become feasible in a variety of applications in engineering. Examples include environmental data gathering [1], industrial process monitoring [2], mobile robots and autonomous vehicles [3] and monitoring of smart electricity grids [4]. A related technology under development is wireless energy collaboration or wireless energy sharing, where energy is transferred between devices using laser beams, resonant electro-magnetic fields, radio signals or sonar signals. While the efficiency of such wireless energy transfer is still relatively low and usually decreases significantly with the distance between the devices, various companies are already developing suitable products. It is reasonable to believe that the technology will improve significantly during the coming years and, hence, become useful in a

Steffi Knorn, Signals and Systems, Department of Engineering Sciences, Uppsala University, Sweden
Daniel E. Quevedo, Department of Electrical Engineering, Paderborn University, Germany

https://doi.org/10.1515/9783110445053-016

growing number of practical applications, [5]. With this, an important question has to be answered, namely, where and in which form should energy harvesting and energy sharing be applied in order to maximize their benefit in engineering applications?

Another relevant engineering application is the use of wireless sensors, which transmit their measurement wirelessly to a remote receiver connected, for instance, to a gateway or fusion centre. Sensor measurements can be transmitted in an analog fashion as well as packets, i.e., digitally. In the case of analog transmissions, the disturbance of the received signal can be used as a performance measure of the signal transmission. When it comes to digital transmissions, a suitable performance measure is the percentage of the correctly received packets at the receiver. In both cases, the performance measure, i.e., the disturbance or the packet arrival rate, depends on the channel gain and the transmission energy. The channel gain describes the quality of the wireless transmission channel and is often assumed to be fading, i.e., changing over time. Depending on the particular channel at hand, the channel gain may or may not be known, might change within a larger or smaller interval, and changes might be observed more or less frequently. When considering energy harvesting devices, the transmission energy cannot be chosen arbitrarily large, and transmissions cannot be repeated arbitrarily often. While energy harvesting provides a potentially everlasting power source, the amount of energy to be harvested during a limited time interval is limited and usually varies over time. This raises the question how the available energy should be used for wireless data transmission in order to achieve the best possible performance.

Our approach to solve such energy management problems is to use systems and control engineering tools. As a first step, suitable abstract models to describe the system have to be chosen. For instance, modeling wireless channel gains as independently, identically distributed (i.i.d.) or as Markov or hidden Markov chains has proven to be particularly useful for system and control engineering purposes. Similar models have also been shown to be suitable to describe harvested energies. Then, tools from the area of optimization can be applied to derive how the available energy should be used to achieve the best possible performance.

2 Modeling

Many suitable models that can be used to describe the optimization problems outlined above can be formulated in discrete time with time slots of a constant length T_{d}. The value of a variable x at the time slot k is then denoted $x(k)$.

2.1 Energy harvesting model

The amount of harvested energies are usually modeled as uncertain, stochastic processes. Consider, for instance, a simple harvesting model where the harvested energy at time k, say H_k, is a discrete value taken from a set of three states describing different harvesting situations. When considering a solar cell, the first state, in which a lot of energy can be harvested, could correspond to sunny weather (denoted "s"), and the associated harvested energy is denoted H_s. A second state, corresponding to cloudy weather, where a moderate amount of energy can be harvested, is then denoted "c", and the amount of energy harvested is denoted as H_c. The third state then corresponds to darkness (denoted "d"), where no energy can be harvested, such that $H_d = 0$.

The second important aspect of such a basic energy harvesting model is to describe with which probability each state is present. The simplest way to describe this probability is to assume that the states are identically, independently distributed (i.i.d.), where the probability of harvesting H_s, H_c or H_d is constant and given by, say, P_s, P_c or $P_d = 1 - P_s - P_c$, respectively. Note that this model is often inaccurate, since the probabilities are assumed independent of time k or the energy harvested during the preceding time slot.

An often more realistic approach is to model the energy harvesting process as a Markov chain, see, for instance, [6]. Consider again the model with three discrete harvested energy values H_s, H_c and H_d. In contrast to the i.i.d. model above, the probability of harvesting a specific amount of energy now depends on the previously harvested energy. This behavior can be illustrated by a graph, as depicted in Figure 1. Since the Markov chain has a finite number of states (in this case, three), it is called a finite-state Markov chain. Each node corresponds to one of the three states with the corresponding discrete harvesting amount. At each time step, the amount of harvested energy depends on the probabilities to change from the corresponding state/node to the next state/node. For instance, $P_{s,c}$ is the probability that after harvesting H_s at time k, the amount of harvested energy at $k+1$ is H_c. Of course, we have $P_{i,s} + P_{i,c} + P_{i,d} = 1$ for all $i \in \{s, c, d\}$.

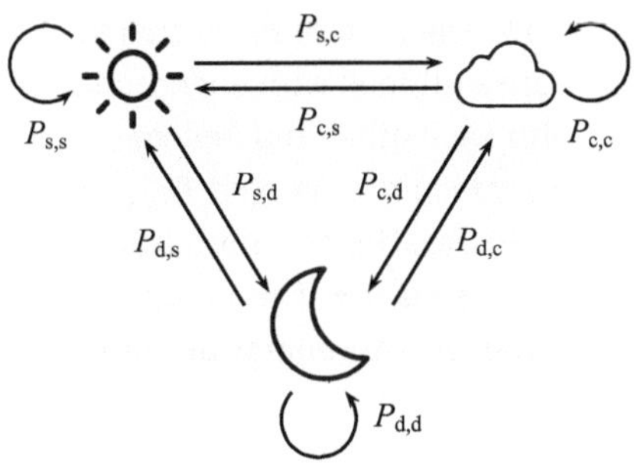

Fig. 1: Modeling the energy harvesting process as a Markov chain.

A significant drawback of the Markov chain model above is the fact that all harvested energies are drawn from a discrete set, which is not realistic in most practical applications. In a more realistic model, harvested energies are drawn from continuous distributions, such as exponential, i.i.d., or other suitable distributions. Combining this approach with the general idea of Markov chains, leads to a hidden Markov chain, where each node does not correspond to a fixed harvested energy amount but to a distinct continuous distribution. A general Markovian model was also considered in [7].

Note that modeling a purely stochastic energy harvesting source is often not accurate when more details about the source are considered. For instance, in the case of solar cells, the model of the harvesting source could be improved by considering the known time of sunrise, solar noon and sunset. Then, a stochastic model could be used to describe variations due to weather conditions.

2.2 Energy sharing model

An important application of energy harvesting are harvesting wireless sensor networks with a large number of sensors located within a common geographical area. Even when assuming that all sensors can be equipped with an energy harvester, the harvested energies will, in general, differ between sensors. Specially, if the immediate environment of the sensors differ, or the sensors harvest energies from different sources, then the differences in harvested energies can be significant over long stretches of time. In such situations, some sensors might not be able to function, e.g., to measure, process or transmit data, while others harvest more than enough energy, which might not even be able to be stored for later use due to limited storage capacities. It is reasonable to assume that such situations could be alleviated by allowing the sensors to share their energies. One possibility to realize this idea is to use wireless energy transfer technologies as described in [5], for instance, by energy transfer between two resonant objects, such as discussed in [8, 9], the use of laser beams, or by the use of beamforming radio waves.

Consider, for instance, a small network of three sensors, as shown in Figure 2. Each sensor $m \in \{1,2,3\}$ draws energy E_m from its local rechargeable battery with battery state B_m, which is described in more detail below. The battery B_m is charged by a source, which harvests H_m. Additionally, sensors are equipped with energy sharing units, that allow Sensors 1 and 2 to share energy with each other and Sensor 3 to receive energy from Sensor 1. The energy sent from sensor m to n is denoted $T_{m,n}$, where $m \neq n$. Note that wireless energy transfer is subject to losses. Hence, only a fraction of the transmitted energy $T_{m,n}$ from sensor m is received at sensor n. This is captured by the energy sharing efficiency $\eta_{m,n} \in [0,1]$, such that the received energy at sensor n is $\eta_{m,n}T_{m,n}$. Note that energy sharing efficiencies depend on the particular wireless energy transfer implementation, as well as the distance between the sensors and other external factors. Thus, the efficiencies depend on the link in the network and might vary over time.

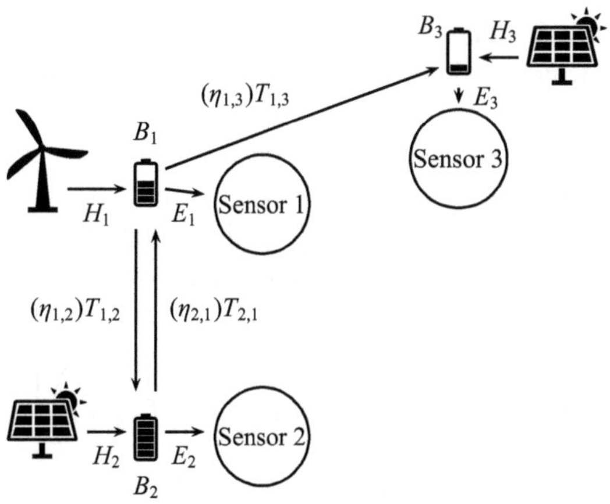

Fig. 2: Network of three energy harvesting and energy sharing sensors.

2.3 Battery model

In order to store harvested energy temporarily for later use, harvesting devices are usually equipped with a rechargeable battery or a capacitor.[1] An important limitation in energy harvesting applications is the so-called energy causality, which refers to the fact that energy can only be used after it has been harvested. Hence, it is important to describe the amount of energy, which is currently available, in order to account for energy causality constraints. To do so, we will derive a suitable, simple model of the battery state of sensor m at time k denoted by $B_m(k)$.

First, assume that battery m is charged by an energy harvester, which harvests $H_m(k)$ at time k. Also, sensor m might use energy $E_m(k)$ at time k for measurements, data processing or data transmission. This leads to the evolution of the battery state according to the following equation:

$$B_m(k+1) = B_m(k) + H_m(k) - E_m(k) , \tag{1}$$

with initial condition $B_m(0)$. To derive more realistic battery models, one must also consider practical limitations of the battery. This includes the maximal battery capacity $\bar{B}_m$, as well as losses due to battery leakage [10, 11]. Including these two limitations leads to the battery model

$$B_m(k+1) = \min\left\{\bar{B}_m; \mu\left(B_m(k) + H_m(k) - E_m(k)\right)\right\} , \tag{2}$$

where $\mu \in (0,1]$ can be interpreted as the battery efficiency describing which fraction of the stored energy at time k is available at time $k+1$. More advanced models might

1 For simplicity, we use the term "battery" to cover both technologies.

also include other limitations of physical batteries such as terms penalizing charging the battery or drawing energy from it, which are known to contribute to ageing of batteries.

When considering energy sharing networks as discussed above, the battery state $B_m(k)$ also depends on the shared energies. Denote the set of neighboring sensors, to which sensor m might transmit energy, by $\mathcal{T}_m$ and the set of neighbouring sensors, from which sensor m can receive energy, by $\mathcal{R}_m$. Then, the battery state of sensor m can be described by the following dynamic model:

$$B_m(k+1) = \min\left\{\bar{B}_m; \mu\left(B_m(k) + H_m(k) - E_m(k) - \sum_{n\in\mathcal{T}_m} T_{m,n}(k) + \sum_{n\in\mathcal{R}_m} \eta_{n,m} T_{n,m}(k)\right)\right\}, \tag{3}$$

which is a suitable abstraction for our current purpose.

2.4 Channel gain model

Wireless sensors often measure and process data of interest, to be transmitted wirelessly to a receiver connected to other sensors or a remote gateway or fusion centre (FC). The reception quality depends on the transmission energy and the current channel gain, see [12]. Hence, in order to choose suitable transmission energies, knowing, measuring or modeling the current channel gain offers a significant advantage.

The channel gain depends on a variety of factors, such as whether there is a direct line of sight between the transmitter or receiver and other stochastic influences. Hence, in general, the channel gain will vary over time. Depending on how fast the gain changes, one denotes the channel as slow fading or fast fading. In some applications, it might also be suitable to simplify the model by assuming that the channel remains constant and is, hence, not fading. This is particularly useful for short time periods or very stable environments. In the case of fading channels, it is usually assumed for simplicity that the channel gain remains constant for the duration of the time slot k and then changes abruptly at the transition between slot k and $k+1$. The value of the modeled channel gain at each time k might be drawn from a variety of different distributions, such as i.i.d. within a continuous interval, i.i.d. from a discrete set, Gaussian distribution, exponential distribution, etc. Further, one usually distinguishes between channel models that are uncorrelated or correlated over time. Correlated models can then be described as Markov chains or hidden Markov models similar to harvesting models as described above.

Which particular model is chosen to describe the channel also depends on the specific application at hand. Usually, a balance has to be found between accuracy of the model and simplicity of the abstraction in order to facilitate an adequate analysis. This modeling decision is often based on measurements of the channel gain of interest over some time or on suitable models for similar applications.

3 Application: Multi-sensor estimation network with energy harvesting and energy sharing

We consider a network with M sensors, where each sensor m measures a signal of interest $\theta_m(k)$, at discrete time instants $k \in \{1,2,3,\dots\}$. The sensors transmit their information to a remote FC, which estimates the vector $\theta(k) = (\theta_1(k), \theta_2(k), \dots, \theta_M(k))^{\mathrm{T}}$ based on the measurements received.

We consider an analog amplify and forward uncoded transmission strategy subject to additive noise [13]. Each sensor is equipped with a local battery, an energy harvester and a unit to transmit and receive energy from other sensors, along with a transceiver for information transmission and reception. A scheme showing a simple system with three sensors is shown in Figure 3.

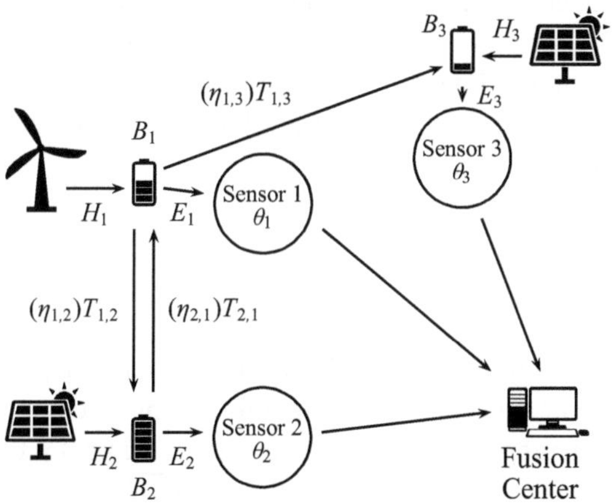

Fig. 3: Network of three energy harvesting and energy sharing sensors.

3.1 Source model and sensor measurements

We consider each $\theta_m(k)$ to be an i.i.d. (regarding time), band-limited Gaussian process with zero mean. The measurements of the sensors are spatially correlated, such that the joint covariance matrix is $R_\theta = \mathbb{E}\left\{\theta(k)\theta^{\mathrm{T}}(k)\right\}$. We assume that R_θ is positive definite. The measurements of sensor m, denoted $x_m(k)$, are subject to measurement noise, $n_m(k)$, such that

$$x_m(k) = \theta_m(k) + n_m(k) \tag{4}$$

for $1 \le m \le M$ and $k \ge 1$. The measurement noises $n_m(k)$ are assumed to be i.i.d. Gaussian, mutually independent and also independent of $\theta(k)$. Further, it is assumed that they all have zero mean and variances σ_m^2.

3.2 Energy harvester, energy sharing and battery dynamics

Each sensor is equipped with an energy harvester to gather energy from its immediate environment. The harvested energy at sensor m at time k, denoted by $H_m(k)$, is described as a finite-state Markov chain, motivated by empirical measurements reported in [7]. It is assumed that the harvested energies are mutually independent and independent of the process $\theta(k)$ and the measurement noise. The energy harvested at time slot k is stored in the battery and can be used for data transmission to the FC or for energy sharing with neighboring sensors in time slot $k + 1$. The energy used to transmit data from sensor m to the FC at time k is denoted $E_m(k)$.

Each sensor can transmit energy to a set of neighboring sensors and can also receive energy from a (possibly different) set of neighboring sensors via directed wireless energy transfer. The set of neighboring sensors from which sensor m can receive energy is denoted by $\mathcal{R}_m$, and the set of neighbouring sensors to which sensor m can transmit energy is denoted by $\mathcal{T}_m$. The energy transferred from sensor m to sensor n at time k is denoted by $T_{m,n}(k)$ and the corresponding efficiency is given by $\eta_{m,n} < 1$.

Further, we assume that during each time interval, some stored energy in the battery is lost due to leakage. Hence, the dynamics of the battery level of sensor m at time $k + 1$ is similar to the model in [14] and is given as in (3).

3.3 Transmission model

Each sensor has a transmitter using an analog amplify and forward uncoded strategy.[2] This scheme is very simple to implement, since it does not require complex coding/decoding and incurs no other delay than propagation delay. Hence, at each time slot k, sensor m transmits its measurement $x_m(k)$ amplified by the factor of $\sqrt{\alpha_m(k)}$. The energy needed for transmission is then given by

$$E_m(k) = \alpha_m(k)\left((R_\theta)_{m,m} + \sigma_m^2\right) \tag{5}$$

where $(R_\theta)_{m,n}$ is element m, n of matrix R_θ. The channel gain of the channel between sensor m and the FC, $g_m(k)$, is modeled as a finite-state Markov chain, [15], as illustrated in Figure 1. We assume that the channel gains are mutually independent and independent of the harvested energies. The signal received at the FC from sensor m at time k is

$$z_m(k) = \sqrt{\alpha_m(k) g_m(k)}\, x_m(k) + \zeta_m(k) \tag{6}$$

where $\zeta_m(k)$ is assumed to be i.i.d. additive white Gaussian noise with variance ξ_m^2.

2 Optimality of analog transmission for multi-sensor estimation of a memoryless Gaussian source over a coherent multi-access channel was shown in [13].

3.4 Distortion measure at the FC

At the FC, the minimum mean-square error (MMSE) estimator (see [16]) provides the vector of estimates $\hat{\theta}(k) = \left(\hat{\theta}_1(k), \hat{\theta}_2(k), \ldots, \hat{\theta}_M(k)\right)^{\mathrm{T}}$ given the vector of received signals gathered in

$$z(k) = (z_1(k), \ldots, z_M(k))^{\mathrm{T}} = \mathbb{H}\theta(k) + v(k) \tag{7}$$

with

$$\mathbb{H} = \operatorname{diag}\left(\sqrt{\alpha_1(k)g_1(k)}, \sqrt{\alpha_2(k)g_2(k)}, \ldots, \sqrt{\alpha_M(k)g_M(k)}\right) \tag{8}$$

and

$$\begin{aligned} v(k) = \Big(& \sqrt{\alpha_1(k)g_1(k)}n_1(k) + \zeta_1(k), \sqrt{\alpha_2(k)g_2(k)}n_2(k) + \zeta_2(k), \\ & \ldots, \sqrt{\alpha_M(k)g_M(k)}n_M(k) + \zeta_M(k)\Big)^{\mathrm{T}} . \end{aligned} \tag{9}$$

Then, the distortion at the FC can be quantified via

$$\begin{aligned} D(E(k), g(k)) &:= \operatorname{trace}\left(\mathbb{E}\left\{\left(\theta(k) - \hat{\theta}(k)\right)\left(\theta(k) - \hat{\theta}(k)\right)^{\mathrm{T}}\right\}\right) \\ &= \operatorname{trace}\left(\left(\mathbb{H}^{\mathrm{T}}R_{\mathrm{v}}^{-1}\mathbb{H} + R_{\theta}^{-1}\right)^{-1}\right) \end{aligned} \tag{10}$$

where $\mathbb{E}\{x\}$ is the expected value of x. Note that the distortion depends on the vector of all transmission energies $E(k) = (E_1(k), E_2(k), \ldots, E_M(k))$, the complete vector of all channel gains $g(k) = (g_1(k), g_2(k), \ldots, g_M(k))$ and $R_{\mathrm{v}} = \operatorname{diag}(\alpha_1 g_1 \sigma_1^2 + \xi_1^2, \ldots, \alpha_M g_M \sigma_M^2 + \xi_M^2)^{\mathrm{T}}$. Note that the distortion $D(E(k), g(k))$ is a random process, since $\theta(k)$ is a random variable. Hence, designing optimal predictive power control strategies is a difficult and challenging task.

3.5 Dynamic energy control

Given the system model described above, we aim to minimize the distortion at the FC, which is described by (10), by choosing the transmission energies and the shared energies at each time k. The optimal solution will depend on the states of the system, such as the channel gains and harvested energies at time k, as well as the available energy at the batteries at the beginning of time slot k. It is important to note here that for decision making, certain causality constraints have to be considered. First, we assume that only causal knowledge is available. Hence, only past and current channel gains and harvested energies are assumed to be known. The channel gains and harvested energies for future time instances are not exactly known, but can be predicted based on the Markov chain model adopted. Also, we will consider energy causality constraints, which describe the fact that only energy that has been harvested previously can be used at a particular time k.

4 Optimization techniques

As discussed in the previous section, we aim to minimize the distortion at the FC by choosing the transmission energies and shared energies for all sensors m and all times k, while relying on causal state information and obeying energy causality constraints. Different techniques have been developed to solve this problem, see for instance, [17]. Which solution is most suitable depends on the particular application at hand.

4.1 Finite time horizon optimization

Consider an application where the wireless sensor network described above is planned to operate for a known, finite time horizon, that is, for $K < \infty$ time slots. In this case, we aim to minimize the sum of the distortion at the FC from time $k = 1$ to time $k = K$. Thus, the optimization problem (see (10)) can be formalized as

$$\min_{E(k),T(k):\,1\le k\le K} \sum_{k=1}^{K} \mathbb{E}\left\{D(E(k), g(k))\right\} , \tag{11}$$

where $T(k)$ is a matrix with entries $(T(k))_{m,n} = T_{m,n}(k)$ for $n \in \mathcal{T}_m$ and $(T(k))_{m,n} = 0$ otherwise, such that all transmission energies and shared energies are nonnegative and the energy causality constraints are met, i.e.,

$$E_m(k), T_{m,n}(k) \ge 0 \quad \text{and} \quad E_m(k) + \sum_{n\in\mathcal{T}_m} T_{m,n}(k) \le B_m(k) \tag{12}$$

almost surely for $1 \le m, n \le M$ and all $1 \le k \le K$, and where $B_m(k)$ satisfies (3).

We assume that complete causal information is available at the FC. Under this information pattern, all sensors report their current battery levels, harvested energies and their estimated channel gains (achieved via pilot transmissions from the FC and channel reciprocity) to the FC via a control channel during the beginning of the time slot k. Hence, the information available at the FC at time k is the vector of battery levels $B(k) = (B_1(k), B_2(k), \ldots, B_M(k))$, the vector of harvested energies $H(k) = (H_1(k), H_2(k), \ldots, H_M(k))$ and the vector of channel gains $g(k)$.

Then, the value of the finite-time horizon minimization problem (11) with causal information is given by $V_1(g(1), H(1), B(1))$, which can be computed recursively from the backward Bellman dynamic programming equation, see [17],

$$\begin{aligned} V_k(g, H, B) = \min_{E(k),T(k)} \Big\{ & D(E(k), g(k)) \\ & + \mathbb{E}\left\{V_{k+1}(g(k+1), H(k+1), B(k+1)) | g(k), H(k), B(k), E(k), T(k)\right\} \Big\} \end{aligned} \tag{13}$$

for $1 \le k \le K-1$ such that $E_m(k) \ge 0$, $T_{m,n}(k) \ge 0$ and $E_m(k) + \sum_{n\in\mathcal{N}_{\mathrm{T},m}} T_{m,n}(k) \le B_m(k)$ with the battery dynamic equation (3) for all m. In (13), the expectation is computed over the random variables g and H. The terminal condition is given by

$$V_K(g, H, B) = D(B(K), g(K)) \tag{14}$$

where all remaining energy is used up for transmission in the final time slot K.

In general, the solution to the dynamic programming recursion (13) and, hence, the optimal energy allocation policy can only be obtained numerically as there is no closed form solution. This is done by recursively (in time, starting at $k = K$) evaluating each feasible energy allocation policy for each possible combination of states and comparing the expected outcome (i.e., expected distortion at the FC). Then, the optimal choices of transmission energies and shared energies are stored in a look-up table for each possible state combination and each time step k to be used when implementing the policy. Since this numerical solution relies on computing the optimal policy for a large number of discretized channel gain and battery level values, we assume that this computation is done off-line at the FC, which has access to potentially unlimited energy and higher computational power. During real-time operation, as the FC receives the channel gains and battery level information of all sensors at the beginning of each transmission phase, the FC looks up the optimal energy allocation policies for the corresponding nearest discretized values of the channel gains and battery levels, and informs all the sensors via a dedicated control channel, which is assumed to be delay free and error free. The sensors subsequently use these optimal decisions for data transmission and energy sharing.

The mayor limitation of the finite time horizon formulation and its solution is that it is only feasible for relatively short horizons, since the numerical solution to the optimization problem otherwise becomes prohibitively large and hence infeasible. Further, wireless sensor networks are often implemented to be used over long times, such that a finite time horizon formulation seems unsuitable.

4.2 Infinite time horizon optimization

While the finite time horizon offers some advantages, it is unsuitable when planning to implement optimal energy allocation policies over long or unknown time horizons. In this case, the optimization problem (11) can be modified into an infinite time horizon problem, which aims to minimize the long-term average distortion at the FC. Hence, we aim to find the stationary energy allocation policy to determine $E(k)$ and $T(k)$ for all k such that the following cost function is minimized

$$\lim_{K\to\infty} \frac{1}{K} \sum_{k=1}^{K} \mathbb{E}\{D(E(k), g(k))\} \ , \tag{15}$$

such that all transmission energies and shared energies are nonnegative and the energy causality constraints (12) are met almost surely for $1 \le m, n \le M$ and $k \ge 1$, and $B_m(k)$ satisfies (3).

The stochastic control problem (15) with centralized information at the FC can be regarded as a Markov decision process (MDP) formulation $\{\mathcal{S}, \mathcal{A}, \mathcal{P}\}$ with state space $\mathcal{S} = \{B, g, H\}$ and action space $\mathcal{A} = \{E, T\}$. The transition probability from state $\mathcal{S}$ to $\mathcal{S}'$ under action $\mathcal{A}$ can be derived from the battery dynamics (3) while considering the Markov chains describing the channel gains and harvested energies, see [6, 17] for more details.

To simplify the notation, the vector of channel gains, harvested energies, battery levels and energy consumption and the matrix of energy shared at time k are henceforth denoted $g = g(k)$, $H = H(k)$, $B = B(k)$, $E = E(k)$ and $T = T(k)$, respectively, and the corresponding vectors of channel gains, harvested energies and battery levels at time $k + 1$ are denoted $\tilde{g} = g(k + 1)$, $\tilde{H} = H(k + 1)$ and $\tilde{B} = B(k + 1)$, respectively.

Then, it can be shown that under suitable assumptions[3], the value of the infinite time horizon stochastic control problem (15) is given by ρ, which is the unique solution of the average-cost optimality Bellman equation

$$\rho + V(g, H, B) = \min_{E,T} \left\{D(E, g) + \mathbb{E}\left\{V\left(\tilde{g}, \tilde{H}, \tilde{B} \middle| g, H, B, E, T\right)\right\}\right\} \tag{16}$$

where E and T satisfy the energy constraints given in (12) almost surely for $1 \le m, n \le M$ and $k \ge 1$, and V is the relative value function. Further, the optimal average cost ρ is independent of the initial conditions $g(0)$, $H(0)$ and $B(0)$.

The Bellman equation (16) can be solved using the relative value iteration algorithm, for which details can be found in [17]. The situation is very similar to that encountered when solving the finite-time horizon problem discussed above, in the sense that the solution consists in testing all possible energy allocation policies for each combination of states and storing the best option for each combination of states in a look-up table. However, in contrast to the finite-time horizon problem, the optimal policy is stationary, i.e., independent of time k, and the algorithm finishes when the value of ρ has converged; in practice, its relative difference between iterations is below a desired tolerance. In order to facilitate the numerical computation, the Bellman equation (16) is solved by discretizing the state and action space, in particular the battery levels and the power level space. (Recall that the state components involving the fading channels and the harvested energy levels are already assumed to be discrete due to the finite-state Markov chain assumption.)

4.3 Suboptimal solutions

The methods proposed to find power control policies in the two preceding subsections, i.e., finding the optimal solution for the finite-time horizon problem or the infinite-time horizon problem, require considerable computational effort. Also, in order to calcu-

3 For instance, suppose that a unichain power control policy exists.

late the look-up tables of the optimal energy allocation policies, knowledge of the underlying stochastic models has to be available. For instance, the finite-state Markov chains describing the channel gains and harvested energies have to be known. This is often impractical. Hence, in practice, it is often beneficial to investigate simple policies that provide suboptimal solutions but require less computational effort and less knowledge of the underlying system dynamics.

4.3.1 Constant level policy

To greatly simplify the implementation of the energy allocation policy, one can, for instance, choose the transmission energies to be a constant, predefined value E in the case that sufficient energy is available in the battery and 0 otherwise. Hence, $E_m(k) = E$ if $B_m(k) \geq E$ and $E_m(k) = 0$ otherwise for all m and k.

4.3.2 Greedy policy

Another very simple policy is the greedy policy, where each sensor just uses all available energy to transmit its data to the FC. Hence, $E_m(k) = B_m(k)$ for all m independently of the channel gain or any other states.

4.3.3 Adapted greedy policy

When implementing the greedy policy, there is a considerable risk of not having any energy available to transmit data from some sensor m to the FC at some time k if no energy has been harvested in the previous step. Thus, the greedy policy is slightly modified such that $E_m(k) = B_m(k)/2$, which ensures that at each time step, some energy is available to transmit data from every sensor to the FC.

5 Simulation study

In this section, we provide a collection of numerical results that illustrate the performance of the dynamic programming based algorithm and the three heuristic policies against important parameters such as the battery capacity and the energy transfer efficiency.

5.1 Varying energy transfer efficiency

A system with two sensors is simulated where $\mu = 1$ (no battery leakage), $\hat{B}_1 = \hat{B}_2 = 4\,\mathrm{mWh}$, $R_\theta = (1,0.2;0.2,1)$, and $\eta = \eta_{1,2} = \eta_{2,1}$ is varied between 0 and 1.

The fading channel gains and harvested energies are modeled as three-level discrete Markov chains with the common transition matrix

$$\mathbb{T} = \begin{bmatrix} 0.2 & 0.3 & 0.5 \\ 0.3 & 0.4 & 0.3 \\ 0.1 & 0.2 & 0.7 \end{bmatrix}. \tag{17}$$

Two cases have been simulated: In the "balanced scenario", the state space for g_1, g_2 is $\{0,0.5,1\}$ and for H_1 and H_2 is $\{0,1,2\}$. In the "unbalanced scenario", the states in the state space of g_2 and H_1 are 4 times lower than g_1 and H_2, respectively. That is, the state space for g_1 and g_2 are $\{0,0.5,1\}$ and $\{0,0.125,0.25\}$, respectively, while the state spaces of H_1 and H_2 are $\{0,0.5,1\}$ and $\{0,2,4\}$, respectively.

To facilitate the implementation of the infinite-time horizon dynamic programming algorithm, the space for the battery levels and the power levels for data transmission or energy transfer to the neighboring sensor were quantized uniformly using 16 levels between 0 and $\hat{B}_1 = \hat{B}_2 = 4\,\mathrm{mWh}$. Despite these discretizations, the dynamic programming (DP) based algorithm can be time consuming for calculating the optimal power control look-up tables, due to the well-known *curse of dimensionality*. The discretization of the decision variables leads to numerical inaccuracies, which can be addressed by averaging the results over a sufficiently long time span. In addition, the three heuristics described in Section 4.3 were implemented, where the adapted greedy policy is denoted "h1", the greedy policy is denoted "h2" and the constant level policy is denoted "h3",

The average distortion and the average energy usages for a simulation time span of 10^4 time steps for the optimal solution based on dynamic programming (DP) and the three heuristics (h1, h2 and h3) are illustrated in the plots in Figure 4.

As can be appreciated, in the balanced case, the average distortion hardly decreases when increasing the energy transfer efficiency despite the increase of average energy transferred between the sensors. In the unbalanced case, the average distortions obtained for the optimal solution (DP) decrease for higher η. The adapted greedy policy (h1) is more suitable for the balanced case, while the greedy policy (h2) achieves better results in the unbalanced case. The constant level policy (h3) does not perform well and performs extremely poorly for the unbalanced case, as one sensor can rarely transmit data.

Note that the average distortion for the optimal solution increases for low values of η. This can be explained by the loss of optimality due to discretization, which is necessary to implement the solution of the Bellman equation on a digital computer resulting in small deviations from the true optimal solution.

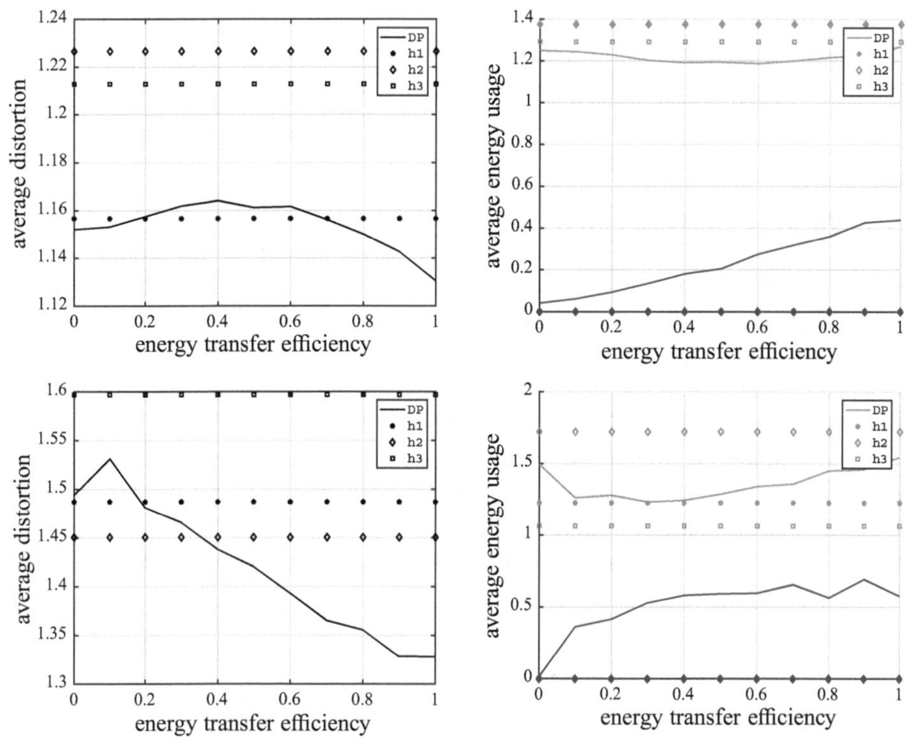

Fig. 4: Average distortion (left) and average energy usage (right, $(E_1 + E_2)/2$ in red, $(T_{1,2} + T_{2,1})/2$ in blue), versus energy transfer efficiency η for the balanced case (top) and the unbalanced case (bottom).

In the unbalanced case, it should also be noted that the optimal shared energy increases when the energy transfer efficiency increases from 0 to approximately 0.3. If the energy transfer efficiency is increased further, the optimal amount of energy shared among the sensors remains roughly the same. Since the measurements from the two sensors carry information about two different sources (although correlated), the FC needs to receive data from both sensors in order to estimate both sources. Hence, in the unbalanced case, one sensor needs to share some energy to allow the other sensor to transmit data that can be received at the FC with an acceptable quality. In the case when wireless energy transfer is possible with a sufficiently high efficiency (such that at least 30% of the transmitted energy is actually received at the receiving sensor), sharing more energy is not beneficial, since the other sensor has enough energy already for information transmission with an acceptable distortion level at the FC.

It can also be observed that the curve of the average energy used for data transmission is "bowl-shaped". Due to the increase in average shared energy when increasing the energy transfer efficiency from 0 to 0.3, on average, less energy is available for data transmission to the FC. Hence, the average energy usage decreases for low energy transfer efficiencies. However, for higher energy transfer efficiencies, the amount of shared energy remains almost the same, leading to an increase in average available energy to be used for data transmission.

5.2 Varying battery capacity

In the second example, instead of varying the energy transfer efficiency, we fix $\eta = \eta_{1,2} = \eta_{2,1} = 0.8$ and examine the effect of varying the battery capacities of the sensors between 1 and 5 mWh, while all other parameters are kept as before. Again, the balanced scenario and the unbalanced scenario with channel gains g_1 and g_2 and harvested energies H_1 and H_2 as above have been simulated.

The simulations are shown in Figure 5. In all cases, the average distortion decreases when increasing the battery capacity. Note that in the balanced case, all heuristic policies perform similarly, where the greedy policy (h2) achieves lower disturbances for low battery capacities, and the adapted greedy policy (h1) outperforms

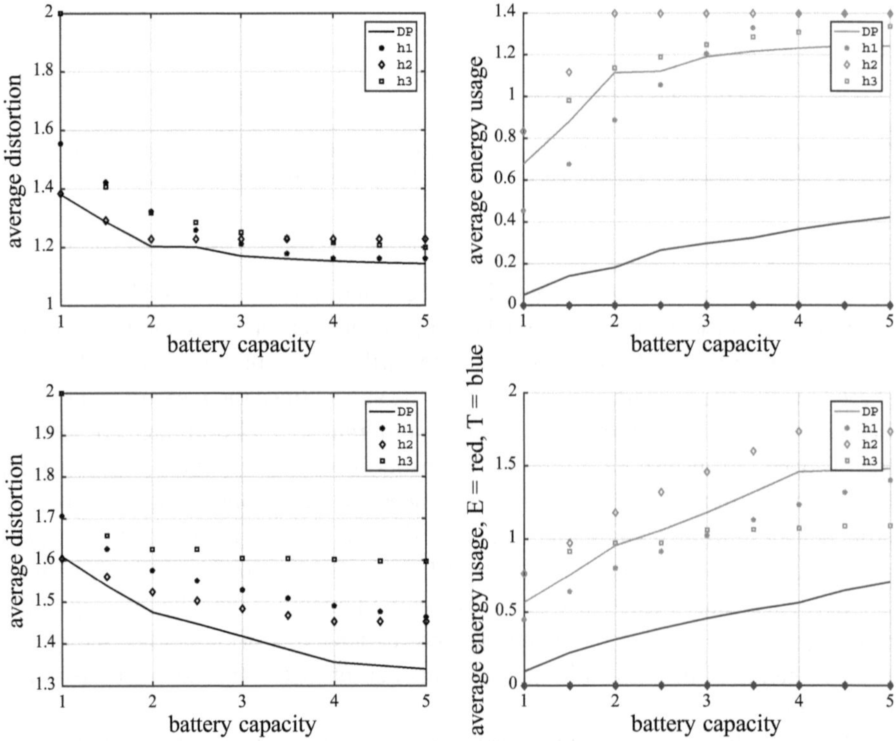

Fig. 5: Average distortion (left) and average energy usage (right, $(E_1 + E_2)/2$ in red, $(T_{1,2} + T_{2,1})/2$ in blue), versus battery capacity $\bar{B}$ for the balanced case (top) and the unbalanced case (bottom).

the other two heuristics for higher battery capacities. In the unbalanced case, where one sensor harvests much more energy but has a much worse channel compared to the other sensor, applying the adapted greedy policy (h2) clearly outperforms the other two heuristic solutions for all battery capacities tested. However, even this heuristic policy performs significantly worse than the optimal energy allocation policy achieved through dynamic programming as presented in Section 4.2.

6 Conclusions

This chapter has discussed how wireless channel gains, and energy harvesting and energy sharing sensors can be modeled using system control engineering methods. These models then enable analysis of the system and apply optimization techniques in order to optimize performance.

To illustrate those techniques, we studied the distortion minimization problem of a multi-sensor system, where each sensor transmits its measurement to a fusion centre, which aims to minimize the average distortion cost. The batteries at the sensors have limited capacity and may be prone to energy leakage, the sensors can harvest energy from their environment and are fitted with transceiver units that allow them to share energy with their neighbors, subject to some loss. Random harvested energies and channel gains are modeled as independent finite-state Markov chains. The fusion center is assumed to have causal information about the sensors' channel gains and harvested energy levels.

The optimal solution is obtained via a stochastic predictive control approach resulting in a Bellman dynamic programming equation. To avoid the computational burden needed to find the optimal solution based on dynamic programming techniques, three heuristic policies are also presented.

Our present results reveal important insights into wireless sensor networks with energy harvesting and energy sharing. It turns out that, even for simple network settings, the optimal energy allocation policy is far from trivial to compute. Indeed, the findings presented here form an important base for further investigation in this area considering more sophisticated network topologies.

Bibliography

[1] I. F. Akyildiz, W. Su, Y. Sankarasubramaniam, and E. Cayirci. A survey on sensor networks. *IEEE Communications Magazine*, 40(8):102–114, 2002.

[2] V. C. Gungor and G. P. Hancke. Industrial wireless sensor networks: Challenges, design, principles and technical approaches. *IEEE Transactions on Industrial Electronics*, 56(10):4258–4265, 2009.

[3] C.-Y. Chong and S. P. Kumar. Sensor networks: Evolution, opportunities and challenges. *Proceedings of the IEEE*, 91(8):1247–1256, 2003.
[4] V. C. Gungor, B. Lu, and G. P. Hancke. Opportunities and challenges of wireless sensor networks in smart grid. *IEEE Transactions on Industrial Electronics*, 57(10):3557–3564, 2010.
[5] W. Lumpkins. Nikola Tesla's dream realized: Wireless power energy harvesting. *IEEE Consumer Electronics Magazine*, 3(1):39–42, 2014.
[6] E. Altman. *Constrained Markov Decision Processes*, volume 7. CRC Press, 1999.
[7] C. K. Ho, P. D. Khoa, and P. C. Ming. Markovian models for harvested energy in wireless communications. In *IEEE International Conference on Communication Systems (ICCS)*, pages 311–315, November 2010.
[8] A. Kurs, A. Karalis, R. Moffatt, J. D. Joannopoulos, P. Fisher, and M. Soljačić. Wireless power transfer via strongly coupled magnetic resonances. *Science*, 317(83):83–86, 2007.
[9] A. Karalis, J. D. Joannopoulos, and M. Soljačić. Efficient wireless non-radiative mid-range energy transfer. *Annals of Physics*, 323(1):34–48, 2008.
[10] T. Zhu, Z. Zhong, Y. Gu, T. He, and Z.-L. Zhang. Leakage-aware energy synchronization for wireless sensor networks. In *Proceedings of the 7th international conference on Mobile systems, applications, and services*, pages 319–332, June 2009.
[11] R. Shigeta, Y. Kawahara, and T. Asami. Demo: Capacitor leakage aware duty cycle control for energy harvesting wireless sensor networks. In *Proceedings of the 9th ACM Conference on Embedded Networked Sensor Systems*, pages 387–388, November 2011.
[12] A. Goldsmith. *Wireless Communications*. Cambridge University Press, 2005.
[13] M. Gastpar. Uncoded transmission is exactly optimal for a simple Gaussian "sensor" network. *IEEE Transactions on Information Theory*, 54(11):5247–5251, 2008.
[14] I. Ahmed, A. Ikhlef, D. W. K. Ng, and R. Schober. Power allocation for an energy harvesting transmitter with hybrid energy sources. *IEEE Transactions on Wireless Communications*, 12(12):6255–6267, 2013.
[15] D. E. Quevedo, A. Ahlén, and K. H. Johansson. State estimation over sensor networks with correlated wireless fading channels. *IEEE Trans. Automatic Control*, 58(3):581–593, 2013.
[16] I. Bahceci and A. K. Khandani. Linear estimation of correlated data in wireless sensor networks with optimum power allocation and analog modulation. *IEEE Transactions on Communications*, 56(7):1146–1156, 2008.
[17] D. P. Bertsekas. *Dynamic Programming and Optimal Control*, volume 1. Athena Scientific, 3rd edition, 1995.

Sabrine Khriji, Dhouha El Houssaini, Ines Kammoun,
and Olfa Kanoun

Energy-efficient techniques in wireless sensor networks

Abstract: Energy constraint is a key feature in wireless sensor networks (WSNs), where sensor nodes are battery powered. In this context, maintenance efforts are frequently required to maintain the lifetime and load balance of the network. In some scenarios, like disaster prevention and chemical process control, changing or recharging nodes batteries is difficult and expensive. This creates an increasing need to develop alternative energy-efficient solutions to power nodes and reduce the energy consumption during network performance. Energy-aware techniques are essential to maintain the sustainability and performance of the network, while reducing the cost as well as the energy consumption. In this chapter, energy dissipation sources are studied to explore new solutions to conserve energy in the network. Also, a detailed description of available energy conservation and optimization techniques for WSNs is presented.

Keywords: Wireless sensor networks, energy management, energy consumption issues, energy supply sources, energy conservation, routing, sensors

1 Introduction

Wireless sensor networks (WSNs) have gained worldwide attention, especially with the proliferation in microelectromechanical systems (MEMS) technology, which enables the development of multi-functional sensor nodes with considerable characteristics like small size, low cost and limited processing and computing resources [1]. These sensors are able to sense both physical quantity and environmental conditions to process information locally, communicate wirelessly and to work cooperatively. The general structure of WSNs is presented in Figure 1. In this context and due to their characteristics, WSNs have emerged in a variety of applications like health, military, smart home, and other commercial applications, such as quality control, inventory management and supervision of non-accessible areas [2]. These applications are classified according to some priorities such as mobility, topology, coverage, heterogeneity and quality of service (QoS).

Sabrine Khriji, Dhouha El Houssaini, Olfa Kanoun, Chair for Measurement and Sensor Technology, Technische Universität Chemnitz, 09126 Chemnitz, Germany
Sabrine Khriji, Ines Kammoun, LETI Laboratory, National Engineering School of Sfax, University of Sfax, 3038 Sfax, Tunisia
Dhouha El Houssaini, Microelectronics and Instrumentation Laboratory, University of Monastir, Tunisia

https://doi.org/10.1515/9783110445053-017

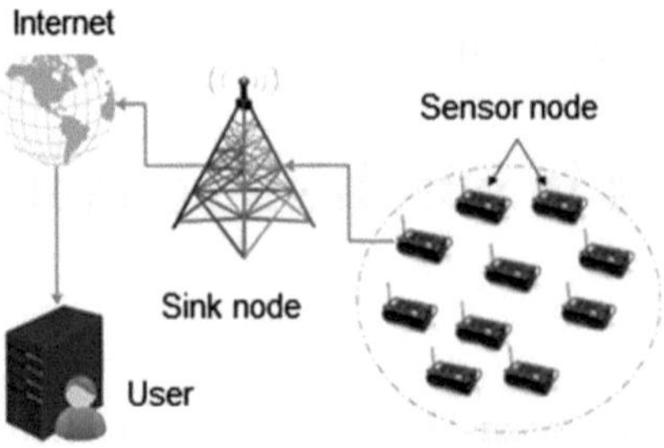

Fig. 1: General structure of WSNs [3].

WSNs are composed of different homogeneous or heterogeneous sensor nodes responsible for collecting and processing sensed data for a defined service. The general architecture of a sensor node is depicted in Figure 2, which presents three main parts: the sensing, processing and transmission subsystems and other complementary and optional components such as the mobilizer and position finding system. In the beginning of each process, the sensors start by collecting physical and environmental information from their surrounding. Once finished, sensed data are transmitted to the processing subsystem for further treatment and calculations. The radio module checks the availability of the medium's access to send these data to the sink or base station, where an adequate decision is made for suitable intervention.

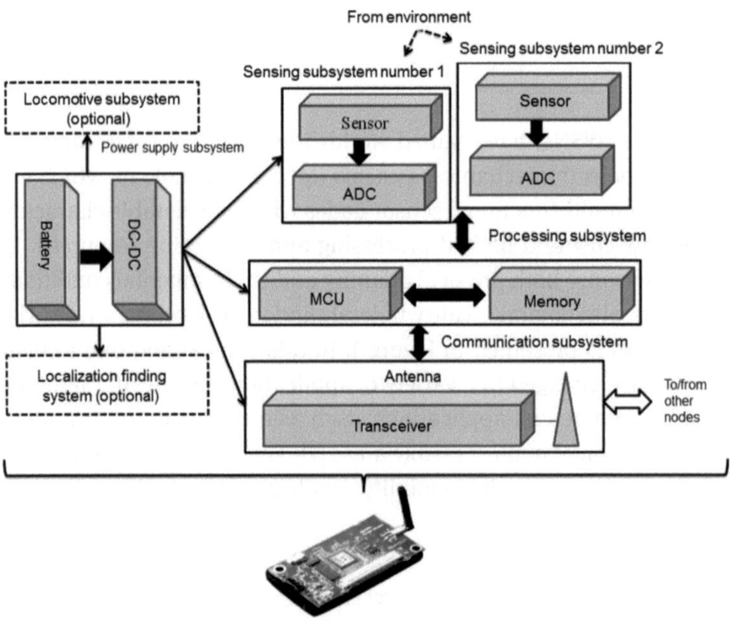

Fig. 2: Typical architecture of a wireless sensor node [3].

Each sensor node deployed in the network is susceptible to follow the same defined process starting by sensing surrounding environment for data collection and finishing by sending this information to the sink and waiting to receive the decision. Indeed, this creates excessive demands to manage the relative available energy sources to increase the lifetime of the node and, consequently, the whole network.

In this chapter, the available energy supplies responsible for powering autonomous sensors are presented. Moreover, a detailed overview of the available energy conservation techniques in WSNs is provided. The remainder of this chapter is organized as follows. Section 2 describes issues related to energy consumption in WSNs. Section 3 presents possible solutions as energy management techniques in WSNs. We close this chapter with a conclusion.

2 Energy consumption in wireless sensors networks

Energy constraint is a prominent feature for WSNs, because sensor nodes consume available energy sources during sensing, processing and transmitting data in order to respond to the application requirements. These services are provided by the radio transceiver, which makes it the most energy consuming component of the node. In this context, defining new, innovative and up-to date solutions for energy consumption during data transmission is a necessity to increase the lifetime of the node and the network as well. Indeed, identifying energy dissipation sources during data transmission is important to develop suitable solutions for the energy consumption problem. Data are transmitted from one node to another through signal propagation, which is susceptible to different energy consuming problems, mainly packet collision, overhearing, interference and idle listening (Figure 3).

The increasing amount of packet density in a channel creates a collision between packets [4]. For example, when two nodes attempt to send a message over the entire network at the same time, the collision phenomenon is very likely to occur (Figure 3(a)). All packets that caused the collision have to be remote and retransmitted again after a certain delay time. For each interrupted packet, retransmissions are processed until they reach the destination node. This is costly in term of power resources, execution time and workload of the node and the network as well. The performance of the network may be affected if the transmitted data are essential for decision making on the desired application and greedy in terms of power dissipation. In this context, defining a chronological plan for transmitting packets is crucial to reduce the number of communicated messages, the sending period and, therefore, the energy consumed. Such an approach can be provided by the media access control layer (MAC), where transmission events are organized with a specific time delay to avoid collision between different packets.

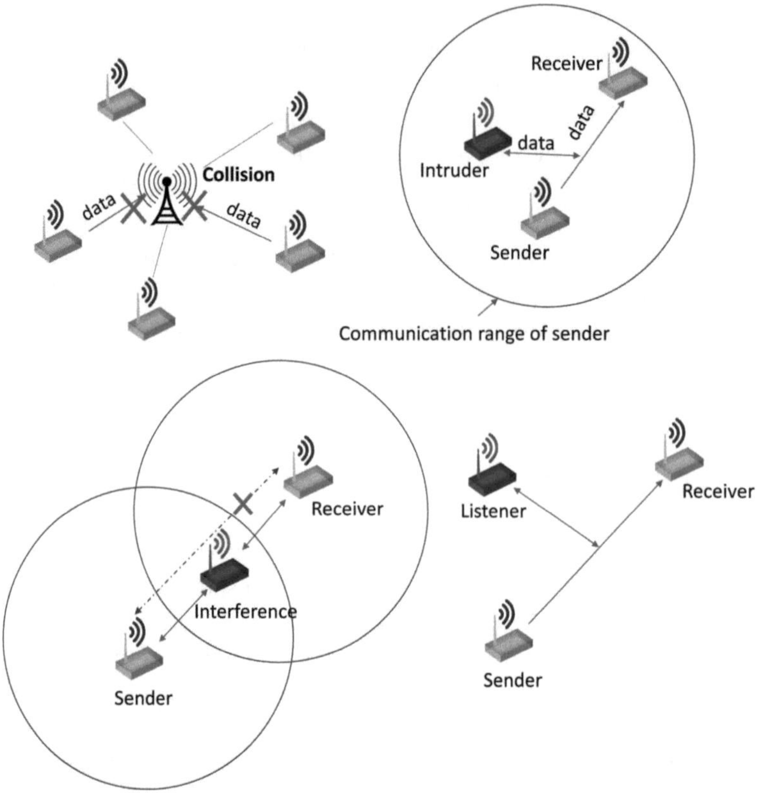

Fig. 3: Energy consumption issues during data transmission: (a) packet collision, (b) overhearing, (c) interference, (d) idle listening.

Each node in the network is capable of listening to other nodes available in its communication range. It is then able to receive messages sent to its neighboring nodes, which is called overhearing [5] (Figure 3(b)). By sending packets to the wrong destination node, neither the corresponding node receives the desired packet nor does the erroneous receiver get coherent data. In fact, when a transmitter sends data packet, all nodes within the same transmission range receive this packet, even if they are not the desired destination. So, the same packet should be resent again until it reaches the desired destination node. This imposes the use of additional energy and computational effort.

Data transmission is based on the radio signal propagation model, which is vulnerable to different problems like reflection, diffraction and scattering [6]. In fact, wave propagation can easily be affected by the surrounding vegetation, as well as the morphological description of the field. For this, the same signal can reach the destination many times via various paths, which creates an excessive use of the available energy. This creates the problem of interference where nodes are to receive the sent message but are not able to decode it to extract the important information [7]. Nodes receiving such information will first try to decode it, and if this is not possible, they will send a message request to the sender again to ask for a new packet. This will be redone continuously until the right information is obtained, which consumes more energy.

Another issue is the idle listening problem, which occurs when a node is listening to an idle channel in order to receive possible traffic [7] (Figure 3(d)). The node will use this available power source for listening to another channel that has no information to circulate. This causes the node to lose its energy for nothing. To avoid such a problem, the node must switch off its transceiver and go to sleep mode to conserve the available power sources for the active mode, where it is responsible for sending and receiving data packets.

New innovative and intelligent solutions are necessary to reduce energy consumption during data transmission in a way as to increase the lifetime of the node and that of the network as well, and also to ensure load balance and sustainability.

3 Energy supply sources in wireless sensors network

WSNs are applied in a variety of challenging and complicated applications. Different low-power wireless sensor platforms have been designed already to run on batteries that have a limited lifetime. Once the energy level drops, the sensor node becomes unable to contribute to the utility of the network as a group. In this context, alternative power sources must be used to increase the lifetime of the node and, therefore, to maintain the sustainability and efficiency of the network. New techniques extract the necessary power supply from different natural sources to harvest energy [8–10] and [11]. An overview of the major energy supply sources for WSNs is provided in this section.

Sensor nodes are battery powered, which presents a limited resource [12]. Based on the application requirements and the desired task, the lifetime of installed batteries varies from one node to another, depending on the computational effort. Once expired, batteries need to be replaced in time so the network does not break down. In some cases, node maintenance becomes complicated or even life threatening, like at military locations or where there is a danger of chemical explosion. This eventually results in permanent failure of the node, and that particular node is then discarded from the network operation. In this context, autonomous, intelligent and easy to use solutions for energy harvesting are necessary to offer an alternative power supply for nodes by exploring environmental resources.

Alternative energy sources are essential to maintain the lifetime of the network instead of only depending on limited batteries. Therefore, exploring harvesting techniques is necessary to generate an efficient power supply for nodes in the network in

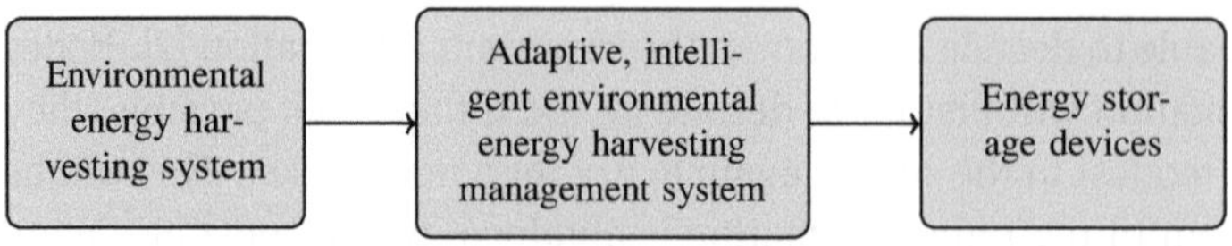

Fig. 4: General diagram of an energy harvesting system.

order to achieve a better sustainability. Renewable energies offer a suitable solution for this approach. An energy harvesting system is composed of three main blocks: an environmental energy harvesting system, an adaptive, intelligent environmental energy harvesting management system and energy storage devices (Figure 4). Different possible energy sources can be found in nature, which offer endless energy supplies like solar energy, mechanical vibration, temperature gradient, wind energy, water flow and magnetic energy. Therefore, harvested energy can be used directly or saved in storage devices. Another alternative is to use wireless energy transfer (WET) technology. In fact, WET has opened up a revolutionary paradigm for the sensor network lifetime. WET is defined as the ability to move across space by transferring electric energy from one storage device to another without any plugs or wires. In addition, WET may distribute available energy from locations with high energy to locations with low energy, in order to balance the overall energy. Many techniques that implement WET in WSN applications, such as lasers [13], or radio transmissions [14], have been published recently.

4 Energy conservation in wireless sensors networks

Sensor nodes are deployed under different scenarios to fulfill the requirements of a certain applications. Therefore, the lifetime of nodes should be managed by adopting energy conservation techniques. Many applications require to operation for many months or even years. In some cases, energy can be saved by exploiting external environment power sources, such as using solar cells or wind as a power source. However, the behavior of external power supplies is discontinuous; thus, batteries are needed as well as a backup load. As network lifetime has become the key characteristic for evaluating WSN, several techniques in the literature aimed to reduce energy consumption and improve network lifetimes. The taxonomy of these techniques will be discussed in this section.

Energy-efficient techniques can be classified into four main categories: data driven approaches [15], duty cycling [16], energy-efficient routing protocol [17] and mobility of nodes or sinks [18] (Figure 5). Data driven approaches are used to reduce the amount of data to be sensed, processed and then delivered to the sink. Duty cycling techniques enable managing sleep and wake-up modes of the transceiver module to

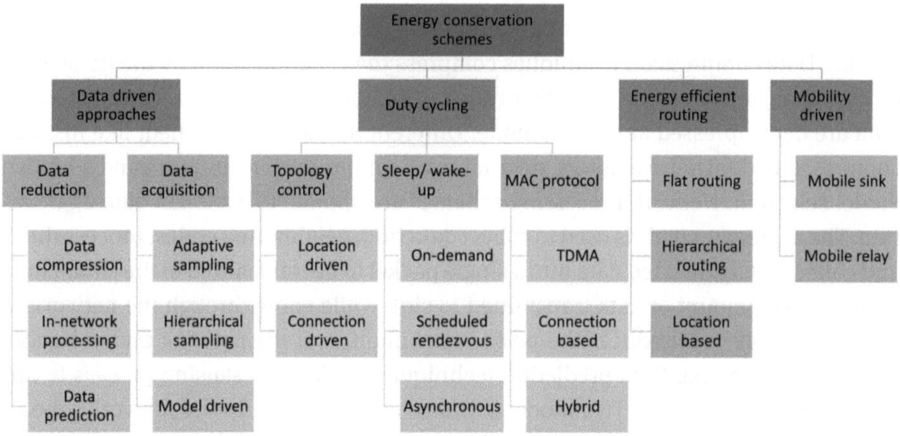

Fig. 5: Taxonomy of approaches for energy conservation in WSNs.

control the energy consumption. A proper routing algorithm enables us to reduce the energy consumed by end-to-end communication and, therefore, increase the lifetime of the network. Also, managing the mobility of nodes in the network enables us to reduce the distance between nodes for communication by assigning specific and energy consuming tasks to mobile nodes. In the next sections more details will be provided about these techniques.

4.1 Data driven approaches

Sampled data from surrounding environment have high temporal or spatial correlation, so there is no need to transmit redundant data to the sink. In this case, transmitting unneeded and redundant data is energy and time consuming. Thus, data driven approaches work as energy-efficient solutions for data redundancy. They are used to reduce the amount of sampled data to be transmitted by maintaining sensing accuracy at an acceptable level imposed by the application requirements. Therefore, data driven approaches are responsible for balancing energy consumption while transmitting sensed data. They are divided basically in two schemes: data reduction and data acquisition.

4.1.1 Data reduction

Data reduction approaches aim to reduce the amount of data to be sensed, processed and delivered to the sink. They convert a large amount of the received data to a smaller useful entity, which can be decoded without any data losses and with reduced energy consumption. Data reduction is based on two main methods: compression and prediction. Data compression techniques compress the generated data by the node itself or by the aggregator node and then transmit only the data needed. Once received, data are decompressed at the sink node. Different methods are applied, like lossless compression, dual Kalman filters, deterministic compression techniques and adaptive model selection. Network processing schemes are applied to establish data aggregation. The aggregator node is used to collect data from neighboring nodes, process them and collect the aggregated data into a single packet to be transmitted to the sink. Thus, reducing the amount of data transmitted to sink while going through the network to the destination node reduces the power consumption of the network and enhances the network lifetime. Data prediction techniques models the sensing process to answer upcoming requests. Two model instances are designed in the network; one in the sink and the other at the corresponding source node. The sink model answers any present requests without the need to communicate with other nodes, which reduces the communication effort and, thus, the energy consumption.

The data prediction method uses three approaches: stochastic approaches, time series forecasting and algorithmic approaches. Stochastic approaches are a combination of random processes, so that probabilistic models can be used in the prediction phase. These approaches offer a high level of data aggregation but with expensive computational efforts. Such approaches are feasible if powerful sensor nodes are available in the network. Time series forecasting presents a set of historical values gathered by periodical sampling, which are helpful to predict the future value in the same series. The most common methods using these approaches are moving average, auto-regressive or auto-regressive moving average methods. Using time series forecasting has been proved to be simpler in implementation and presents an acceptable accuracy. Algorithmic approaches depend on the desired application. Nevertheless, one common factor that they share is the algorithmic approach used to obtain a prediction, beginning with a heuristic or behavioral characterization of the phenomena.

4.1.2 Data acquisition

Recent research considers that the sensing process is the most energy consuming service in WSNs. Several causes create such a problem, like the use of power hungry transducers, which require high power resources to perform sampling tasks, the use of power hungry analogs/digital converters, which require a high rate and high resolution, or the use of active sensors, which they need to send out a probing signal to acquire any necessary information of the sample or the use of long acquisition time, which can be in the order of hundreds of milliseconds or sometimes in seconds. To control the energy dissipation during the sensing process, energy-efficient data acquisition approaches are required. These approaches are classified into three main classes: adaptive sampling, hierarchical sampling and model-based active sampling.

Adaptive sampling reduces the amount of data required from the transducer based on either spatial or temporal correlation. These methods are implemented on centralized architectures, which requires high computational effort to treat all received packets in the same sink. The hierarchical sampling method assumes that the network is equipped with different types of sensors. Each sensor has a specific resolution and is associated with a given energy consumption. To get the tradeoff between accuracy and energy saving, this approach dynamically selects which class to activate. Finally, model-based approaches aim to reduce the number of sampled data by using computed models, meanwhile, saving energy through the use of data acquisition protocols. A model used to sense phenomenon is built upon sampled data to predict the next possible sample with a certain accuracy.

4.2 Duty cycling

4.2.1 Topology control protocols

Topology control protocols are basically associated to network redundancy. They search for the minimum subset of nodes that is able to ensure network connectivity by exploiting the network redundancy. In most cases, nodes are randomly deployed in the network, where the number of nodes is not managed, which may overweigh the network by deploying additional and unnecessary nodes. Thus, managing the number of deployed nodes is critical to ensure the balance and sustainability of the network and to eliminate node redundancy [19]. In this context, topology control protocols aim to adapt the network topology based on the application requirement and the number of active nodes. These approaches consider the state of the nodes, active/passive, to establish an energy-efficient managing system, which is classified into two categories: location driven protocols and connectivity driven protocols. Location driven protocols define the state of node (active/ sleep), which node to turn on and when based on the node's location, which is assumed to be previously known. Connectivity driven protocols are responsible for activating/deactivating a node dynamically in such a way as to achieve a total network connectivity and full coverage. For example, geographical adaptive fidelity (GAF) [20] is location driven. This protocol reduces energy consumption while keeping a constant level of routing fidelity. The working area is divided into small virtual grids, which are defined, such that for any two adjacent grids A and B, all nodes in A are able to communicate with nodes in B and vice versa. Also, nodes in the same grid are equivalent for routing, so that just one node is activated at a time. This will need the nodes to cooperate together. For connectivity driven approaches, a well-known protocol is Span [21], which selects coordinators of all nodes in the network. These coordinators stay awake continuously and apply a multi-hop routing, while others stay in sleeping mode and periodically check for any updates.

4.2.2 Sleep/wake-up protocols

Sleep/wake-up protocols are defined through the radio subsystem of the sensor node, which reduce the time that a node stays in the idle state. Sleep/wake-up protocols can be classified mainly into three categories: on demand protocols, scheduled rendezvous and asynchronous schemes. On demand protocols are considered as the most intuitive approach to reduce power consumption. A sensor node should be awakened only when it receives a packet from neighbors that is suitable for low duty cycle applications. To inform the sleeping node about possible neighbors that are trying to communicate with, sleep/wake-up protocols use multiple radios with low power and rate. Indeed, a relay node is required wait for the node to wake up and receive the data. This makes on demand schemes unsuitable for disruption tolerant networks (DTN) [22]. An alternative solution to the duty cycle technique consists in using a scheduled rendezvous approach, which ensures that all nodes wake up at the same time. According to a wake-up schedule, sensor nodes wake up and are still active for a short time interval to communicate with their neighbors and then, they go to sleep until the next rendezvous time. All sensors should be synchronized in order to wake up and work at the same time. The third category used for the duty cycle approach is asynchronous schemes, which allow sensor nodes to manage their sleep and wake-up schedules. Nodes select the time to wake up when they are able to communicate with their neighbors. To achieve this, neighbors should have an overlap between their wake-up periods. The main objectives of this scheme is that the nodes do not have to be synchronized together; it allows each node in the WSN to set its own sleep/wake-up schedule independently, and accordingly the contention and workload are reduced.

4.2.3 MAC protocols with low duty cycle

Embed a duty cycle within the channel access. The low duty cycle scheme is implemented by most of these protocols in order to manage the power of the network. This way, the MAC protocol can be classified into three categories: TDMA-based protocols, contention-based protocols and hybrid protocols. In the TDMA (time division multiple access)-based MAC protocol, time is divided into time frames, and each time frame is further split into a fixed number of slots having an equal length called time slots, as shown in Figure 6. Each node is allocated a time slot in a time frame and is allowed to transmit only in the allocated time slot. Furthermore, a node depending on the schedules of its neighboring nodes may remain in sleep mode when it is neither to transmit nor to receive, i.e., as nodes activate their radio only during their own slots, the energy consumption is significantly decreased to the minimum required for transmitting/receiving data. The schedule is performed by a specific node, the base station (BS) node, which broadcasts data to the rest of the network via a management frame referred to beacon. Each node is just required to respect the instructions coming from

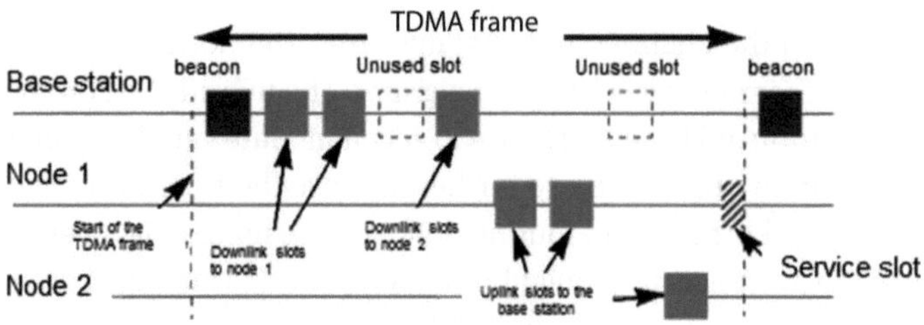

Fig. 6: TDMA channel access mechanism.

the BS by following its time plan blindly. Generally, the structure of the frame is organized as follows: downlink slots going from the BS to the rest of the network and backwards, uplink slots [23].

Another type of MAC protocol is based on integrating channel access functionalities with a sleep/wake-up scheme and are referred to as contention-based protocols, which are based on carrier sense multiple access (CSMA) or carrier sense multiple access/collision avoidance (CSMA/CA) [24]. This protocol serves to manage the collision between nodes, where colliding nodes will back off for a random duration of time before attempting to access the channel again, which offers a good robustness and scalability. However, high energy is consumed because of collisions and multiple access. Some of the popular contention-based MAC protocols are Sensor-MAC (S-MAC), Berkeley- MAC (B-MAC) and Time-out-MAC (T-MAC) [25]. S-MAC is based on sleep-listen schedules. It periodically, wakes up for short time intervals to look for available packets in the channel. The schedules are exchanged within the neighboring nodes in order to provide synchronization. During the listening interval, every node needs to stay awake and wait for upcoming messages, which consumes more energy. Moreover, listen and sleep periods of this protocol are fixed from the beginning of the deployment; this limits the scalability of the network and makes it complex to implement. Regarding the B-MAC protocol, it implements a back-off scheme, an accurate channel estimation facility and optional acknowledgments as basic channel access control components [26]. Then, B-MAC involves an asynchronous sleep/wake-up scheme based on low-power listening (LPL) and periodic listening in order to perform a low duty cycle. Nodes wake up for a short period and check for channel activity to detect eventual ongoing transmissions. They return to sleep if no activity is detected. This process is repeated periodically. The period between successive wake ups is called the check interval. When the wake-up time is specified, the check interval is fixed by the application. If a sender wants to transmit a message, it sends a long preamble to make sure that the receiver is listening for the packet. The preamble has the size of a sleep interval. The T-MAC protocol is introduced to enhance S-MAC protocol by allowing nodes to sleep again if no message has been received for a defined delay time [27]. The basic idea behind T-MAC is to adaptively adjust the sleep and wake-up periods based on the estimated traffic flow. The main advantage of this protocol is getting into sleep mode when nothing is heard in the channel, but in some cases, the node may enter sleep mode while it still has other messages from its neighbor. The hybrid MAC protocol presents a perfect combination of TDMA and CSMA to accommodate variable traffic patterns. It uses information from upper layers to further improve their performance, like information about the neighborhood and routing policies. The main objectives of this protocol are to reduce the overhearing of neighboring nodes and to managed their channel sniffing patterns like Z-MAC, which executes CSMA in low traffic and switches to TDMA in high traffic conditions.

5 Energy-efficient routing

Routing is a key service in WSNs that is responsible for routing transmitted data from the source node to the destination across the network. Transmission is done in one single hop or via an intermediate node in multi-hop communication. Implementing routing protocols in WSNs is very challenging due to several characteristics specific to the application. These protocols can be classified as flat routing (data centric), hierarchical routing (clustering) and location-based routing (geographic). They depend on the network structure and protocol operation and can be classified into multi-path-based, query-based, negotiation-based, QoS-based, and coherent-based protocols.

5.1 Flat routing

Flat routing protocols are also called data centric protocols, where all nodes have the same role in the network and communicate together to perform the sensing task. The node source transmits its data packet to a specific node called the sink. The sink node sends queries to nodes within the same sensing area and then waits for data from the sensors located in the selected area. The intermediate nodes aggregate data from multiple sources within the deployment region and forward the collected data toward the sink (Figure 7). This results in less data transmission and redundant data elimination, so, the energy of the whole network is successfully saved. There are a plenty of protocols implemented in the literature, like SPIN (sensor protocol for information via negotiation) [28], direct diffusion [29] and rumor routing [30].

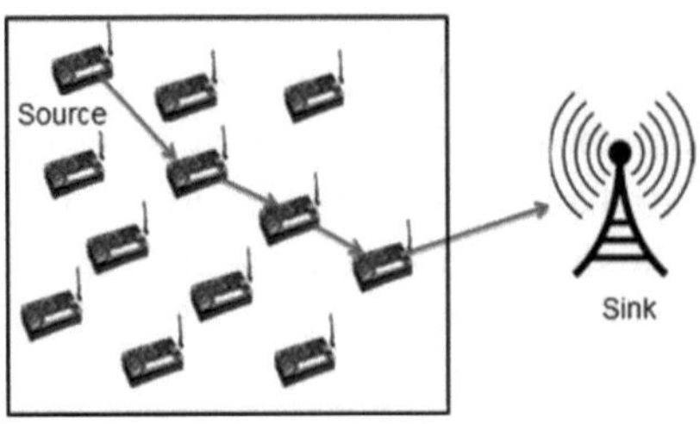

Fig. 7: The flat routing protocol.

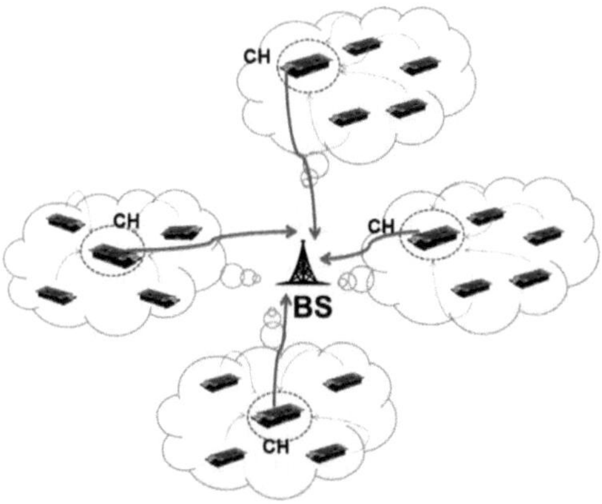

Fig. 8: The hierarchical routing protocol.

5.2 Hierarchical routing

Traditional routing protocols implemented for WSNs are not optimal in terms of energy efficiency and load balancing [31]. Clustering is introduced as an energy-efficient communication protocol that can be used by sensor nodes to forward their sensed data to the sink, resulting in load balancing and network lifetime extension (Figure 8). Moreover, clustering techniques have emerged as a popular choice to achieve energy efficiency and scalable performance in large-scale sensor networks. Clustering techniques are based on three kinds of nodes, defined as follows: normal nodes (NN), cluster heads (CH) defined as an aggregator of data received from normal nodes within the same cluster and a base station (BS). In fact, NNs sense the environment, obtain data and forward them to their CH, which, in turn, aggregates its data from all cluster members by turns. The main objective of hierarchical routing is to efficiently manage the energy consumption of the whole network. To do so, sensor nodes should be involved in multi-hop communication inside a particular cluster. Clustering enables reduction of the amount packets transmitted from the cluster head to the base station, which includes data aggregation and fusion techniques, which also reduces the overload and packets loss. To enhance the performance of clustering algorithms, specific clustering characteristics should be considered, like changeability of clusters, uniformity of cluster size, and intra-cluster and inter-cluster connectivity (Figure 9). Clusters' definition could be fixed, where the choice of CH is defined from the beginning of the process or is variable, where CHs are selected randomly, which varies the number of CHs for each iteration. The size of the cluster itself can be fixed or modified during the clustering. Clusters having the same size present a uniform distribution while clusters with different sizes show a non-uniform distribution. Once the clusters are created and the CHs are defined, communication protocols should be defined. Two possible connectivities are available: single-hop and multi-hop. For single-hop communication, normal

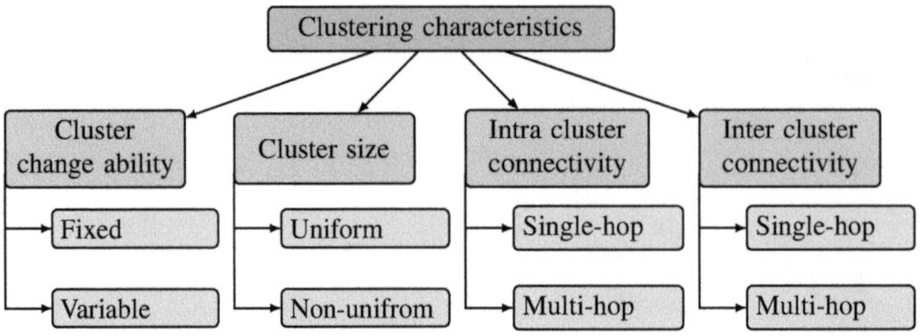

Fig. 9: Taxonomy of cluster characteristics.

nodes in cluster transmit their data directly to the corresponding CH. In multi-hop scenarios, a relay node is used to transmit data from one sensor node to the corresponding CH and then to the BS.

In general, cluster-based algorithms often include four stages: CH selection, cluster formation, data aggregation and data transmission. For many clustering protocols, the network schedule is split into different rounds of fixed period. As can be seen in Figure 10, each round includes two phases, namely, the setup phase and the steady-state phase. The cluster head selection process is done during setup phase. The steady-state phase attempts to aggregate data and assures data communication. For this reason, this phase is divided into different frames. In each frame, the sensor nodes transmit their data to a CH within same cluster range and this CH forwards the data collected to BS. The timing diagram of these two phases is explained further in Figure 10.

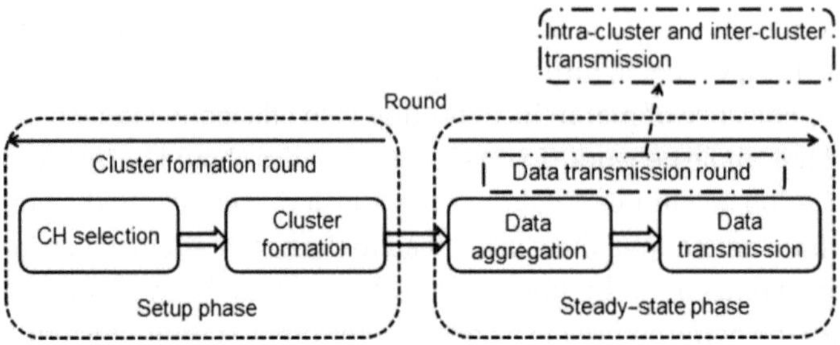

Fig. 10: Phases of clustering techniques in one round.

5.3 Mobility-based energy conservation schemes

Node mobility has come to be considered as a recent solution for energy-efficient data collection in WSNs. Actually, deploying mobile nodes can be done in two ways: either with the use of a mobilizer installed on the node itself or by placing the node on a mobile element like an animal or a car. In fact, mobilizers are energy hungry, and therefore, the mobility can be limited to powerful nodes. Once the mobility has been applied in WSNs, several issues concerning the connectivity may surface. To start with, a sparse architecture is needed during the sensor network design to ensure that all application requirements are satisfied. So, the number of nodes deployed will be reduced as the connectivity constraint is relaxed, since mobile elements can reach all parts of the network, which reduces the energy consumption. Actually, communicated packets traverse the network to the sink in a multi-hop path. Thus, when the sink is static, some paths are loaded more than others. Therefore, nodes closer to the sink are subject to energy depletion. So, mobility should be alternated between nodes to simplify the data collection process by affecting some mobile data collector. Now, ordinary nodes stay idle waiting for mobile data collector to route messages towards it. Consequently, link error and communication costs are reduced.

6 Conclusion

In this chapter, energy management concepts in WSNs are studied. First, we presented the main energy consumption issues in WSNs responsible for reducing the load balance and lifetimes of nodes, as well as the network. Data transmission is considered as one of the most power consuming services, where different communication limitations create a disturbance in the energy balance in the network, like packet collision, overhearing and interference problems. Therefore, new innovative and energy-efficient solutions have been developed to maintain efficiency as well as the lifetime of the network. Energy conservation main schemes were presented and classified into four main approaches: data driven, duty cycling, routing and mobility driven. An overview of these approaches is presented, which focuses on their ability to reduce the energy consumption during data transmission. By implementing these methods, energy dissipation in sensor networks can be reduced. However, methods like data reduction and prediction may affect the robustness of transmitted information. For this, effects of theses approaches on link quality and data feasibility should be considered for further investigations.

Bibliography

[1] A. Rathee, R. Singh, and A. Nandini. Wireless sensor network-challenges and possibilities. *International Journal of Computer Applications*, 140(2), 2016.

[2] S. Khriji, D. El Houssaini, M. W. Jmal, C. Viehweger, M. Abid, and O. Kanoun. Precision irrigation based on wireless sensor network. *IET Science, Measurement & Technology*, 8(3):98–106, 2014.

[3] A. Tzounis, Katsoulas, T. Bartzanas, and C. Kittas. Internet of things in agriculture, recent advances and future challenges. *Biosystems Engineering*, 164:31–48, 2017.

[4] I. Dbibih, I. Iala, D. Aboutajdine, and O. Zytoune. Collision avoidance and service differentiation at the mac layer of wsn designed for multi-purpose applications. In *Cloud Computing Technologies and Applications (CloudTech), 2016 2nd International Conference on*, pages 277–282. IEEE, 2016.

[5] R. Sett and I. Banerjee. An overhearing based routing scheme for wireless sensor networks. In *Advances in Computing, Communications and Informatics (ICACCI), 2015 International Conference on*, pages 2076–2082. IEEE, 2015.

[6] A. Razaque and K. M. Elleithy. Low duty cycle, energy-efficient and mobility-based boarder node—mac hybrid protocol for wireless sensor networks. *Journal of Signal Processing Systems*, 81(2):265–284, 2015.

[7] M. A. Kafi, J. Ben-Othman, A. Ouadjaout, M. Bagaa, and N. Badache. Refiacc: Reliable, efficient, fair and interference-aware congestion control protocol for wireless sensor networks. *Computer Communications*, 101:1–11, 2017.

[8] W. Han, B. Zhu, N. Wang, and J. Xu. Development and evaluation of a wireless sensor network monitoring system in various agricultural environments. In *2013 Kansas City, Missouri, July 21-July 24, 2013*, page 1. American Society of Agricultural and Biological Engineers, 2013.

[9] V. Raghunathan, S. Ganeriwal, and M. Srivastava. Emerging techniques for long lived wireless sensor networks. *IEEE Communications Magazine*, 44(4):108–114, 2006.

[10] T. Rault, A. Bouabdallah, and Y. Challal. Energy efficiency in wireless sensor networks: A top-down survey. *Computer Networks*, 67:104–122, 2014.

[11] J. A. Khan, H. K. Qureshi, and A. Iqbal. Energy management in wireless sensor networks: A survey. *Computers & Electrical Engineering*, 41:159–176, 2015.

[12] C. Antonopoulos, F. Kerasiotis, C. Koulamas, G. Papadopoulos, and S. Koubias. Experimental evaluation of the waspmote platform power consumption targeting ambient sensing. In *Embedded Computing (MECO), 2015 4th Mediterranean Conference on*, pages 124–128. IEEE, 2015.

[13] N. A. Bhatti, A. A. Syed, and M. H. Alizai. Sensors with lasers: Building a wsn power grid. In *Proceedings of the 13th international symposium on Information processing in sensor networks*, pages 261–272. IEEE Press, 2014.

[14] M. Erol-Kantarci and H. T. Mouftah. Radio-frequency-based wireless energy transfer in lte-a heterogenous networks. In *Computers and Communication (ISCC), 2014 IEEE Symposium on*, pages 1–6. IEEE, 2014.

[15] Q. A. Bakhtiar, K. Makki, and N. Pissinou. Data reduction in low powered wireless sensor networks. In *Wireless Sensor Networks-Technology and Applications*. InTech, 2012.

[16] S. Lai. *Duty-cycled wireless sensor networks: wakeup scheduling, routing, and broadcasting*. PhD thesis, Virginia Polytechnic Institute and State University, 2010.

[17] S. A. Nikolidakis, D. Kandris, D. D. Vergados, and C. Douligeris. Energy efficient routing in wireless sensor networks through balanced clustering. *Algorithms*, 6(1):29–42, 2013.

[18] M. Bouaziz and A. Rachedi. A survey on mobility management protocols in wireless sensor networks based on 6lowpan technology. *Computer Communications*, 74:3–15, 2016.

[19] L. P. Damuut and D. Gu. On redundancy identification in randomly deployed wsns, another perspective. In *Sensors Applications Symposium (SAS), 2013 IEEE*, pages 27–32. IEEE, 2013.

[20] S. Roychowdhury and C. Patra. Geographic adaptive fidelity and geographic energy aware routing in ad hoc routing. In *International Conference*, volume 1, pages 309–313, 2010.

[21] B. Chen, K. Jamieson, H. Balakrishnan, and R. Morris. Span: An energy-efficient coordination algorithm for topology maintenance in ad hoc wireless networks. *Wireless Networks*, 8(5):481–494, 2002.

[22] C. Velásquez-Villada and Y. Donoso. Delay/disruption tolerant network-based message forwarding for a river pollution monitoring wireless sensor network application. *Sensors*, 16(4):436, 2016.

[23] S. Zhuo, Z. Wang, Y.-Q. Song, Z. Wang, and L. Almeida. A traffic adaptive multi-channel mac protocol with dynamic slot allocation for wsns. *IEEE Transactions on Mobile Computing*, 15(7):1600–1613, 2016.

[24] M. U. H. Al Rasyid, I. U. Nadhori, A. Sudarsono, and R. Luberski. Analysis of slotted and unslotted csma/ca wireless sensor network for e-healthcare system. In *Computer, Control, Informatics and Its Applications (IC3INA), 2014 International Conference on*, pages 53–57. IEEE, 2014.

[25] A. Bachir, T. W. M. Dohler, and K. Leung. Mac essentials for wireless sensor networks. *IEEE Commun. Surveys Tutorials*, 12(2):222–248, 2010.

[26] S. Bdiri, F. Derbel, and O. Kanoun. Wireless sensor nodes using energy harvesting and b-mac protocol. In *Systems, Signals & Devices (SSD), 2013 10th International Multi-Conference on*, pages 1–5. IEEE, 2013.

[27] W. Lee, Y. Lee, S. Lee, and D. Kim. Analysis of s-mac/t-mac protocols for wireless sensor networks. In *Proceedings of the 10th WSEAS international conference on Communications*, pages 262–267. World Scientific and Engineering Academy and Society (WSEAS), 2006.

[28] E. Woodrow and W. Heinzelman. Spin-it: a data centric routing protocol for image retrieval in wireless networks. In *Image Processing. 2002. Proceedings. 2002 International Conference on*, volume 3, pages 913–916. IEEE, 2002.

[29] C. Intanagonwiwat, R. Govindan, and D. Estrin. Directed diffusion: A scalable and robust communication paradigm for sensor networks. In *Proceedings of the 6th Annual International Conference on Mobile Computing and Networking*, pages 56–67. ACM, 2000.

[30] D. Braginsky and D. Estrin. Rumor routing algorthim for sensor networks. In *Proceedings of the 1st ACM International Workshop on Wireless Sensor Networks and Applications*, pages 22–31. ACM, 2002.

[31] Z. Manap, B. M. Ali, C. K. Ng, N. K. Noordin, and A. Sali. A review on hierarchical routing protocols for wireless sensor networks. *Wireless Personal Communications*, 72(2):1077–1104, 2013.

Sadok Bdiri, Faouzi Derbel, and Olfa Kanoun

A wake-up receiver for online energy harvesting enabled wireless sensor networks

Abstract: Supplying power-hungry radio receivers with an energy harvester is challenging. A wake-up receiver (WuRx) for wireless sensor networks consumes much less power. The introduced WuRx operates with a modified medium access protocol (MAC), allowing low-energy consumption and practical latency. The sensitivity of the WuRx is optimized with minimally integrated active parts, emphasizing −61 dBm. The idle power consumption of the WuRx reaches down to 7.2 μW and 500 μW during decoding. The WuRx can handle a 32-bit long pattern at a bit rate of 4 kbit/s with Manchester encoding. The design blocks are discussed in this paper. A prototype is realized for evaluation purposes.

Keywords: WuRx, wake-up receiver, event-triggered, on demand, ultra-low power, WSN, IoT, energy harvesting

1 Introduction

Internet of Things (IoT) has promoted the needs of wireless sensor network (WSN) applications. WSN is implemented on devices referred to as wireless sensor nodes (WSn). They rely on stand-alone power sources like batteries or energy harvesters, as to disassociate them from any sort of maintenance. The energy consumed is a highly critical feature to be considered in WSn architecture. A traditional sensor node embeds mainly a processing unit, generally a microcontroller (MCU), a radio chip, sensor(s) and the power source.

Today, commercially available MCUs feature low-power modes consuming less than 1 μW in the standby state. This drastically reduces energy usage when a WSn switches to the idle state. The same applies to sensors when their duties have been accomplished. However, radio receivers still require a large amount of energy to either transmit or receive radio packets, taking more than 70% of power out of the battery. It is worth considering that the physical dimensions of a WSn should be kept to a minimum. This forces the batteries to also be relatively small. For such reasons, batteries (i.e., coin cells) have limited capacities ranging between 80 mAh to 1000 mAh. An always-on radio chip consuming an order of mW will deplete those batteries in

Sadok Bdiri, Faouzi Derbel, Faculty of Electrical Engineering and Information Technology, Leipzig University of Applied Sciences, Wächter Str. 13, 04107 Leipzig, Germany
Olfa Kanoun, Faculty of Electrical Engineering and Information Technology, Chemnitz University of Technology, Reichenhainer Str. 70, 09126 Chemnitz, Germany

https://doi.org/10.1515/9783110445053-018

only few months. Unlike the MCU, the radio chip cannot be completely turned off, as it needs to be in the receiving state waiting for incoming radio packets. Otherwise, the WSn is disconnected from the network and will not be reached by external events. Radio duty cycling is a compromise between the always-on and off states.

A certain protocol should be agreed upon between a transmitter and a duty cycling radio receiver for them to be able to communicate adequately. This is done by different MAC protocols, either in asynchronous or synchronous schemes. While energy usage is reduced, other problems arise, including latency and overhearing. In event-triggered communication, WSN traffic is very low. Applying MAC protocols with very low duty cycles to reduce energy induces a large end-to-end latency. The latter is more of a concern in real-time applications. To better handle this issue, a trending mechanism suggests using an ultra-low-power radio receiver to serve for permanent idle listening, while the rest of the WSn's peripherals are in sleep state. This radio is known as the WuRx. It is mainly characterized by the power demand (order of μW), sensitivity and latency. Several tradeoffs exist between these parameters, and they set the final performance. A WuRx is meant to continuously monitor the channel, waiting to receive a triggering event referred to as a wake-up packet (WuPt). The latter contains information about the transmitter and mainly the destination address. It is sent only when the transmitter has to communicate with the receiving node. When a WuPt is received, the WuRx checks if the destination address correlates and then issues an interrupt command to the MCU. The receiver radio is activated to establish the link with the transmitter. For a duty-cycled radio to have a similar energy consumption of a WuRx, the duty cycle should be very low (i.e., 0.01%). As a result, the latency is significantly increased.

The challenging part of the design of a WuRx lies in enhancing its sensitivity while preserving the same power budget. In the literature, WuRxs based on several architectures have been introduced. The design diversity can be classified in terms of complexity of architecture and the applications targeted applications. Some WuRx are designed to consume a few nW, featuring low sensitivity so as to be deployed in low-range applications (i.e., WBAN). Highly sensitive WuRx are also proposed and demand more power, since more active parts and more complex architecture is required. In the following chapter, different WuRx designs introduced in the literature are elaborated and compared together.

2 System implementation

The design process of the WuRx depends on the architecture chosen, power consumption awareness and sensitivity improvement. The introduced WuRx is based on passive detection. The different blocks are illustrated in Figure 1. A baseband amplifier

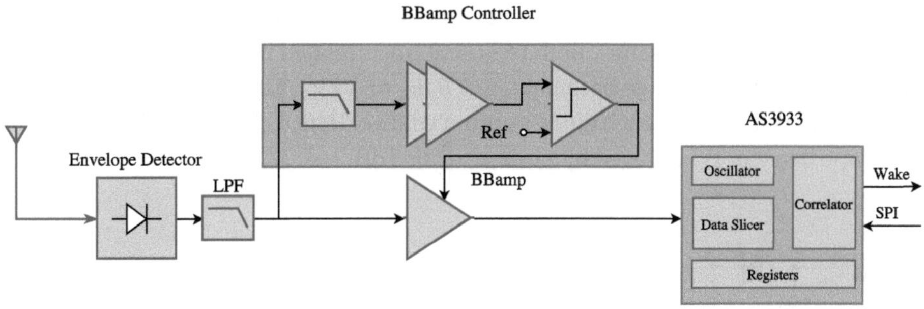

Fig. 1: Wake-up receiver block diagram.

(BBamp) amplifier boosts the output voltage of envelope detector and feeds it to the decoder. The BBamp controller sets the duty of the BBamp with dependance on the WuPt arrival. In this section, the design analysis of all blocks of WuRx are discussed.

2.1 Envelope detector

The performance of the WuRx is limited by the envelope detector characteristics. The latter relies on the zero-bias Schottky diodes HSMS-2852 [1]. They provide fast switching and are optimized for small-signal handling of less than −20 dBm with an input signal frequency below 1.5 GHz. Detector diodes act as square law detectors for low-level signals. This means that the output voltage is proportional to the square of the RF voltage at the diode junction. The tangential signal sensitivity (TSS) is the lowest input signal power level P_{TSS} in watt, for which the detector will have an 8 dB signal-to-noise (SNR) ratio at the output V_{det} of a single diode detector; P_{TSS} can be calculated as follows:

$$P_{TSS} = \frac{2.5\sqrt{4kTR_V B_V}}{\gamma}\,, \quad (1)$$

where T is the temperature in kelvin, k is Boltzmann's constant, R_V is the video resistance in Ω, B_V is the bandwidth in Hz and γ is the voltage sensitivity in V/W. At 2 MHz of signal bandwidth B_V, TSS = −57 dBm at room temperature. From (1), it is clear that a lower signal B_V results in a lower detected power [2]. The detector performs an RF-to-DC conversion from an on-off-keying (OOK) modulated signal. TSS degrades eventually with the increase of the detector's noise floor. The root-mean-square (RMS) noise V_n [2] generated by a single diode is given by:

$$V_n = \sqrt{4kTB_V R_V}\,, \quad (2)$$

At the square-law region, the detection law obeys the relation in (3).

$$V_{det} = \gamma P_{in} \quad (3)$$

A voltage detector with two diodes, where the output voltage is $V_{out} = 2V_{det}$, can be seen as two resistors in series. Both represent uncorrelated noise sources. Therefore, the total RMS noise voltage becomes $\sqrt{8kTB_vR_v}$ or $\sqrt{2}V_n$. Coherently, the final rectified voltage is the sum of each diode's output voltage. Hence, the SNR of the two-diode envelope detector is improved by $2/\sqrt{2} = \sqrt{2}$ or 3 dB.

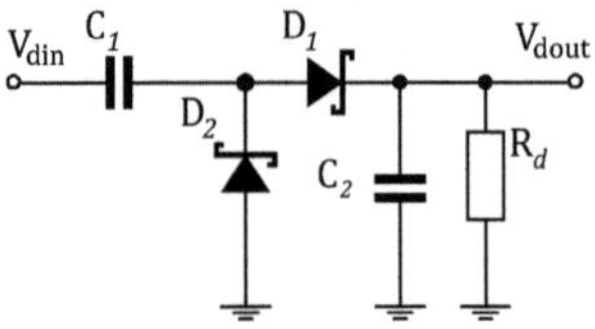

Fig. 2: Envelope detector in a voltage multiplier configuration.

A Greinacher voltage multiplier configuration is employed for the proposed WuRx. The chosen band is the license-free European 868 MHz. The two diodes are in shunt at RF, so the equivalent input impedance Z_{in} is halved. Impedance matching circuits will, thus, be easier to design. The input impedance is simulated $Z_{in} = 31.8 - j358.2\,\Omega$. An LC matching network precedes the diodes for maximum power transfer with the antenna. The resultant input return loss S_{11} using scattering parameter simulation is shown in Figure 3.

Placing a low noise video amplifier after the output of the envelope detector boosts the voltage level of the extracted envelope.

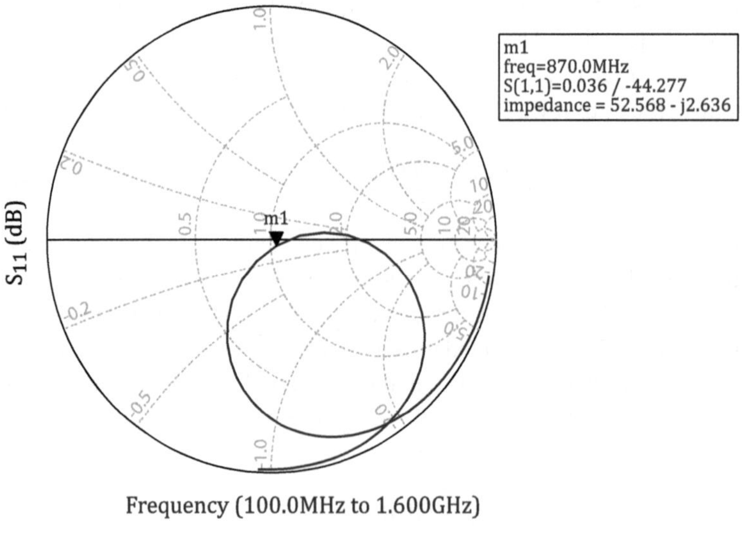

Fig. 3: Simulated envelope detector reflection coefficient.

2.2 Baseband amplifier

In this WuRx, the digital baseband is handled by the low frequency WuRx AS3933 [3]. It performs within the range of 15 kHz to 150 kHz. The minimum voltage sensitivity is $80\,\mu V_{RMS}$. For the envelope detector to output an equivalent voltage of $V_{dout} > 80\sqrt{2}$, the input power P_{in} must be $P_{in} > -58.5$ dBm. The BBamp is necessary to boost the voltage V_{dout} with enough gain G_v. From (1) and (2), G_v should satisfy the following relation:

$$G_v \geq \frac{32\sqrt{2}}{\sqrt{V_n^2 + V_a^2}}, \tag{4}$$

where V_a denotes the RMS noise voltage of the total circuit that follows the envelope detector. Excessive gain must be avoided not to saturate the amplifier at high input RF power. Moreover, a low noise operational amplifier (opAmp) and low offset voltage V_{os} must be used to achieve an adequate performance. While an instrumentation amplifier offers better noise immunity because of the differential inputs, it generally consumes more power than a single opAmp. The signal frequency of V_{dout} is $f_{bb} = 17$ kHz. For an opAmp to handle such a signal with the required G_v, a gain-width product (GBP) of at least $\text{GBP} \geq f_{bb} G_v$ is necessary.

Another requirement is a very low input-referred RMS voltage noise e_{n1}. The selected opAmp is the LMV641 [4]. It features GBP = 12 MHz, a low spectral noise density of $17\,\text{nV}/\sqrt{\text{Hz}}$ and a typical voltage offset $V_{os} = 30$ mV at room temperature. The BBamp is set in a non-inverting configuration. One critical downside of the selected opamp is its current demand of $I_{bb} = 158$ mA at 3 V. Lower power amplifiers will either have a higher input noise or lower GBP. To reduce the current consumption, the BBamp is only activated in an on-demand fashion. This is controlled by a different block of the WuRx that enables the BBAmp upon reception of a WuPt. The BBamp controller is discussed in the next section.

2.3 BBamp controller

The BBamp consumes more energy than required by the WuRx. As an energy saving procedure, it is initially disabled and activates only upon the reception of a WuPt. This is maintained by means of a logic block that extracts a high/low signal from the incident WuPt, which is fed to the VCC/enable pin of the BBamp. The triggering signal should have a time duration t_{Active} to allow the BBamp enough time to process the WuPt. The circuitry mainly consists of a nano-power opAmp to add gain to the received signal plus a comparator to digitize it. The idea is similar to what is presented in [5], where the amplifying stage is based on two opAmps, consuming 330 nA per channel. The latter emphasizes a noise spectral density of $265\,\text{nV}/\sqrt{\text{Hz}}$ at an input frequency of 100 Hz [6]. In this work, the 3 dB bandwidth is limited to 100 Hz. This translates to an equivalent input-referred RMS noise voltage of $e_{n2} = 2.5$ mV. The first stage will be an active low-pass filter, while the second stage is the gain block to further boost the signal. Unlike the BBamp, a much higher gain is needed to overcome the threshold of the comparator input. The ISL28194 has a GBP of 3.5 kHz, which sets the gain to be 35 for an input signal frequency of 100 Hz. A nano-power comparator decides to issue a digital signal by comparing the output signal from the opamps to a certain threshold. It depletes a current of only $I_{Cmp} = 75$ nA at 3 V.

The structure of the WuPt is essential, as it is dynamically correlated with the BBamp controller. It is chosen according to the requirement set by the decoder. A WuPt sample waveform is shown in Figure 4. The transient simulation illustrated in Figure 4 shows the different slope obtained at the output of the BBamp controller opamps.

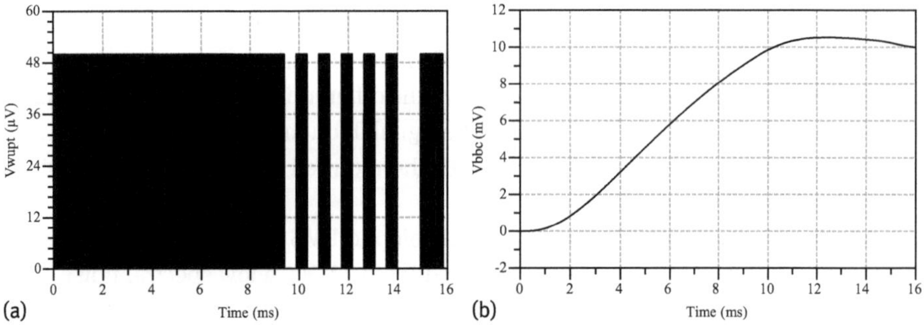

Fig. 4: The BBamp controller waveform during the reception of a WuPt. (a) WuPt sample at 17 kHz carrier frequency; (b) output of the BBamp controller opamps.

2.4 The decoder

In this WuRx, AS3933 is chosen to perform WuRx's identity (ID) correlation. It handles three channels with a carrier frequency f_{bb} ranging between 15 kHz and 150 kHz. The decoder features a duty cycle (on/off) mode to save more energy at the cost of a slightly longer preamble. Within this mode, the AS3933 enables all three channels simultaneously for a duration of 1 ms. Afterwards, it turns them off for a maximum delay of 8 ms. The decoder is able to decode a 16/32 bit OOK modulated wake-up pattern. The WuPt structure should comply with that of the decoder, as described by the manufacturer. The sequence contains a carrier burst, separation bits, a preamble and a user-defined pattern. In the case of 32-bit ID, the WuPt must be Manchester encoded. The AS3933 incorporates two configurable registers with a total capacity of 16 bit to store the ID pattern. When those bits represent Manchester symbols, the pattern is 32-bit in length, since 1 Manchester symbol is a 2-bit representation. With this configuration, there are $2^{16} = 65{,}536$ unique patterns. For a carrier frequency $f_{bb} = 17$ kHz, the two following relations are mainly used to determine the exact timings of the WuPt sequence, as to fulfill the AS3933 protocol rules,

$$f_{\text{RCClk}} = 1.75 f_{\text{bb}} \quad (5)$$

$$\frac{92}{f_{\text{RCClk}}} + \frac{8}{f_{\text{bb}}} < t_{\text{C}} < \frac{155}{f_{\text{RCClk}}}, \quad (6)$$

where f_{RCClk} is the clock frequency and t_{c} is the carrier burst time slot. From (5) and (6), the required value $t_{\text{C}} = 5$ ms is chosen in the current design with $f_{\text{RCClk}} = 31.5$ kHz. Moreover, a 1-bit representation is a successive number N_{C_T} of on/off periods of the carrier signal, which has to last a user-defined number N_{CLK_T} of clock periods. The same principle applies for a 0-bit, but only with the complete absence of carrier (off periods). The internal RC oscillator is used to generate the clock signal. The AS3933 supports bit rates within the range of [1, 8] kbit/s. According to the design requirements, AS3933 is programmed to perform an address decoding at a bit rate of 4 kbit/s. The decoder uses clock signals to help frequency and bit detection. The number of clock periods varies depending on the bit rate chosen, and in this case, $N_{\text{CLK}\_T} = 7$. Conversely, the carrier has to last a number of on/off periods N_{C_T} equal to seven clock periods. The relation is calculated in the following expression:

$$N_{\text{C}\_T} = \frac{f_{\text{bb}}}{f_{\text{RCClk}}} N_{\text{CLK}\_T} \quad (7)$$

A WuPt, which complies with the above settings, is illustrated in Figure 5 where $f_{\text{bb}} = 17$ kHz, $N_{\text{C}\_T} = 4$, and the pattern is `0xD669`. The total WuPt duration is $t_{\text{WuPt}} \approx 25$ ms. The current consumption of AS3933 is measured as 1.7 µA.

The WuPt should be preceded with additional carrier burst to compensate against the slow rising time of BBamp controller.

2.5 Wireless module

The host used is the CC430 System-on-Chip [7]. The chip integrates an MCU and a radio core. It emphasizes different low power modes (LPM), where the lowest, LPM3.5, consumes around 1 µA with memory retention. The radio can perform in either 433 MHz band or 868 MHz to 915 MHz. The radio core is used as a wake-up transmitter (WuTx), since the CC430 is able to handle amplitude-shift keying (ASK) and OOK modulation schemes. The packet transmission through a 64-byte first in first out (FIFO) buffer is not enough to handle the transmission of the complete WuPt. Alternatively, an infinite packet length mode is necessary to be able to send a complete WuPt. This needs to work on the FIFO by refilling it byte by byte until the WuPt is transmitted. The standard packet structure is shown in Figure 6.

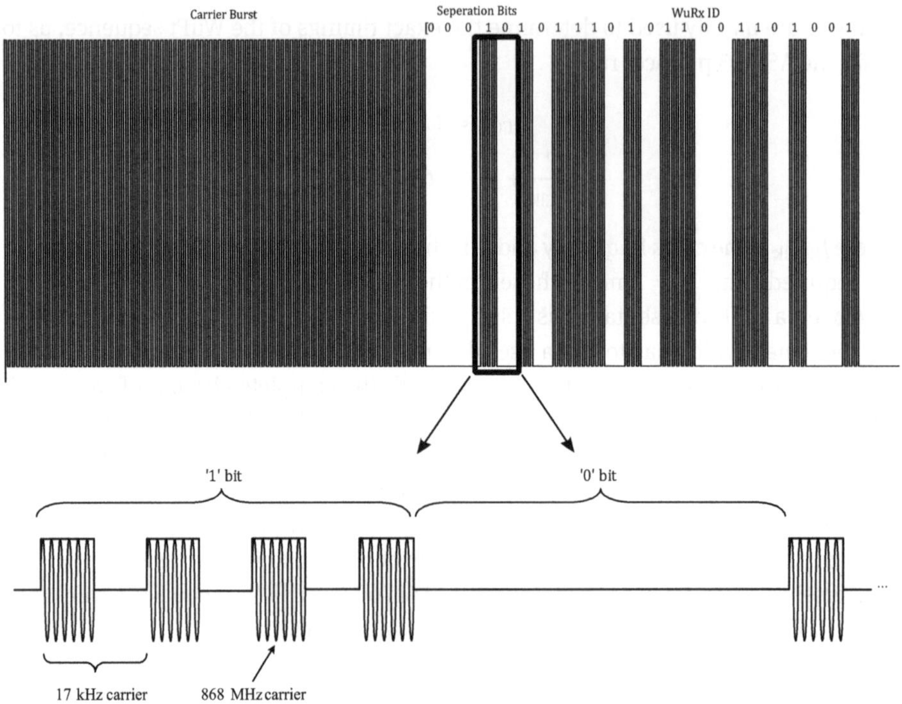

Fig. 5: AS3933 Wake-up packet structure.

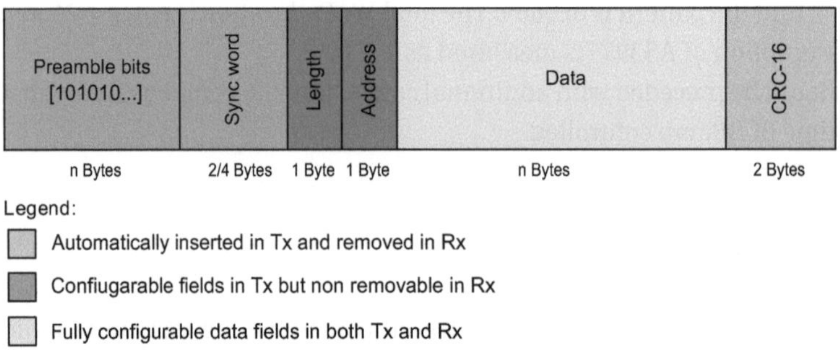

Fig. 6: CC430 packet structure.

Preamble, synchronization, length and address fields are set to a value of 0xAA to represent the carrier burst of WuPt. The data field contains the rest of the carrier burst, separation bits, preamble and pattern bytes. The code redundancy check (CRC) is disabled, and its corresponding field contains 0x0 bytes. The CC430 is able to transmit a packet with transmission power up to 11 dBm.

3 System evaluation

In this section, a proof of concept, for the presented WuRx design is carried out for experimental evaluation. The CC430 and the WuRx are assembled on a single printed circuit board (PCB), as shown in Figure 7.

Fig. 7: Wireless sensor node embedded with the proposed WuRx (33.5 × 20.5 mm).

3.1 Measurements

The manufactured PCB is a four-layer 0.8 mm FR4 substrate. The PCB characteristics directly affects the impedance of the transmission lines. The effects are taken into account to realize the circuitry. To validate that, the WuRx is matched to 50 Ω; the reflection coefficient S_{11} is measured using the Agilent N9914A. A matching network trimming was necessary to counter the realization imperfections. Figure 8 shows the corresponding S_{11}. $S_{11} < 10$ dB means that less than 10% is reflected back to the antenna.

Changing the operation frequency of the WuRx within 867 MHz to 870 MHz does not alter the performance of the circuit. Nevertheless, it is found to be of benefit, as the frequency can be changed freely within a certain bandwidth. An example could be the use of one channel for WuRx and another channel for normal communication between nodes to avoid collision or interference.

Furthermore, to test the successful reception of WuPt, the WuTx is configured to send 100 WuPts. A visualization of the received WuPt waveform can be seen on an oscilloscope (Figure 9).

Measuring the sensitivity of the WuRx is done by connecting an attenuator between two sensor modules (Figure 10), where one is the WuTx and second has the

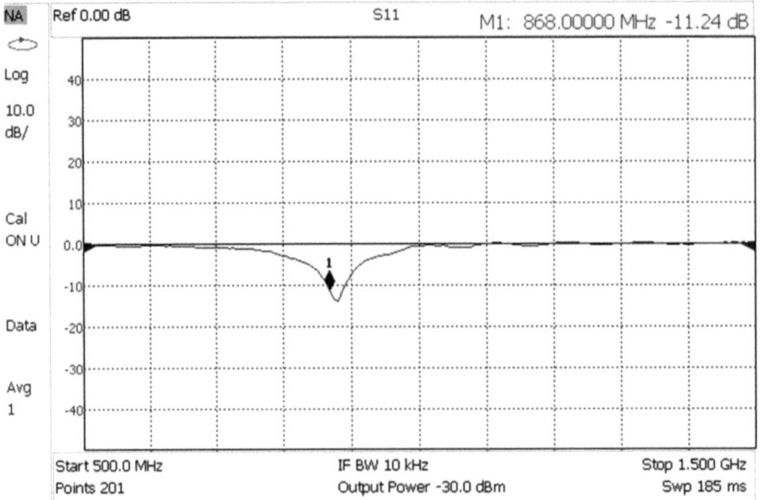

Fig. 8: Measurement of reflection coefficient (S_{11}) at the input of the WuRx.

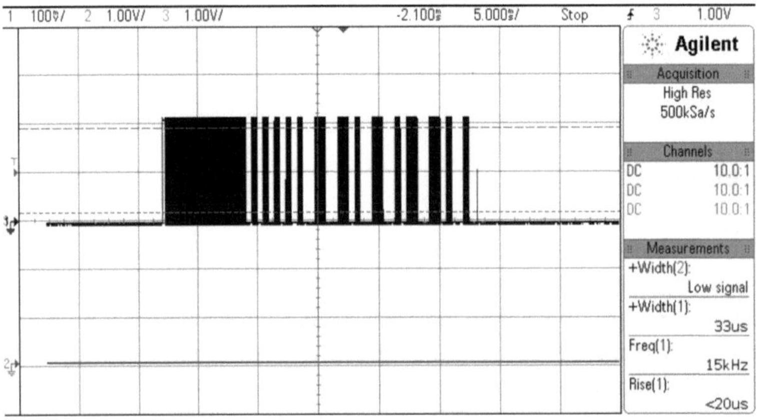

Fig. 9: Oscilloscope capture of WuPt at the output of the envelope detector.

WuRx active. The WuTx is set with a transmission power of –20 dBm. The decoder issues an interrupt when WuPt is received and the address matches the WuRx's ID. With every attenuation step, the number of successful interrupts are counted. At 41 dB of attenuation, the packet error rate (PER) starts to increase. At –61 dBm, PER = 10^{-2}, that is one missed packet out 100 transmitted ones.

Furthermore, Table 1 summarizes the current consumption of the WuRx parts for different operation modes, including the host node. The WuPt power consumption is crucial when decoding a WuPt.

The current WuRx design has dynamic energy consumption that depends on the number of received WuPt. The highest consumption is dominated by the BBamp. Let

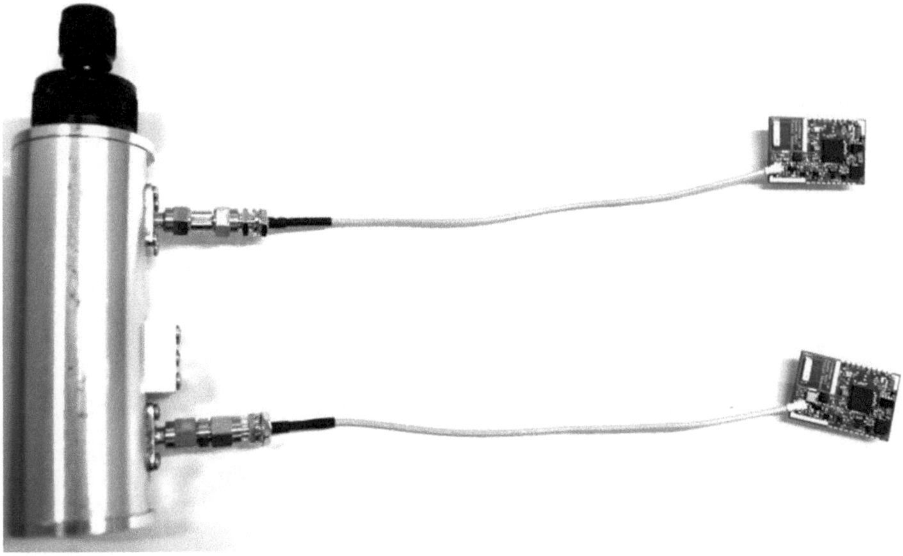

Fig. 10: Assembled WuRx PCB attached to a host node.

Tab. 1: Current consumption of WuRx and Panstamp in different states.

Components	Channel monitoring [μW]	WuPt decoding [μW]
Envelope detector	0	0
BBamp controller	2.19	2.2
BBamp	0	474
AS3933	5.1	24
Total	7.29	500.2

β be the duration of the communication cycle within the network. The average power consumption P_{av} of the WuRx can be calculated as follows:

$$P_{av} = \frac{(\beta - N_{WUPT}t_{dec})P_{lis} + t_{dec}N_{WUPT}P_{dec}}{\beta}, \tag{8}$$

where N_{WUPT} denotes the number of WuPt received in β, t_{dec} is the time needed for decoding, P_{lis} and P_{dec} are the power consumption during listening and decoding, respectively. Assuming all WuRx peripherals have no delay in between when switching from the listening to the decoding state, the corresponding power parameters are $P_{dec} = 500.2\ \mu W$ and $P_{lis} = 7.29\ \mu W$. From the relation (8), P_{av} can be plotted against N_{WUPT}, where $\beta = 1$ d is chosen arbitrarily.

In Figure 11, P_{av} is plotted with different t_{dec} values. For a 10^3 received WuPts packets in one day, the average power consumption is less than 10 μW. The interferences of unwanted bands can trigger BBamp, resulting in a higher energy consumption. A sur-

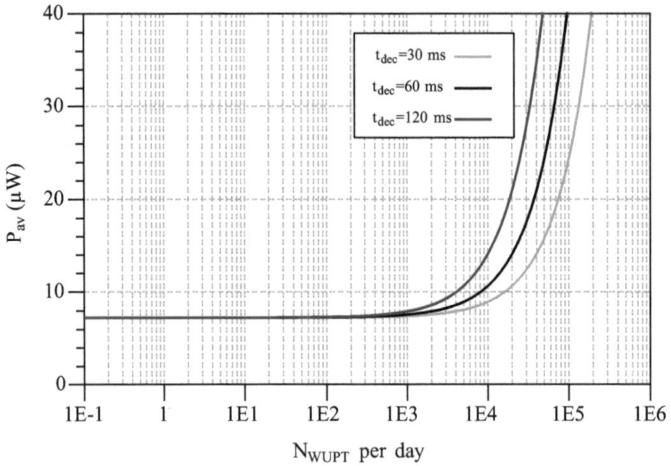

Fig. 11: Average WuRx power consumption against the number of received packets per day and for different decoding periods t_{dec}.

face acoustic wave (SAW) filter helps rejecting those bands at the price of increased insertion loss. The sensitivity of the WuRx plays a key role in extending the wireless coverage of the network. However, antenna gains and path losses also impact the link distance R. The relation between these parameters is expressed in the Friis equation (equation (9)),

$$\frac{P_r}{P_t} = G_r G_t \left(\frac{\lambda}{4\pi R} \right)^2 \tag{9}$$

where P_r, P_t are, respectively, the received and transmitted signal power; G_r and G_t represent the antenna gains of the receiver and the transmitter, respectively. However, such a relation stands in an ideal condition where reflections, diffractions, atmospheric absorption are not present. Therefore, the communication range is significantly different between indoor and outdoor applications. A practical range test of the network is realized in an indoor line-of-sight to measure the maximum possible coverage. The test was performed inside the Electrical Engineering (EE) Department of HTWK Leipzig to witness the impact of different wave propagation effects. As shown in Figure 12, WuTx is fixed at a height of 2 m, and WuRx is reallocated until no WuPt is received. The distances are measured with an accuracy of ±1 mm, due to the laser rangefinder.

The antennas used are omnidirectional monopole antennas with a gain of 1 dBi. At 8 dBm a received WuPt is observed at a maximum distance of 74 m. In an open space environment, longer distances can be achieved. Accordingly, this is lower than the equivalent distance calculated from equation (9). Furthermore, characteristic WuRx parameters are summarized in Table 2.

Fig. 12: Range test at the EE Department of HTWK.

Tab. 2: Summary of wake-up receiver specifications.

Specification	Value	Condition
Sensitivity	−61 dBm	4 kbit/s, PER=1%, 10 ms latency
Input impedance	(31.8−358.2i) Ω	Simulated
Wake-up pattern	32-bit	Manchester
Bit rate	1–4 kbit/s	Decoder bit rate
Supply voltage	2.85–83.6 V	
Current supply	2.43 μA	Channel monitoring
	166.73 μA	WuPt decoding

4 Comparison with the literature

The WuRxs with the lowest power consumption are based on a passive RF front end. For instance, passive radio frequency identifiers (RFID) serve as WuRxs with very low sensitivity and zero power consumption. They rely on the energy extracted from the incident wave received from the antenna. An RFID-based WuRx is presented in [8]. Using seven stages of voltage multiplier circuit by means of Schottky diodes [1], the WuRx is completely passive. The resulting range is less than 10 m at 13 dBm transmission power. Here, the WuRx also acts as an RF energy harvester, so that it powers the MCU and transmits back the tag's information. Therefore, the received WuPt should be packed with more power. Yet, the diodes feature a sensitivity of −57 dBm at 2 MHz bandwidth, which enables them to detect a signal at more than 30 m distance with the same transmission power. This coverage is considered insufficient for most applications of WSN. The same diodes construct a WuRx by means of active elements, so as to enhance the overall sensitivity.

The authors of [9] introduced a WuRx with a digital baseband consuming 1.2 μW when monitoring for the wake-up packet (WuPt) while providing −55 dBm of sensitivity. The decoding process is done by an MCU. The latter switches from sleep state to active state, consuming 63 μW. The MCU is only triggered when a WuPt with more than a pre-configured length is received. Although it reduces the WuRx's vulnerability against interferences, it adds more latency. Another passive front-end-based WuRx is proposed by [5]. The sensitivity is enhanced by placing a post-amplifier right after the envelope detector. The WuRx incorporates a mechanism that extracts a logic signal from the received packet. This is used to turn on/off the amplifiers. The digital backend is handled with a low-frequency wake-up receiver AS3933 [3] to correlate a 16-bit address pattern. The total power consumption in listening is only 7.5 dBm and features a sensitivity of −60 dBm. For practical considerations, the WuRx is susceptible to interferences, which trigger the post-amplifiers, causing increased energy consumption.

In different WuRx designs, a low noise amplifier (LNA) can be placed before the envelope detector to boost the incoming signal. This architecture is classically known as tuned-RF (TRF). It is still used in many radio systems due to its simplicity. However, increasing sensitivity will require large gain to overcome the noise figure (NF) of the following envelope detector. This requires a significant amount of power. In [10], the authors applied a duty cycle of 0.6% on the TRF-based WuRx. It consumes an average power of 8.5 μW for 8.1 ms and features a sensitivity of −75 dBm. When the WuRx switches to decoding mode, the power demand rises up to 1.1 mW. The monitoring mechanism used to detect a WuPt is a simple preamble sequence (i.e., '0101 ...') detector. Unfortunately, this can be easily triggered from surrounding interferences and result in higher energy usage. With TRF architecture, the effort is mainly spent on reducing the consumption of gain blocks.

For super-heterodyne (SH) architecture, the amplifier, mixer and post amplifiers contribute in boosting the signal. A SH receiver features an increased selectivity and performance comparing to TRF architecture. Yet, it requires accurate local oscillator (LO) with high current demands. For WuRx, SH can probably be the most complex architecture. To name a few, the authors of [11] apply 270 ns of duty to the power-hungry blocks of the SH WuRx. This results in an average power demand of 3 μW at 64 bit/s. At 100% duty, the receiver demands 27.5 mW. [12] proposed an RF front end of a similar WuRx, consuming 8.25 mW. The reduction of current of those SH WuRxs relies mainly on the duty cycle and fast sampling. In [11], the sensitivity is reported to be −83 dBm. A different approach to enhancing sensitivity is employed in super-regenerative radio (SRR) receivers by means of positive feedback. The more interesting features about SRR lies in its energy consumption (generally less than 1 mW). An SRR-based WuRx is introduced by the authors of [13]. Featuring an excellent sensitivity of −97 dBm, the WuRx incorporates a complex programmable logic device (CPLD) for the decoding process. While the RF front end consumes only 20 µW, the CPLD demands as much power, which is considered higher than other reported digital baseband circuits [14]. In [15], a duty-cycled WuRx based on SRR is presented. With a reported sensitivity of −90 dBm and a power demand of only 1 µW at 1 kbit/s, the WuRx probably features the most enhanced power/sensitivity tradeoff. However, the evaluation omits crucial evaluation details of the decoding mechanism. Therefore, the WuRx lacks empirical measurements to prove the expected performance. The designed WuRx has a minimally increased complexity. Its features are compared to those of recently published ultra-low-power radios and provided in Table 3.

Tab. 3: Comparison of WuRx prototypes.

	This work	[9]	[13]	[11]	[16]
Frequency [GHz]	0.868	0.868	0.868	0.868	2
Power [µW]	7.2^{α}	1.2	40	3–86.7	52
Sensitivity [dBm]	−61	−55	−97	−83	−72
Data rate [kbps]	4	–	1.25	0.06–08	100
Addressing	√	√	√	√	–
Implementation	OtS*	OtS	OtS	130 nm	90 nm

* Off-the-Shelf
$^{\alpha}$ N_{WuPt}/day < 10^4

5 Conclusion

An ultra-low power wake-up receiver for wireless sensor networks is presented. The chapter details the different design steps. The architecture is based on an on-demand duty, where the power-hungry part within the WuRx activates at the reception of the WuPt. This makes it possible to enhance the sensitivity of the WuRx while remaining in the micro-power budget as long as the it operates in low traffic event-triggered applications. The sensitivity achieved is −61 dBm, while consuming 7.2 µW in channel monitoring mode. There are 65,535 possible unique patterns that are configurable with Manchester encoding. At 8 dBm of WuPt transmission power, a communication range of nearly 74 m in the line-of-sight scheme is observed. The WuRx allows a low latency and an always reachable wireless sensor node whilst preserving the battery's lifetime.

Bibliography

[1] Avago Technologies. HSMS285x-S., datasheet. http://www.avagotech.com/docs/AV02-1377EN, 2009.

[2] SkyWorks Inc. Mixer and Detector Diodes. http://www.skyworksinc.com/uploads/documents/200826A.pdf, 2008.

[3] Austrian Mikro Systeme. 15–150-KHz AS3933 3D Low Frequency Wakeup Receiver, datasheet. http://ams.com/eng/Wake-up-receiver/AS3933, 2010 [Revised 2015].

[4] Texas Instruments. LMV641, 10 MHz, 12 V, Low Power Amplifier, datasheet. http://www.ti.com/product/lmv641, 2007 [Revised 2016].

[5] S. Bdiri, F. Derbel, and O. Kanoun. An 868 MHz 7.5 μW Wake-up Receiver with −60 dBm Sensitivity. *Journal of Sensors and Sensor Systems*, 5(2):433–446, 2016.

[6] Intersil. ISL28194, Ultra-Small, 330nA and 1μA Single Supply, Rail-to-Rail Input/Output (RRIO) Op Amps, datasheet. http://www.intersil.com/content/dam/Intersil/documents/isl2/isl28194-95.pdf, 2009.

[7] Texas Instruments. CC430F5137, Ultra-Low-Power MCU, datasheet. http://www.ti.com/lit/ds/symlink/cc430f5137.pdf, 2012 [Revised 2016].

[8] K. Kaushik, D. Mishra, S. De, K. R. Chowdhury, and W. Heinzelman. Low-cost wake-up receiver for RF energy harvesting wireless sensor networks. *IEEE Sensors Journal*, 16(16):6270–6278, 2016.

[9] M. Magno, V. Jelicic, B. Srbinovski, V. Bilas, E. M. Popovici, and L. Benini. Design, implementation, and performance evaluation of a flexible low-latency nanowatt wake-up radio receiver. *IEEE Trans. Industrial Informatics*, 12(2):633–644, 2016.

[10] D. Y. Yoon, C. J. Jeong, J. Cartwright, H. Y. Kang, S. K. Han, N. S. Kim, D. S. Ha, and S. G. Lee. A new approach to low-power and low-latency wake-up receiver system for wireless sensor nodes. *IEEE Journal of Solid-State Circuits*, 47(10):2405–2419, 2012.

[11] H. Milosiu, F. Oehler, M. Eppel, D. Fruhsorger, S. Lensing, G. Popken, and T. Thones. A 3-μW 868-MHz wake-up receiver with −83 dBm sensitivity and scalable data rate. In *2013 Proceedings of the ESSCIRC (ESSCIRC)*. Institute of Electrical & Electronics Engineers (IEEE), sep 2013.

[12] C. Tzschoppe, R. Kostack, and F. Ellinger. A 2.4 GHz fast settling wake-up receiver frontend. In *2014 10th Conference on Ph.D. Research in Microelectronics and Electronics (PRIME)*, pages 1–4, 2014.

[13] J. Petäjäjärvi, K. Mikhaylov, R. Vuohtoniemi, H. Karvonen, and J. Iinatti. On the human body communications: wake-up receiver design and channel characterization. *EURASIP Journal on Wireless Communications and Networking*, 2016(1):1–17, 2016.

[14] N. S. Mazloum, J. N. Rodrigues, O. Andersson, A. Nejdel, and O. Edfors. Improving practical sensitivity of energy optimized wake-up receivers: proof of concept in 65 nm CMOS. *CoRR*, abs/1605.00113, 2016.

[15] M. Eppel, H. Milosiu, and F. Oehler. A novel 1 μw super-regenerative receiver with reduced spurious emissions and improved co-channel interferer tolerance. In *2016 IEEE Topical Conference on Wireless Sensors and Sensor Networks (WiSNet)*, pages 85–88, Jan 2016.

[16] N. M. Pletcher, S. Gambini, and J. M. Rabaey. A 2 GHz 52 μW wake-up receiver with −72 dBm sensitivity using uncertain-IF architecture. *Digest of Technical Papers - IEEE International Solid-State Circuits Conference*, 51:524–526, 2008.

Part V: **System design and applications**

Dhouha El Houssaini, Sabrine Khriji, Kamel Besbes,
and Olfa Kanoun

Wireless sensor networks in agricultural applications

Abstract: Technological development in wireless sensor networks (WSNs) has made it possible to build up decision support systems for many real world problems. Agricultural applications are considered as one of the most interesting fields, which presents an increasing demand for decision making and optimization systems. WAN-based solutions are used in agriculture to reduce cost and improve productivity by enabling real-time control of changes in the field, animals and crops states and machinery maintenance possibilities. This chapter highlights the importance of WSNs in agriculture and their use in controlling agricultural fields for farmers. As an example of a WAN application in agriculture, an automated irrigation system is presented. The system developed is based on the use of specific sensors and actuators to enable a real-time control of the irrigation task.

Keywords: Wireless sensor network, agricultural monitoring, deployment scenarios, wireless communication, platforms, irrigation system, soil analysis, weather control, sensors

1 Introduction

Nowadays, the agricultural domain is becoming more dependent on new technologies in a concern to improve the traditional methods of farming [1]. Different new technologies have been adopted in agricultural applications to improve productivity. Technologies such as geographical positioning system and WSNs are based essentially on the use of micro-mechanical components to control input parameters. WSNs enable a real-time monitoring system based on assembling and processing all kinds of data with a non-habitual spatial and temporal resolution. Basically, WSNs appeared in the 1980s and were defined as a collection of spatially-distributed, autonomous and specialized transducers with a communications infrastructure for monitoring and recording the physical conditions of the environment at different locations [2]. The wireless solution enables the use of small and cheap sensors, making the de-

Dhouha El Houssaini, Sabrine Khriji, Olfa Kanoun, Chair for Measurement and Sensor Technology, Technische Universität Chemnitz, 09126, Germany
Dhouha El Houssaini, Kamel Besbes, Centre for Research on Microelectronics and Nanotechnology, Technopark of Sousse, 4050, Tunisia
Kamel Besbes, Microelectronics and Instrumentation Laboratory, Faculty of Science of Monastir, 5019, Tunisia

https://doi.org/10.1515/9783110445053-019

ployment phase much easier in different scenarios, such as animal tracking, forest control, crop and yield monitoring, etc. Moreover, WAN offers self-controllable tools able to maintain full and complete control of the targeted network. Also, the use of different sensor nodes for different tasks reduces the number of employed staff, which implies a reduction in expenses.

Generally, a network contains several detection stations, called sensor nodes. Every sensor node is composed of a transducer, micro-controller, transceiver and power source. The transducer converts the physical effects and phenomena to electrical signals, while, the micro-controller processes the data and store it. The transceiver receives and transmits the data from and to the base station. All components are battery powered with a limited energy. Wireless communication protocols are used to send and collect sensed data instead of the usual wired solutions. This provides an 80% reduction in communication and connectivity expenses compared to wired solutions. Expenses can also be reduced by the simplicity of deployment, usage and maintenance [3].

WAN employs different and heterogeneous sensor nodes, which enables a large range of applications due to the easier deployment and better flexibility. Several monitoring systems are available and based on WAN designed for healthcare, environmental, industrial and military applications, etc. One of the most important fields of application is agricultural applications.

2 Applications of wireless sensor networks

Sensor nodes are used for continuous data sensing, event detection, location determining, and local monitoring of actuators, which ensure a wide range of applications for sensor networks. Actually, WSNs can be deployed in various environments where a wired network is impractical or even impossible to employ, like in dense forests, volcanoes, mountains, moving cars, etc. [1]. Features such as reliability, flexibility, self-organization, small size, energy efficiency and ease of deployment, mean that sensor nodes can be used for a wide range of applications such as military tasks, industrial process monitoring, environmental control, precision agriculture, healthcare, home automation, vehicle networks and intelligent transportation systems, etc.

Environmental monitoring: WSNs are useful for environmental monitoring, such as the movement of animals and monitoring environmental conditions such as landslides, temperature, humidity, etc. Also, it is useful for the atmospheric context, e.g., forest fire detection, meteorological research, flood detection and pollution. One of the most important examples of sensor nodes in environmental monitoring is precision agriculture. In agriculture, WSNs could be employed to monitor, for example, pesticide levels in drinking water, the level of soil erosion and that of air pollution in real time.

Military applications: WSNs have been an essential part of the military field in areas such as command, control, communications, computing, intelligence, surveillance, reconnaissance and tracking systems. Several examples of military applications of sensor networks may be cited. Some of the most important are monitoring friendly forces, equipment and ammunition, reconnaissance of opposing forces and terrain and tracking of enemies.

Healthcare monitoring: Sensor nodes have also invaded the medical field by providing interfaces for the disabled, integrated patient monitoring, diagnostics, telemonitoring of human physiological data and tracking and monitoring of doctors and patients inside a hospital. The most important functionality of the wireless node in this domain is the telemonitoring of human physiological data, since it facilitates access to the collected data of patients and stores it in real time. It also allows the monitoring and detection of patient behavior, which, in turn, allows doctors to identify pre-defined symptoms earlier.

Other applications: The features of WSNs allow them to be used in other domains, for example, home applications such as smart homes, industrial monitoring such as control of material fatigue, robot control and guidance, local control of actuators, monitoring product quality. Other applications may be cited, e.g., environmental control in office buildings, monitoring of disaster area such as nuclear and chemical areas, detection and monitoring of car thefts, vehicle tracking and detection, etc.

3 Wireless sensor network deployment scenarios

According to the specificity and requirements of the system to be implemented, the deployment follows certain scenarios. A scenario itself presents many features, such as the field's dimensions, the parameters to be monitored, the distance between one node and another, parameter thresholds, the platform used in the system, the number of nodes to be used and the topology applied. All the terms enumerated above contribute in one way or another in the communication process when dat is transmitted and received data to and from nodes. Indeed, the dimensions of the field, as well as the distance between nodes and the choice of deployment topology intervene, on the one hand, in the data transfer time and latency, and, on the other hand, in the energy consumed to transmit packets from a node to another or to the base station. Therefore, in some cases, there is no need to transmit data frequently, but only if the monitored parameter exceeds certain threshold value. This fact promotes the energy management process during communication to reduce the amount of power consumed. Moreover, the kind of parameter to be measured has an influence, as there are factors that should be supervised during short periods of time, like humidity and others that can be controlled once a week such as the pH level of soil. Accordingly, wireless sensor deployment in agriculture could be classified into two widely used variants [4]: the terrestrial wireless sensor network (TWSN) and the underground wireless sensor network (UWSN).

3.1 Terrestrial wireless sensor networks

WSNs are battery powered systems interconnected and deployed for a specific application purpose [5]. In TWSN, nodes are deployed on the surface of the field, where smart, tiny and low-cost sensors are installed in the network. Such energy-efficient devices empower the sensor nodes to precisely collect the surrounding data. Based on the data sensed, these nodes communicate between each other in a way so as to satisfy the application specification. The nodes communicate among themselves by using radio-frequency (RF), ISM band frequency [6] (such as 902–928 MHz and 2.4–2.5 GHz). Typically, a gateway node is used to collect sensed values in the network in order to analyze the observed variables. Such nodes are powered with both RF and GSM communication modules. The end user, the farmer, is able to monitor the state of the field and execute any necessary intervention by means of in-field sensors and actuators. For example, a user can switch on/off a valve when the water level applied to the field reaches some predefined threshold value. A typical TWSN is presented in Figure 1(a).

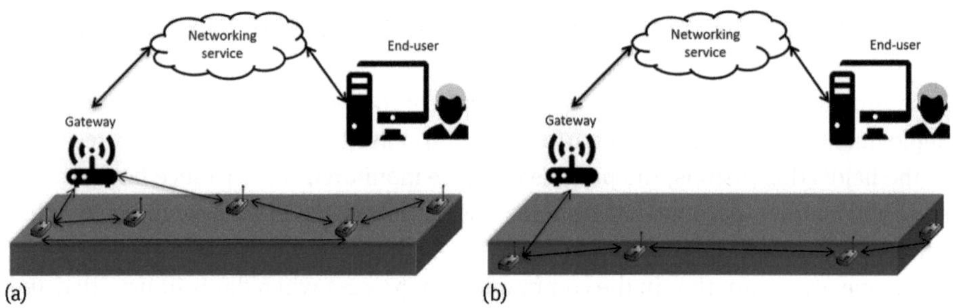

Fig. 1: Deployment scenarios in WSNs: (a) terrestrial and (b) underground.

3.2 Underground wireless sensor networks

Another variant of the WAN is the wireless underground sensor network (WUSN) [7]. In this version, wireless sensors are buried inside soil. Thus, some additional constraints should be considered. In such scenarios, higher frequencies suffer severe attenuation, and comparatively lower frequencies are able to penetrate through the soil [8]. This phenomenon will limit the communication radius of nodes and increase the number of nodes needed in the network to cover a larger area. In such a way, wired sensors are used to increases the network coverage. However, in such a scenario, both sensors and wires used may be sensitive to farming activities such as ploughing. A typical agricultural application based on underground sensor networks is shown in Figure 2(b). Sensor nodes are placed under the ground, and one gateway is deployed in order to transmit the information collected. However, due to the comparatively shorter communication distance, a larger number of nodes is required to be deployed.

To ensure a complete and full control of deployed sensor nodes, different communication technologies are emerged in the agricultural field in order to improve the outcomes and preserve the available resources. A detailed description of wireless communication technologies and the standards used in various agricultural applications is presented in the next section.

4 Wireless technologies and standards

This section describes the different wireless technologies and standards used in agricultural applications. A comparison between these technologies is provided to identify the adequate technology in terms of energy consumption, data rate, communication range and cost.

Zigbee: [9] is a network and application layer protocol based on the IEEE 802.15.4 standard [10]. It is judged to be energy efficient, low cost and reliable, which makes it conform to agricultural and farming application specifications. It supports a small range of about 10 to 20 meters of data communication, decentralized, ad-hoc and mesh networks. Devices supporting ZigBee enable a low-duty cycle, which makes them suitable for agricultural applications, where periodic information updates are required. Nevertheless, ZigBee applications produce a low data rate of 20 to 40 kbps for 868/915 MHz and a rate of 250 kbps for a 2.4 GHz ISM band. Thus, a low hardware specification is required.

Wi-Fi: is a wireless local area network (WLAN) standard for information exchange or wireless connection to the Internet based on the IEEE 802.11 standard [11]. This technology provides a communication range in the order of 20 meters indoors and about 100 meters outdoors with a data transmission range of 2 to 54 Mbps at a frequency of 2.4 GHz. With the possibility to connect heterogeneous architectures over an ad-hoc network it is very useful for agricultural applications where multiple devices can be used.

Bluetooth: is based on the IEEE 802.15.1 standard [12]. This technology is generally known for being a low-power, low cost wireless technology used for communication between portable devices and desktops over a short range (8–10 m). The data rate achieved in various versions of Bluetooth ranges from 1 to 24 Mbps. One advantage of using Bluetooth is that it needs only a very small amount of radio power to work, which makes it perfect for battery powered devices. Another advantage is that it can reduce interference from any other Bluetooth device located nearby.

WIMAX: stands for Worldwide Interoperability for Microwave Access, a wireless communication standard referring to the inter-operable implementations of the IEEE 802.16 standard (IEEE Standard for Local and Metropolitan Area Networks, 2011) family. WiMAX is targeted to achieve a 0.4–1 Gbps data rate on fixed stations, and the maximum transmission range with this technology is 50 km. The Mobile WiMAX (IEEE 802.16e standard) provides data rates in the order of 50–100 Mbps. Also, WiMAX is stated to be energy efficient over the pre-4G long-term evaluation (LTE) and the evolved High-Speed Packet Access (HSPA+) (Deruyck et al., 2010; Louta et al., 2014). The long-range support together with high-speed communication features place WiMAX as the best suitable technology for agricultural applications involving asset monitoring, such as farming system monitoring, crop area border monitoring and real-time diagnostics, such as the remote control of water pumps, lights, gates and remote diagnosis of farming systems.

GPRS/3G/4G: The General Packet Radio Service (GPRS) is a packet service based on GSM communication, and it complements some existing services such as circuit switching cellular phone connections. The GPRS network is often called the 2.5 G network and has been phased out in favor of newer 3G/4G installations. GPRS technology was standardized by the European Telecommunications Standard Institute (ETSI). It supports the Internet (IP) protocol, Point-to-Point Protocol (PPP) and X2.5 connections. GPRS offers a high data transfer, which can rate up to 172 Kbps. It is ten times faster than existing GSM network services. It uses packet switching for data transmission. The data packets are distributed from several terminals in the system across multiple channels. Moreover, GPRS facilitates the usage of Internet applications over mobile networks. It is used to gain access for machinery control and monitoring in some agricultural applications.

LoRa: The Long-Range Radio protocol was introduced by the LoRa alliance as a protocol stack for low power consumption, wide area communication between remote nodes and gateways and secure data transmission. It uses an unlicensed radio spectrum in the industrial, scientific and medical (ISM) bands. This technology presents mainly two layers. A physical layer, also called LoRa RF, maintains direct contact with the outside world over the radio interface, and it includes many parameters such as frequencies, bands, power levels and modulation. It uses worldwide unlicensed frequencies, 868 MHz 915 MHz and 433 MHz. A medium access control (MAC) layer called LoRaWAN is used to manage communication between LPWAN gateways and end node devices. LoRaWAN defines the communication protocol and system architecture for the network, while the physical layer of LoRa enables the long-range communication link. Moreover, LoRaWAN is responsible for the management of data rate, communication frequencies and the power of all connected devices. The basic architecture of LoRaWAN consists of LoRa end devices, LoRa gateways and LoRa servers. A data packet is transmitted by an end-node device and received by the LoRa gateways to ensure the transmission of packets from the LoRa end devices to a centralized network server responsible for packet duplication, security checking and network management. The data received is then forwarded to the application servers.

Tab. 1: Comparison of different wireless communication technologies.

Parameters	Zigbee	WiFi	Bluetooth	WiMAX	GPRS/3G/4G	LoRA
Communication standard	IEEE 802.15.4	IEEE 802.11a,b,g,n	IEEE 802.15.1	IEEE 802.16.a,e	NA	IEEE 802.15.4.g
Frequency band	868/915 MHz, 2.4 GHz	2.4 GHz	2.4 GHz	2–66 GHz	900–1800 MHz	868/915 MHz
Data rate	20–250 Kbps	2–54 Mbps	1–24 Mbps	0.4–1 Gbps 50–100 Mbps	Up to 172 kbps	50 kbps
Transmission range	10–20 m	10–200 m	8–10 m	≥ 50 km	1–10 km	5 km

5 Wireless sensor platforms for agricultural applications

Agriculture is a demanding field, where applied technology should fulfill the requirement for low-power conception, precision, real-time synchronization, reduced man power, reduced environmental impact, small size and low cost. Thus, the choice of the working platform becomes challenging. In this context, many WSNs were designed especially for agricultural applications. Basically, they can be classified according to different features and characteristics. Table 2 presents a comparison of some WAN platforms used in agricultural applications.

6 Taxonomy of agricultural monitoring applications

Agriculture is a growing field where farmers are asked to follow new technological trends in order to improve benefits while reducing expenses. Agriculture is very demanding, since it is a combination of multiple tasks such as planting, cultivating, animal care and tool maintenance. Thus, associating wireless sensor technologies in agricultural tasks will help farmers manage the agricultural system by offering a flexible, real time and easy to use solution.

Based on the targeted application, the deployment of a WAN in an agricultural system should be deeply studied. In fact, many factors should be considered, such as the degradation of weather conditions like rain or fog, the morphological structure of fields or the presence of obstacles. Therefore, a taxonomy of agricultural applications based on WSNs can be deduced (Figure 2). Wireless technology is used to help farmers control their agricultural fields any time and from anywhere in an accurate and source-efficient manner.

Tab. 2: Comparison of some existing wireless sensor platforms used in agriculture.

Features	MICAz [13]	TelosB [14]	IRIS [15]	Imote2 [2]	Waspmote [16]
Processor	ATmega 128L	TIMSP 430	ATmega 128L	Marvell XScalePXA271	ATmega 128L
Processor performance					
System memory (kB)	4	10	4	256	8
I/O connectivity	UART, I2C, SPI, DIO	ART, I2C, SPI, DIO	ART, I2C, SPI, DIO	ART, I2C, SPI, DIO	ART, I2C
Current draw					
Active (mA)	8	1.8	8	31	9
Sleep (μA)	15	5.1	8	390	62
Radio specifications					
Operating frequency (MHz)	2400	2400	2400	2400	2400
Number of channels	programmable	programmable	programmable	steps of 5 MHz	programmable
Data rate (kbps)	250	250	250	250	NA
RF power (dBm)	−240 to 0	−24 to 0	+3	–	1 mW
Sensitivity (dBm)	-90 to −94	-90 to −94	−101	-94	-92
Range (m)					
outdoor	75 to 100	75 to 100	>300	–	90
indoor	20 to 30	20 to 30	>50	–	Up to 30
Additional features					
Embedded sensors	–	Temperature, light and humidity	Light	–	Temperature, luminosity

6.1 Soil preparation system

Soil presents a crucial element for growing plants, since it contains important nutrients responsible for the plant's growth, and it helps in absorbing water. Also, the choice of the adequate spot to be prepared, as well as the suitable type of crop to plant, necessarily depends on the physical characteristics of the soil. Thus, soil preparation offers a detailed study of field productivity, which will help farmers create a strategy to exploit the agricultural field. Kotamäki et al. [17] used a 2000 m^2 covered field to measure, via 70 sensors, soil moisture, air humidity and temperature, wind direction and speed, etc. This WAN tends to collect spatially and temporally accurate data as a real-time application for water and soil monitoring. In the same context, Corke et al. deployed 30 sensor nodes spaced 10 m from each other. The transmission occurs every half an hour in a clustering topology, providing information about the pH of the soil,

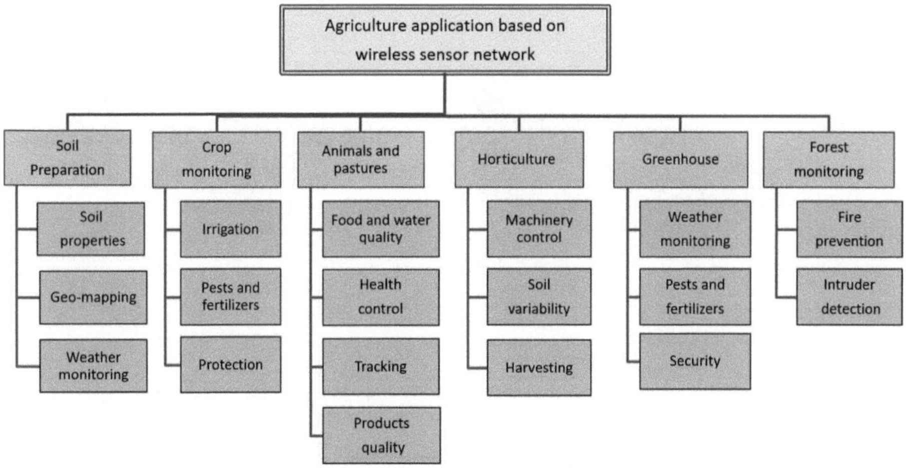

Fig. 2: Taxonomy of agricultural applications based on WSNs.

temperature, moisture, salinity, ambient light, etc. This network aims to manage outdoor environmental applications that need a long lifetime and range without recourse to maintenance.

6.2 Crop yield system

The main aim of agricultural applications is to increase the crop yield while improving the quality of products. To guarantee better quality, many factors intervene all along the lifetime of the plant, starting while it is a seed until it becomes a fruit. Therefore, multiple sensors can be deployed in the field to help farmers keeping track of the well-being of their plants by applying any necessary interventions such as irrigation, fertilizers or pesticides (Figure 3(a)). Abd-Elkader et al. monitored a field of potatoes in Egypt [18]. Mica nodes were placed at a distance of 10 m to keep an eye on the damage of pests on crops. This helped to improve their quality as well as their quantity.

6.3 Livestock monitoring

Livestock is very crucial for the welfare of man, as it provides essential food supplies like eggs, milk and meat, and helps in some agricultural tasks. The fact that humans eat and use animal products is the main reason most livestock animals are alive in the first place. Therefore, it is very important to establish favorable and healthy conditions for raising them to guarantee their well-being, so improving the quality of livestock products. Another issue may occur in big farms where supervising and tracking animals become tiring and time consuming. Runaway animals could cause important

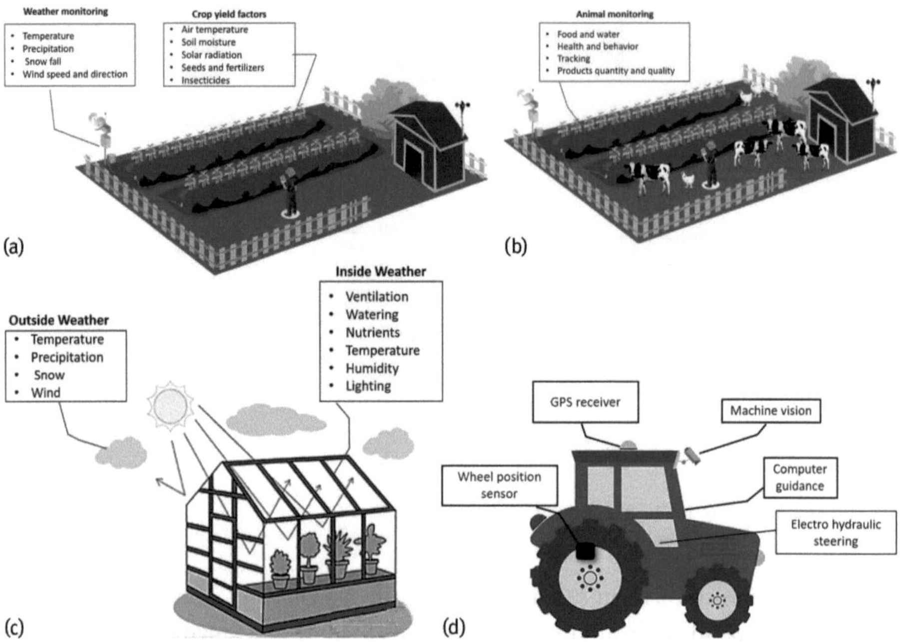

Fig. 3: Agricultural monitoring systems for(a) crop yield, (b) livestock and farm products, (c) greenhouse and (d) agricultural machinery.

losses for farmer. So, it is very necessary to control animal movements and behavior. Thus, many factors should be controlled to help farmers maintain the quality of their products and organize their tasks (Figure 3(b)). This fact has led to Zhang et al. [19] work on controlling the position of zebra. A GPS is used to locate the position of the animals every 8 minutes. The distances considered in this study are 2 m to 3.1 m for grazing and for walking from 10 to 20 m. Multi-hop transmission is topology employed. The sensors are attached to collars around the necks of the animals.

6.4 Greenhouse monitoring

Agricultural technology research and application has been paid more and more attention, especially the greenhouse, which has become more and more important as a way to provide agricultural products all through the year. Optimal greenhouse climate adjustment enables users to attain noticeable energy savings and achieve efficient agricultural production especially during the winter season. What brings better management and control is the deployment of various sensors in open spaces as well as in the soil (Figure 3(c)). In [20], a real-time monitoring system for greenhouse control is designed. It aims to transmit temperature, humidity, light, air quality and other environmental parameters from sensor nodes to a host computer for the specific operations. It is based on the ZigBee WAN.

6.5 Horticulture monitoring

In agricultural applications, many tasks use different kinds of machineries such as tractors, cultivators, ploughs and harvesters. Actually, the use of machinery in agricultural application helps reduce the time required by farmers to execute tasks like tilling, planting cultivating or harvesting and also reduce the number of workers. Many parameters can be controlled with WAN to improve the results obtained by using machinery such as soil characteristics, crops features, machinery emplacement, etc. Tractors are generally equipped with a global positioning receiver to locate the exact position of the machine. The use of an installed camera helps to collect instantaneous pictures of the field that are useful for creating schematic maps. Some sensors may be integrated on the machine, such as soil property, moisture, mass flow, etc. (Figure 3(d)). Wireless capabilities can be used to control electrical motors [21], to optimize energy [22] or to monitor machine tools [23].

6.6 Forest monitoring

WSNs have become an important tool in environmental monitoring. The relatively low cost of the devices allows the installation of a huge range of nodes that can adequately represent the variability in the environment. They can provide risk assessment information, like alerting farmers of frost damage and providing better micro-climate awareness. Johnson and Margalho [24] monitored the agro-climate in the Amazon, analyzing short-range WAN transmissions. They found that more distant nodes suffered a performance loss, while nodes closer to the base station maintained their throughput levels. Another example of climate supervision is flood prediction by means of wireless sensors that can detect rainfall, water levels and weather conditions. The sensors supply information to a centralized database system.

7 Application-specific sensors in agriculture

Actually, sensor networks allow collecting multiple in-situ information, necessary for exploiting and controlling agricultural systems by analyzing different environmental variables such as light, temperature, humidity and soil moisture. This information can be acquired by different sensors deployed in the network. Hence, a variety of specific sensors have been adopted in agricultural applications to control and monitor fields [25].

7.1 Soil analysis

Soil is a basic element in agriculture, as it is responsible for crop development, yield and product quality. Many characteristics of soil can be tested and analyzed for better agricultural outputs, such as moisture, salinity, temperature and electrical conductivity. Soil moisture, temperature and electrical conductivity were analyzed in [26] by applying time domain refractometry (TDR). The device is designed with eight TDR channels, two-rod probes and a GPRS modem for collecting data. Also, frequency domain refractometry (FDR) can be adopted to evaluate soil salinity assessment for sandy minerals and bulk conductivity, as presented in [27]. Some other work used simple low-cost capacitance resistance in order to collect information about soil water content and salinity [28]. In Table 3 a simple comparison between available soil sensors is provided.

Tab. 3: Comparison of some soil related sensors.

Sensors	soilmoisture	Soiltemperature	Watercontent	Conductivity	Salinity
Pogo portable sensor [29]	√	√	√	√	√
Hydra probe II [30]	√	√	√	√	√
ECH2 EC-5 [31]	√	×	√	×	×
VH-400 [32]	√	×	√	×	×
EC-250 [33]	√	√	√	√	√

7.2 Detection and classification of crops, weeds and fruits

Many elements are essential for the growth and good quality of crops and fruits. In [34, 35], red peaches in orchard images were detected through an RGB camera based on linear color models through a distance-based classification approach. Then, a combination of an ultrasonic sensor and a camera were used for weed detection in a cereal plantation [36]; the ultrasonic sensor identifies the height of the plants, and the camera determines the weed and crop coverage. In [37], a low cost smart camera was adapted with selected filters to pass red and near infra-red spectral bands and the normalized difference vegetation index (NDVI) was obtained for plant detection. Fruit grading is addressed in [38] on the basis of a mobile sensing platform mounted on a glove that integrates several sensors, such as touch pressure, imaging, inertial measurements, localization and an RF identification (RFID) reader.

7.3 Weather and climate-related sensors

Actually, weather is a critical factor affecting the production of crops, the welfare of animals and all agricultural practices, starting from soil preparation to the harvesting operation. It presents a direct input in the agricultural system. Many factors that have a direct or indirect impact on the agricultural applications can be included in weather such as precipitation, temperature, humidity, sunshine, wind velocity, storms, snowfall, etc.; for instance, the forecast of weather has an important impact on the planning of the agricultural process. It helps to launch or withhold the sowing or seeding operation and soil preparation tasks. Also, in the case of rainfall, it helps farmers to reduce the dependency on stored water sources as a way to save them. In this context, environmental sensors such as humidity, ambient temperature and wind speed are deployed with application-specific soil and plant sensors in various agricultural applications. Such kinds of heterogeneous placement ensures intelligent and improved decision making. Table 4 lists some sensors specific to the measurement of environmental parameters used in some agricultural applications.

Tab. 4: Comparison of some weather-related sensors.

Sensors	Temp	Hum	Wind-S	Wind-D	S-R	Rain	Pres
CM-100 compact weather station [39]	√	√	√	√	×	×	√
WXT520 compact weather station [40]	√	√	√	√	×	√	√
All-in-one (AIO) weather sensor [41]	√	√	√	√	×	×	√
SHT71[42]	√	√	×	×	×	×	√
CI-340 hand-held photosynthesis [43]	√	√	×	×	×	×	×
RG13/RG13H [44]	×	×	×	×	×	√	×
CS300-L pyranometer [45]	×	×	×	×	√	×	×

Where Temp: temperature; Hum: humidity; Wind-S: wind speed;
Wind-D: wind direction; S-R: solar radiation; Pres: pressure

8 Solutions for irrigation systems based on WAN in the measurement and sensor technology laboratory

A precision irrigation solution based on WSNs is proposed. The main challenges of the work is to design an automated irrigation system that could reduce manpower, water waste, time and money, and also offers reliable data communication between sensor nodes and fulfills the requirements for low-power consumption and low cost. The

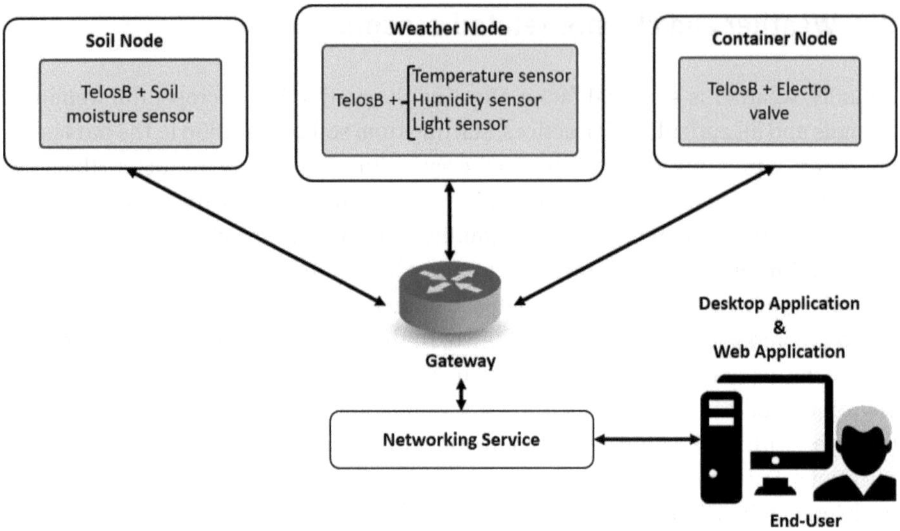

Fig. 4: Architecture of the proposed automated irrigation system based on WAN.

automated irrigation system developed uses multiple sensors to control water quantity in real time. The system is proposed to detect soil and environmental parameters through a sensor node and display the measured parameters in real time via a Java interface or via a web site. These two interfaces must be simple to use and receptive to the user's needs. The prototype consists of three parts: the base station, the wireless sensor module and the software part. The base station should be attached to a computer that receives logs and displays all the sensor readings. The sensor module will transmit sensing data to the base station. The software part consists of a Java application and a website to log and display sensor data. The base station is often thought of as just a central component that is used to gather data from distributed nodes. This node is the heart of the whole system; its failure is very critical and constitutes a single point of failure phenomenon.

For the system's implementation, different sensor nodes were compared in terms of energy consumption, data rate and cost. Based on the results obtained, TelosB mote was chosen, since it is an ultra-low-power wireless module for monitoring applications, a user-friendly product and allows rapid application prototyping. TelosB works properly within minus 40 to 123.8 °C, which is suitable for extreme weather conditions. Specifications are illustrated in Table 5.

Tab. 5: Specification of TelosB mote.

Features	TelosB [14]
Processor	TIMSP430
System memory (kB)	10
I/O connectivity	ART, I2C, SPI, DIO
Current	Active: 1.8 Sleep: 15 (μA)
Operating frequency (MHz)	2400
Number of channels	programmable
Data rate (kbps)	250
RF power (dBm)	−24 to 0
Sensitivity (dBm)	−90 to −94
Communication range	Outdoor: 75 to 100 Indoor: 20 to 30
Embedded sensors	Temperature, light and humidity sensor

8.1 The nodes developed

A large number of sensor nodes are spread over an area controlling environmental and soil factors that affect the irrigation process. Further details for each node will be given in the following sections.

Soil node: Water is required for the basic growth of plants. When a sufficient amount of water is not present for plant needs, then stress can occur, which ultimately leads to reduced quality or death. Several soil parameters should be controlled to have an efficient irrigation system. Measuring soil moisture is important to estimate the exact quantity of water needed for each plant in each field. Therefore, farmers can control the water supply properly and avoid waste. Measuring temperature helps farmers to know when the water container should be opened, which is very important in hot and dry seasons.

The VH400 soil moisture sensor by Vegetronix [32] was used for soil monitoring; it has a small size and lower power consumption. Since it measures the dielectric constant of the soil using transmission line techniques, it is insensitive to water salinity and will not corrode over time. Soil temperature can influence the germination rate and survival of woody species that grow in the environment. Soil temperature also affects how quickly plants take up water and nutrients.

Weather node: The environment monitoring part consists of various types of sensors to collect environmental factors such as luminance level, temperature, relative humidity and wind speed. Air temperature has a crucial role in plant growth; low temperature causes a decrease in the absorption and movement of water in plants. Humidity also plays a major role in plant growth; low humidity causes fast transpiration, which means that the plants take up and use large amounts of water (and nutrients). Normally, the range of healthy relative humidity for plants is from 50 to 70%. High air moisture reduces the required plant watering frequency. With the help of a sensor (SHT11 [46]) integrated in the TelosB, we can control the ambient temperature and humidity. The small size and low-power consumption makes SHT11 the ultimate choice for even the most demanding applications, including sensor networks.

The application pattern of water is susceptible to distortion by wind. Although wind speed and direction are not controlled variables, their effect on irrigation uniformity is significant, so farmers should take wind speed into consideration when irrigating. So, to obtain more information about the weather changes that affect plant growth, the mote is accomplished with a wind speed sensor. Using this sensor is helpful to detect whether or not irrigation is good for this period, because when the wind speed is high, the water will not persist in the soil. For this node, we used the Vortex anemometer wind speed sensor [47].

Container node: In the container, an actuator is attached to the TelosB mote. When urgent conditions are detected on other nodes, a simple order will be given to the actuator to open or not open the water valve for irrigation. To control the irrigation process, we used the solenoid valve provided by Rain Bird [48]. A solenoid valve is an electromechanical valve that is controlled by an electric current. The electric valve is a low-power consumption device, and it is powered with 24 V for 50/60 Hz. A relay is used to switch a high power system with a small operating power. Then, the relay is connected in series with the electro-valve and an external 9 V battery. This way, when the TelosB mote triggers the relay, the circuit closes, the electro-valve opens and the irrigation process begins.

8.2 Mote deployment and network architecture

The motes used are placed in a star topology, where sensors are mounted on different motes, as shown in Figure 3. This topology consists of a central node (base station), to which all other nodes (soil node, weather node and container node) are connected; this central node provides a common connection point for all nodes, and it acts as a conduit to transmit messages. The star topology reduces the damage of the failure caused by line failure by connecting all the systems to a central node. The failure of a transmission line linking any peripheral node to the central node will result in the isolation of that node from all the others, but the rest of the system will be unaffected. With this topology there is the potential of battery power saving, because all the nodes connected to the central node spend their time in sleep mode, only waking up to take measurements and send the data to the base node.

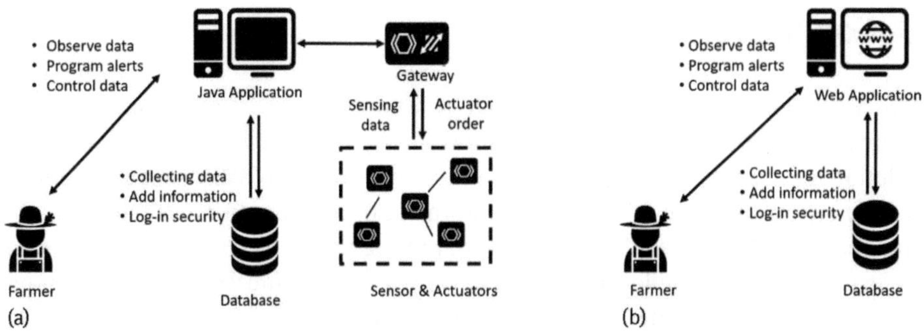

Fig. 5: Architecture of proposed Java application (a) and web application (b).

Each mote is responsible for retrieving the appropriate parameters (moisture, temperature, wind speed, etc.). The data from each node will be transmitted to the base station, where it is recorded and sent to the farmer's personal computer (PC) in time to take the proper action. The base station is a mote connected to a PC on a universal serial bus (USB) port that acts as a gateway for the rest of the motes. When it receives measurement values, it forwards it to the PC where a Java application receives the data and stores them in a MySQL database [49] for post-processing. The displayed results contributed by the Java application or by the website will allow the user to view maps, tables and graphs in a comprehensible form to ease decision making, like where the irrigation amount will be delivered as the season progresses. With this solution, the decision whether to or not irrigate is no longer a matter of guessing and just looking at the crop. With soil sensors, it becomes possible to know exactly what is happening in the zone where it matters: inside the soil where the roots from the crop should take up their water.

8.3 Software application

Throughout the operation of the system, the levels of moisture and temperature of the soil node and the levels of some environmental parameters of the weather node are forwarded to the base station by the corresponding motes. When the base station receives a measurement value, it forwards it to the PC, where a Java application receives the data and stores it in a MySQL database for post-processing. The implementation of the system has been handled successfully, so the farmer can control the amount of water needed in the field in two ways: by displaying data onto the Java application if he is inside the farm or from a website if he is outside.

9 Conclusion

WSNs combine sensor technology, modern network and wireless communication technology. They present a flexible, autonomous, cost effective and energy-efficient solution for a variety of applications. In agriculture, the use of WSNs enables farmers to fully control and maintain their agricultural tasks, and so to improve the profits. In this chapter, WSNs were briefly introduced, putting emphasis on their application in agriculture. Some example applications of WSNs in agriculture were given to focus on the benefits of integrating WSNs in agricultural applications. One important task in agriculture is irrigation; thereby, an automated irrigation system based on WSNs was developed.

Bibliography

[1] F. C. Oliveira, Á. C. Collado, and L. F. C. Leite. Autonomy and sustainability: An integrated analysis of the development of new approaches to agrosystem management in family-based farming in Carnaubais Territory. *Piauí, Brazil, Agricultural Systems*, 115:1–9, 2013. ISSN 0308-521X.

[2] Memsic. *Imote2 High-Performance Wireless Sensor Network Node*, datasheet 6020-0117-01 Rev A. [Online]. Available: http://thermofisher.com.au/Uploads/file/Environmental-Industrial/Process-Monitoring-Industrial-Instruments/Data-Acquisition/Wireless-Sensors/Memsic/IMOTE2-high-performance-wireless-sensor-network-node.pdf, (accessed on Sep 18, 2018).

[3] G. Merrett. Wireless sensor networks for process monitoring. Lecturer in intelligent sensing and pervasive systems at the University of Southampton, 2011.

[4] Ojha et al. Wireless sensor networks for agriculture: the state-of-the-art in practice and future challenges. *Computers and Electronics in Agriculture*, 118:66–84, 2015.

[5] I. F. Akyildiz and I. H. Kasimoglu. Wireless sensor and actor networks: research challenges. *Ad Hoc Network*, 2(4):351–367, 2004.

[6] J. Gutiérrez, J. F. Villa-Medina, A. Nieto-Garibay, and M. Á. Porta-Gándara. Automated irrigation system using a wireless sensor network and GPRS module. *IEEE Transactions on Instrumentation and Measurement*, 63(1):166–176, 2014.

[7] Vuran and Akyildiz. Cross-layer packet size optimization for wireless terrestrial, underwater and underground sensor networks. In *Proceedings of IEEE INFOCOM, Phoenix, AZ, USA*, pages 780–788, 2008.

[8] X. Yu, P. Wu, W. Han, and Z. Zhang. A survey on wireless sensor network infrastructure for agriculture Computer Standards and Interfaces. *35(1), 59–64*, 2013.

[9] ZigBee Specifications, ZigBee Alliance Std. [Online]. Available: http://www.zigbee.org/.

[10] *IEEE Standard for Information technology, Telecommunications and Information Exchange Between Systems – Local and Metropolitan Area Networks – Specific Requirements Part 15.4: Wireless Medium Access Control (MAC) and Physical Layer (PHY) Specifications for Low-Rate Wireless Personal Area Networks (WPANs)*, EEE Std 802.15.4a-2007 (Amendment to IEEE Std 802.15.4-2006), 1–203, 2007.

[11] *IEEE Standard for Information technology–Local and metropolitan area networks–Specific requirements–Part 11: Wireless LAN Medium Access Control (MAC) and Physical Layer (PHY) Specifications – Amendment 8: Medium Access Control (MAC) Quality of Service Enhancements*, IEEE Std 802.11e-2005 (Amendment to IEEE Std 802.11, 1999 Edition (Reaff 2003), 1–22, 2005.

[12] Bluetooth Technology Special Interest Grou [Online]. Available: https://www.bluetooth.org/.
[13] MICAz wireless Measurement System, MEMSIC. [Online]. Available: http://www.memsic.com/userfiles/files/Datasheets/WSN/micaz_datasheet-t.pdf, (accessed on May 16, 2017).
[14] Memsic. *TELOSB Mote Platform*, datasheet 6020-0094-03 Rev A. [Online]. Available: http://www.memsic.com/userfiles/files/Datasheets/WSN/telosb_datasheet.pdf, (accessed on Sep 18, 2018).
[15] Memsic. *IRIS Measurement System*, datasheet 6020-0124-02 Rev A. [Online]. Available: http://www.memsic.com/userfiles/files/Datasheets/WSN/IRIS_Datasheet.pdf, (accessed on Sep 18, 2018).
[16] Waspmote, *Libelium*, Libelium.com, 2018. [Online]. Available: http://www.libelium.com/libeliumworld/waspmote/, (accessed on Sep 18, 2018).
[17] Kotamäki et al. Wireless in-situ Sensor Network for Agriculture and Water Monitoring on a River Basin Scale in Southern Finland: Evaluation from a Data User's Perspective. *Sensors*, 9:2862–2883, 2009. doi:10.3390/s90402862.
[18] S. M. Abd El-kader and B. M. M. El-Basioni. Precision farming solution in Egypt using the wireless sensor network technology. *Egyptian Informatics J*, 2013. doi:10.1016/j.eij.2013.06.004.
[19] Zhang et al. Hardware Design Experiences in ZebraNet. *SenSys'04, November 3–5*, 2004.
[20] The Design of Greenhouse Monitoring System Based on ZigBee WSNs. In *Computational Science and Engineering (CSE) and Embedded and Ubiquitous Computing (EUC), 2017 IEEE International Conference on, Guangzhou, China, 21–24 July 2017*, 2017.
[21] Lima-Filho et al. Embedded system integrated into a wireless sensor network for online dynamic torque and efficiency monitoring in induction motors. *M. A., IEEE/ASME Transactions on Mechatronics*, 17(3):404–414, 2012.
[22] F. Salvadori et al. Monitoring in industrial systems using wireless sensor network with dynamic power management. *A. C., IEEE Transactions on Instrumentation and Measurement*, 58(9):3104–3111, 2009.
[23] K. K. Tan et al. Distributed fault detection in industrial system based on sensor wireless network. *Computer Standards and Interfaces*, 31(3):573–578, 2009.
[24] T. M. Johnson and M. Margalho. Wireless Sensor Networks for Agroclimatology Monitoring in the Brazilian Amazon. In *International Conference on Communication Technology, November 2006*, 2006. doi:10.1109/ICCT.2006.341876.
[25] G. Pajares, A. Peruzzi, and P. G. de Santos. Sensors in Agriculture and Forestry. *Sensors (Basel, Switzerland)*, 13(9):12132–12139, 2013. doi:10.3390/s130912132.
[26] W. Skierucha, A. Wilczek, A. Szypłowska, C. Sławiński, and K. Lamorski. A TDR-Based Soil Moisture Monitoring System with Simultaneous Measurement of Soil Temperature and Electrical Conductivity. *Sensors*, 12:13545–13566, 2012.
[27] A. Wilczek, A. Szypłowska, W. Skierucha, J. Cieśla, V. Pichler, and G. Janik. Determination of Soil Pore Water Salinity Using an FDR Sensor Working at Various Frequencies up to 500 MHz. *Sensors*, 12:10890–10905, 2012.
[28] E. Scudiero, A. Berti, P. Teatini, and F. Morari. Simultaneous Monitoring of Soil Water Content and Salinity with a Low-Cost Capacitance-Resistance Probe. *Sensors*, 12:17588–17607, 2012.
[29] POGO Portable soil moisture, salinity and temperature sensor, STEVENS water monitoring systems, http://www.easybib.com/reference/guide/apa/website (accessed on May 16, 2017).
[30] Hydra Probe II Soil Sensor, STEVENS water monitoring systems, https://www.licor.com/documents/hoy85pj515o0rc4uw9q3 (accessed on May 16, 2017).
[31] EC-5 SMALL SOIL MOISTURE SENSOR, Decagon devices, http://manuals.decagon.com/Manuals/13876_EC-5_Web.pdf (accessed on May 16, 2017).
[32] VH400 Soil Moisture Sensor Probes, Vegetronix. [Online]. Available: http://www.vegetronix.com/Products/VH400/, (accessed on May 16, 2017).

[33] EC250 ELECTRICAL CONDUCTIVITY SENSOR, Pentair environemental systems. [Online]. Available: http://www.pentairenvironmental.com/products/ec250-electrical-conductivity-sensor.html, (accessed on May 16, 2017).
[34] M. Teixidó, D. Font, T. Pallejà, M. Tresanchez, M. Nogués, and J. Palacín. Definition of Linear Color Models in the RGB Vector Color Space to Detect Red Peaches in Orchard Images Taken under Natural Illumination. *Sensors*, 12:7701–7718, 2012.
[35] M. Teixidó et al. An Embedded Real-Time Red Peach Detection System Based on an OV7670 Camera, ARM Cortex-M4 Processor and 3D Look-Up Tables. *Sensors*, 12:14129–14143, 2012.
[36] D. Andújar, M. Weis, and R. Gerhards. An Ultrasonic System for Weed Detection in Cereal Crops. *Sensors*, 12:17343–17357, 2012.
[37] V. Dworak et al. Strategy for the Development of a Smart NDVI Camera System for Outdoor Plant Detection and Agricultural Embedded Systems. *Sensors*, 13:1523–1538, 2013.
[38] R. V. Aroca et al. A Wearable Mobile Sensor Platform to Assist Fruit Grading. *Sensors*, 13:6109–6140, 2013.
[39] CM-100 COMPACT FIVE-PARAMETER WEATHER SENSOR, Sensor experts. [Online]. Available: https://sensorexperts.com/node/222596, (accessed on May 16, 2017).
[40] WXT520 Weather Sensor, Campbell Scientific. [Online]. Available: https://www.campbellsci.co.uk/wxt520, (accessed on May 16, 2017).
[41] All-In-One (AIO) Weather Sensor, Climatronix corporation. [Online]. Available: http://www.climatronics.com/Products/Weather-Station-Systems/AIO_compact_weather_station.php, (accessed on May 16, 2017).
[42] Datasheet SHT7x (SHT71, SHT75), Humidity and Temperature Sensor IC, Sensirion. [Online]. Available: http://www.mouser.com/ds/2/682/Sensirion_Humidity_SHT7x_Datasheet_V5-469726.pdf, (accessed on May 16, 2017).
[43] CI-340 Handheld Photosynthesis System, CID Bio-Science. [Online]. Available: https://cid-inc.com/plant-science-tools/photosynthesis-measurement/ci-340-handheld-photosynthesis-system/, (accessed on May 16, 2017).
[44] Vaisala Rain Gauge RG13, RG13H, Vaisala. [Online]. Available: http://www.vaisala.com/Vaisala%20Documents/Brochures%20and%20Datasheets/WEA-MET-G-RG13-RG13H-datasheet-B010195EN-F-LOW-v1.pdf, (accessed on May 16, 2017).
[45] CS300-L Pyranometer, Campbell Scientific. [Online]. Available: https://www.campbellsci.de/cs300-pyranometer, (accessed on May 16, 2017).
[46] *Datasheet SHT1x (SHT10, SHT11, SHT15)*. Humidity and Temperature Sensor IC, Sensirion.
[47] Inspeed Vortex Wind Sensor, Inspeed. [Online]. Available: http://www.inspeed.com/anemometers/vortex_wind_sensor.asp, (accessed on May 16, 2017).
[48] Solenoid irrigation Valves, Rain Bird. [Online]. Available: http://www.rainbird.com/Landscape/products/valves/index.htm, (accessed on May 16, 2017).
[49] Database, MySQL. [Online]. Available: https://www.mysql.com/, (accessed on May 16, 2017).

Paul Cahill and Vikram Pakrashi

Piezoelectric energy harvesting for monitoring of rail bridge infrastructure

Abstract: The use of structural health monitoring systems for civil structures is ever expanding, and by assessing the dynamical condition of structures, informed maintenance management can be conducted at both individual and network levels. With the continued growth of information age technology, the potential arises for smart monitoring systems to be integrated with civil infrastructure to provide efficient information on the condition of a structure. In this regard, energy harvesting techniques have immense potential to form an integral part of wireless sensor networks (WSN) through the harvesting of energy from the host civil structure and provide a continuous, renewable energy supply for the WSN node. This chapter focuses on this problem, investigating the integration of energy harvesting devices with civil infrastructure. While there has been considerable focus on the development and optimization of such devices using steady state loading conditions, their applications for civil infrastructure are less known. Although research is still in the initial stages, studies into the uses associated with such applications are very promising. A model bridge is considered, and the through the use of the dynamical response of the bridge to a variety of train loading conditions, the energy harvesting outputs from such devices is established and the potential energy output determined.

Keywords: Energy harvesting, piezoelectric, bridge, train, vibration, finite element analysis

1 Introduction

The harnessing of vibrations from large-scale civil infrastructure is an attractive application for energy harvesting devices, with bridges, wind turbines and tall buildings being identified as potential host structures for such technology [1]. Utilizing the high dynamic responses of such structures under operational conditions for energy harvesting can offer a wide range of applications, including forming the basis of power independent WSNs [2] and their use for structural health monitoring (SHM) of civil infrastructure [3].

Of all civil infrastructure, the harnessing of vibrations resulting from bridge–vehicle interaction is an attractive application for energy harvesting devices [4]. This

Paul Cahill, MaREI Centre, Environmental Research Institute, University College Cork, Ireland
Vikram Pakrashi, Dynamical Systems & Risk Laboratory, School of Mechanical and Materials Engineering, University College Dublin, Ireland

https://doi.org/10.1515/9783110445053-020

is due in part to the number of bridge structures present in society, along with the relatively large dynamic amplitudes and repeatability of the loadings that they experience. The modeling of bridge–vehicle interaction is subject matter that has received intensive investigation, with models of increasing complexity being proposed. When investigating such interaction, the model chosen depends very much on the nature of the problem being considered and the degree to which the intrinsic details of the dynamic behavior of the train is required. Such models can range from the most basic, that of a point load in the form of a direct delta function traversing a beam [5], to train loadings with a singular degree of freedom (1-DOF) [6], three DOFs [7] and multiple degrees of freedom (MDOF) [8].

Energy harvesting utilizing the response of a bridge structure to the induced vibrations of vehicular traffic was recently formulated for a cantilever piezoelectric energy harvesting device [9]. The use of piezoelectric patch devices for energy harvesting directly from the strain fluctuations in a bridge due to vehicle passage has also been established [10]. Efforts to quantify the amount of energy that can be harvested from bridges have also been ongoing, including the amount of energy that can be harvested from bridge–vehicle interaction for a highway structure using a cantilever piezoelectric harvester [11], including the use of a laboratory experiment to experimentally verify the energy harvesting potential [12]. Initial experimental studies into the application of cantilever-based energy harvesters with a highway bridge [13] and a train bridge [14, 15] have also been conducted.

The use of piezoelectric devices for the SHM of bridge infrastructure has also received attention in recent times, including the monitoring of damage evolution in a concrete beam using lead zirconate titanate (PZT) sensors [16], as well as polyvinylidene fluoride (PVDF) sensors [17]. Similarly, piezoelectric energy harvesters have also been employed for an experimental model bridge structure, of 5 m span, undergoing cyclic loading of varying frequency and amplitude [18]. A parametric frequency-increased generator (PFIG), based on electromagnetic principles, has also been proposed for energy harvesting from bridge infrastructure [19]. The use of train tracks as the location for energy harvesting from train passages has been studied, with a SDOF harvester being proposed that has a natural frequency matching that of the train passage [20]. Energy harvesting using a piezoelectric patch device has been proposed for train track-based harvesting [21]. Studies have also investigated the monitoring the response of a truss bridge due to train passage using piezoelectric sensors compared against traditional strain gauges and ultrasonic techniques [22]. Other applications arising from piezoelectric harvesters coupled with bridge infrastructure is the monitoring of traffic loadings through a piezoelectric weigh-in-motion (WIM) system [23].

Bridge infrastructure offers great possibilities for energy harvesting applications due to the magnitudes of the vibrations, the comprehensive knowledge available regarding bridge–vehicle interaction and the number of bridges at national and international levels. Utilizing such harvested energy and energy harvesting technology for applications such as powering wireless nodes and the monitoring of the health of structure to which it has been applied is a concept that has far reaching implications for SHM. This chapter investigates energy harvesting applications from bridge vibrations due to train passage. Two piezoelectric energy harvester configurations are presented in this regard. The creation of a finite element model bridge is detailed, and the dynamic response due to the passage of an international train fleet obtained. The energy harvesting output from the two piezoelectric harvester configurations is finally presented.

2 Piezoelectric energy harvesting

Piezoelectric energy harvesters convert strain variations acting on the piezoelectric element within the harvester into electrical energy. Utilizing the operational responses of civil infrastructure elements as the dynamic excitation sources for such harvesters is exceptionally attractive due to the relevant sizes associated with such infrastructure and the corresponding loadings and responses, which can be substantial [24]. Two piezoelectric device configurations can be used to exploit the dynamic response of such structures, namely cantilever devices and patch devices (Figure 1), which allows for either the acceleration response of the host structure to be used as the base excitation or the strain response to be utilized freely by imposing strain fluctuations directly on the piezoelectric harvester.

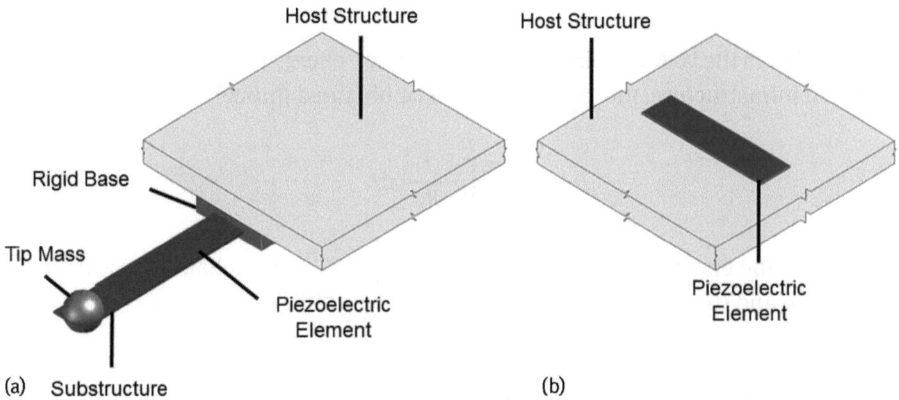

Fig. 1: Illustration of piezoelectric device configurations attached to host structure, including (a) cantilever device configuration and (b) patch device.

2.1 Piezoelectric cantilever energy harvester

Piezoelectric cantilever energy harvesting devices most commonly consist of a cantilever beam substrate onto which piezoelectric materials are bonded (Figure 1(a)). The device is subsequently embedded within a rigid base and attached to the designated host structure, the acceleration response of which under operational conditions provides the base excitation to the harvester. The optimization of cantilever devices has received significant attention in recent times [25], with the addition of a tip mass at the free end of the cantilever beam now frequently adopted to allow for frequency optimization. The electromechanical behavior of a cantilever energy harvester can be expressed by the coupled linear single-degree-of-freedom equations as [26]

$$m_c\ddot{z} + c_c\dot{z} + k_c z - \theta V = -m_c\ddot{y}_b \tag{1}$$

$$\theta\dot{z} + C_p\dot{V} + \frac{1}{R_l}V = 0 \tag{2}$$

where m_c, c_c and k_c are the mass, damping and stiffness of the energy harvester, respectively, and z is the relative displacement of m_c, with over-dots denoting differentiation with respect to time and y_b is the base acceleration from the host structure. The electromechanical coupling coefficient is given as θ, V is the voltage and C_p and R_l are the capacitance and load resistance, respectively. The natural frequency, ω_c and the damping, ξ_c, of the harvester are defined as

$$\omega_c = \sqrt{\frac{k_c}{m_c}} \quad \text{and} \quad \xi_c = \frac{c_c}{2m_c\omega_c} \tag{3}$$

Following from the voltage being calculated from the energy harvester when deployed with civil infrastructure, the total energy can be obtained through:

$$E_h = \int_0^T \frac{V(t)^2}{R_l}\,dt \tag{4}$$

where E_h is the energy harvested by the harvester, $V(t)$ is the voltage across the load resistor, R_l, and the total time considered for the harvester is given by T.

2.2 Piezoelectric patch energy harvester

Unlike cantilever energy harvesters, piezoelectric patch energy harvesters are designed so as to be bonded directly to the surface of a host structure (Figure 1(b)). Consequently, the surface strain fluctuations resulting from the host structure undergoing operational loadings are transferred to the bonded patch energy harvester, resulting in an electrical response from the device [4]. The electrical output from such harvesters can be described by [27]

$$\frac{dv(t)}{dt} + \frac{v(t)}{\tau} = \frac{e_{31}A}{C_p}\frac{d}{dt}[S_1(t) + S_2(t)] \tag{5}$$

where $v(t)$ is the voltage, τ is a time constant, e_31, A and C_p are a piezoelectric stress constant and the area and capacitance of the material, respectively; $S_1(t)$ and $S_2(t)$ are the strain vectors, with respect to time t, acting on the material in the longitudinal and translational directions, respectively, due to strain fluctuations from the host structure. The time constant is given as $\tau = C_p R_l$, where R_l is the external resistive load, and C_p of the piezoelectric material can be calculated by

$$C_p = \frac{\epsilon_{33}^S A}{t_m} \tag{6}$$

where ϵ_{33}^S is the permittivity constant at constant strain and t_m is the thickness of the material. The amount of energy that can be harvested from the patch energy harvester is determined from the generated voltage, as determined in equation (4). Following from this, our attention turns to the host structure to which the piezoelectric energy harvesters are to be deployed.

3 Bridge–vehicle interaction model

3.1 Modeling of bridge and train loadings

To determine the energy harvesting potential due to the induced vibrations in a bridge that arise due to the passage of the international train fleet, a three-dimensional finite element model of the bridge was created. A double-track sectional model was created using 20-node hexahedral bricks that has the dimensions of 10.6 m in length and 10 m in breadth (Figure 2). Two sets of concrete sleepers were modeled on the deck of the bridge at distances of 0.8 m, over which the track was modeled.

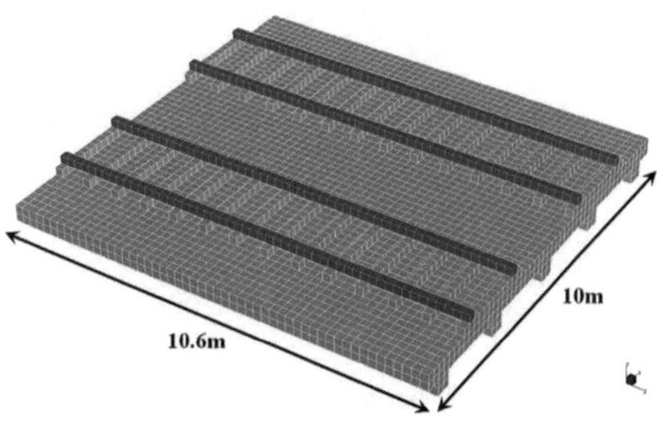

Fig. 2: Finite element bridge model.

Applied to the model tracks are load paths, with the individual axle loads being applied at distances and magnitudes as determined by the axle spacings and loadings, respectively. These axle loads being applied are modeled as point loads. Of note is the possibility of modeling such train loadings with greater degrees of complexity, as previously discussed. Since the primary focus of this study is to establish the feasibility of energy harvesting from the vibrational response of a bridge to the fleet of trains as a whole, these conservative point loads are sufficient in this regard.

3.2 Attributes of an international train fleet

Five trains from an international stock were chosen for the purposes of comparing the energy harvesting potential from a bridge structure from traffic induced vibrations (Figure 3). Of the five trains considered, the Irish 071Loco and 201Loco are diesel powered locomotives and the remaining three, the French TGV, German ICE and Japanese Shinkansen, are electric trains. Each train was modeled with the same loading, axle spacings and carriage configuration as it would have under operational conditions, as outlined (Table 1).

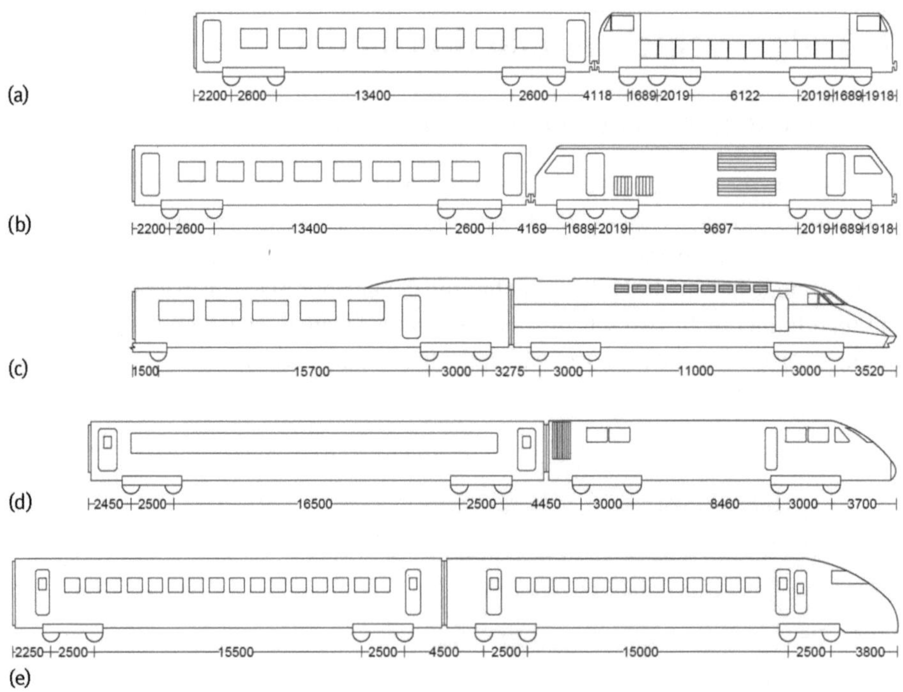

Fig. 3: International locomotives and carriages, including axle spacings for (a) 071Loco, (b) 201Loco, (c) TGV, (d) ICE, (e) Shinkansen.

Tab. 1: Characteristics of the international train fleet.

Parameter		071Loco	201Loco	TGV	ICE	Shinkansen
Locomotive length	m	17.4	21.0	22.2	20.2	26.1
Number of locomotives		1	1	2	2	2
Carriage length	m	23.0	23.0	18.7–21.9	26.4	25.0
Number of carriages		7	7	10	12	14
Total train length	m	178.4	182.1	237.6	357.1	402.1
Locomotive axle load	kN	161.9	182.5	158.3	190	107.3
Carriage axle load	kN	117.7	117.7	158.3	140.0	107.3
Total train load	kN	4,267	4,391	4,748	8,240	6,867

The 071Loco and 201Loco trains consist of a single locomotive hauling seven carriages, with the carriages of both being similar in nature as they originate from the same rail stock. The TGV, ICE and Shinkansen trains consist of two locomotives, located on opposite ends of the train, and haul 10, 12 and 14 carriages, respectively. While the three electric trains can operate at high speeds of up to 300 km/h, the two diesel trains are limited in their operational speed, with a maximum speed of 160 km/h for both trains. Taking into account these operational speed constraints, for the 071Loco and 201Loco, the range of train speeds considered ranged between 40 km/h and 160 km/h at 20 km/h intervals and for the TGV, Shinkansen and ICE, the range considered was between 40 km/h and 300 km/h, again with 20 km/h spacing.

3.3 Train-induced vibrations in the bridge model

The response of the train-induced vibrations was obtained using a linear transient solver, with two loading cases investigated for each speed for the different trains, namely that of a single train passage and that of a double train passage. For the double passages, the trains are modeled to travel at equal speeds in opposite directions and to enter the bridge at the same time. For all train passages, a sampling rate of 100 Hz was applied and an additional time period was added after each train left the bridge model to account for damping, applied to the model at 2.158%. For all train passages, the position of maximum dynamic response was determined and the acceleration and strain profiles obtained for determining the energy harvesting potential of different devices.

The computed acceleration responses of the bridge can be used as the base excitation of a cantilever energy harvesting device, as has previously been described. A double passage results in a greater acceleration response when compared against a single train passage at similar speeds (Figure 4), while the duration of the response is dependent on the length of the train considered. The influence of trains consisting of locomotives with greater loadings compared to the carriage loadings can be identified. This is true for the 201Loco and 071Loco trains, whereby the acceleration response is

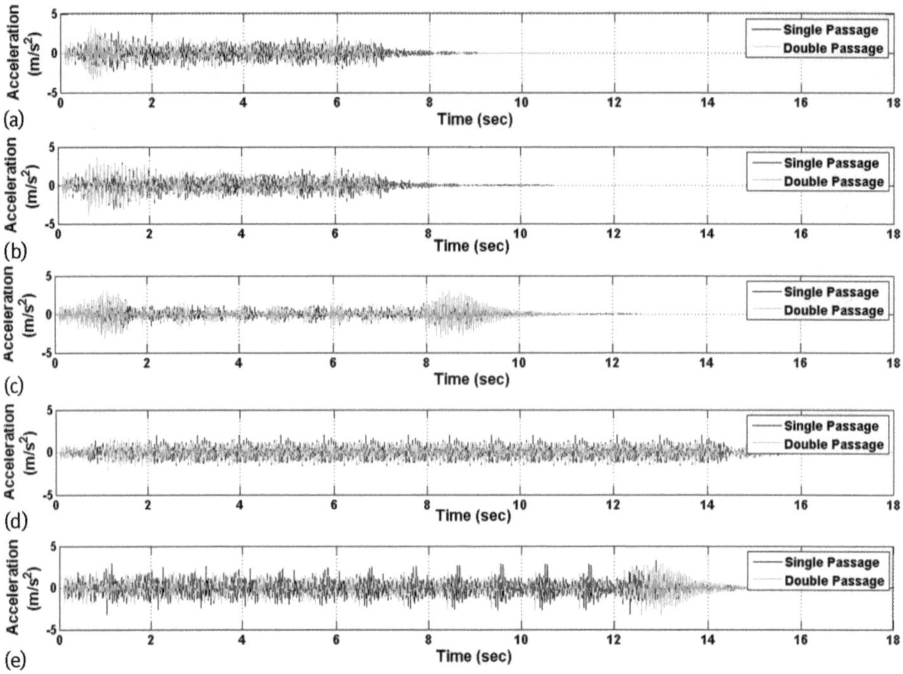

Fig. 4: Example acceleration response for single and double train passages at 100 km/h for (a) 071Loco, (b) 201Loco, (c) TGV, (d) Shinkansen, (e) ICE.

greatest in magnitude at the beginning of the train passage, and also for the TGV, both at the beginning and at the end of the passages. This can also be seen in the analysis of the strain profiles for similar train passages (Figure 5). The loadings from the individual axles can be more easily detected in the strain output, while the disparity in the overall train lengths of the fleet is recognizable in both profiles. The strains obtained for the train passages can be subsequently utilized to determine the energy harvesting potential of a patch harvester.

4 Energy output from a bridge structure using a cantilever harvester

4.1 Properties of the cantilever energy harvester

To determine the energy harvesting potential from a cantilever energy harvester, an energy harvester chosen for the modeling from previous literature [13], due in part to its high efficiency rates at low natural frequency, which will maximize the energy harvesting output. The properties used for the cantilever harvester are outlined in Ta-

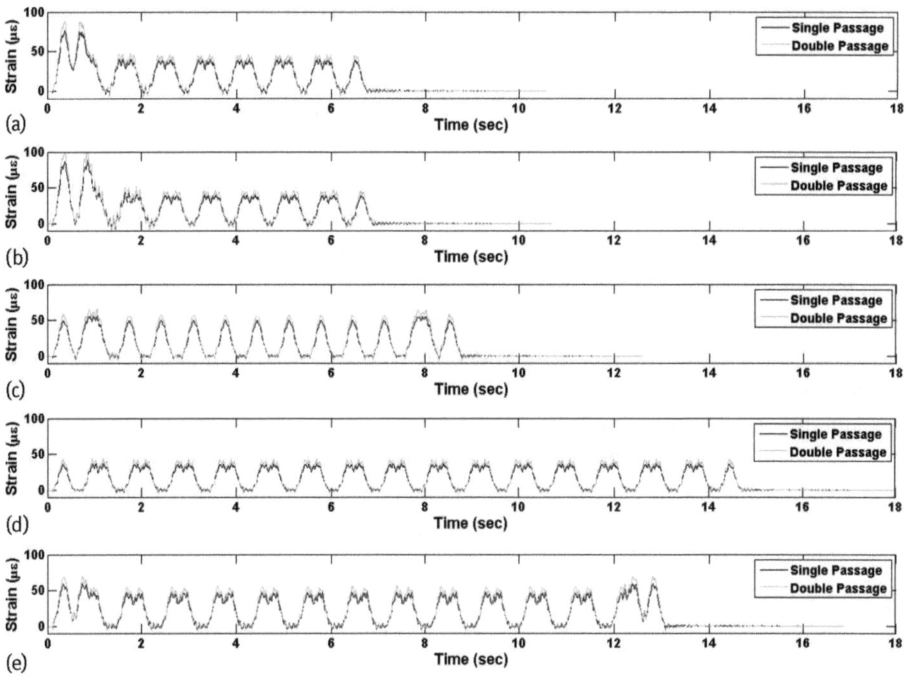

Fig. 5: Example strain response for single and double train passages at 100 km/h for (a) 071Loco, (b) 201Loco, (c) TGV, (d) Shinkansen, (e) ICE.

ble 2, with the piezoelectric material PZT forming the basis of the harvester. Utilizing these parameters and Equations (1) and (2), the energy harvester can be subsequently investigated for civil infrastructure applications. Utilizing the acceleration response of the bridge vehicle, as detailed in the preceding section, the amount of energy that can be harvested due to passages of the international fleet can be determined, including for a wide range of speeds and for different loading conditions, including passing trains.

Tab. 2: Properties of piezoelectric cantilever energy harvester for train–bridge interaction applications.

Parameter		Value	Unit
Mass	m_c	0.023	kg
Equivalent viscous damping ratio	ξ_c	0.02	
Capacitance	C_p	2.24	nF
Electromechanical coupling	θ	$2.34 \cdot 10^{-4}$	C/m
Resistive load	R_l	364	kΩ
Natural frequency	ω_c	14.4	Hz

4.2 Energy harvesting output from the cantilever energy harvester

Using the acceleration responses of the model bridge, the potential output of the cantilever energy harvesting device can be determined. In this study, a single location is considered so as to determine the energy harvesting potential at the position of maximum acceleration response, the mid-span of the bridge. It was observed that the voltage outputs of the devices were dependent on the train loading characteristics and the speed at which it was traveling. Both the length of the train and the speed at which it was traveling determined the time period over which the cantilever energy harvester generated voltage, as can be seen in a typical voltage response for the five train vehicles, all traveling at 100 km/h illustrated by Figure 6, with both single and double train passages for the fleet illustrated. Following from the voltages being determined, the amount of energy obtained from each passage was determined using equation (4).

For single passages of the five trains considered, it was found the speed of the train had a significant impact on the amount of energy harvested by the cantilever device (Figure 7). The maximum energy harvested was found to occur during the passage of

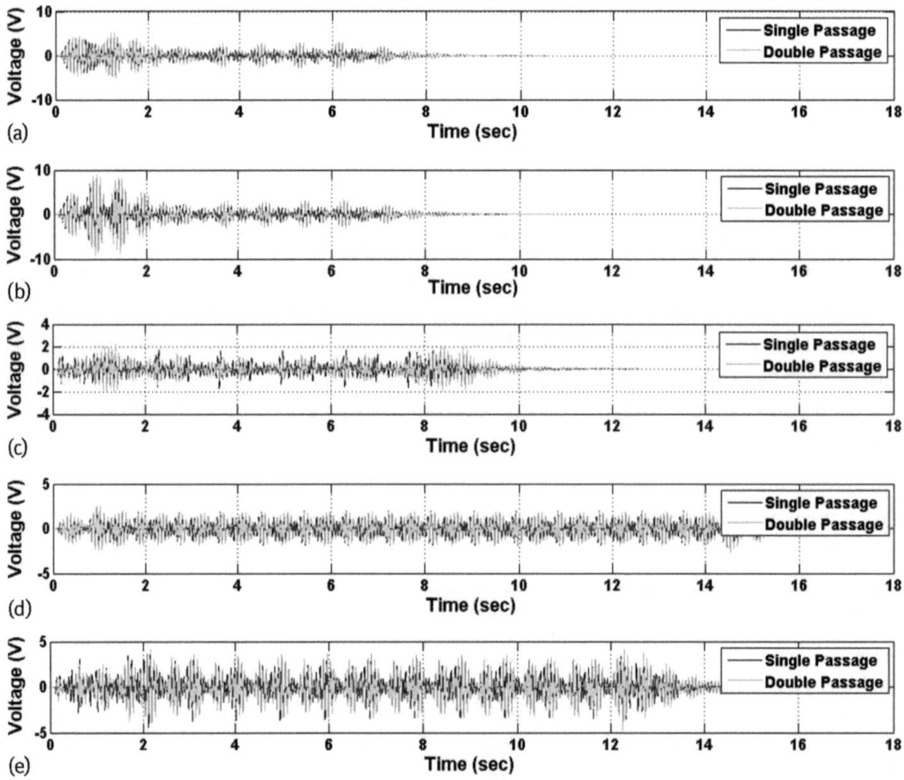

Fig. 6: Example voltage output from cantilever energy harvester for single and double train passages at 100 km/h for (a) 071Loco, (b) 201Loco, (c) TGV, (d) Shinkansen, (e) ICE.

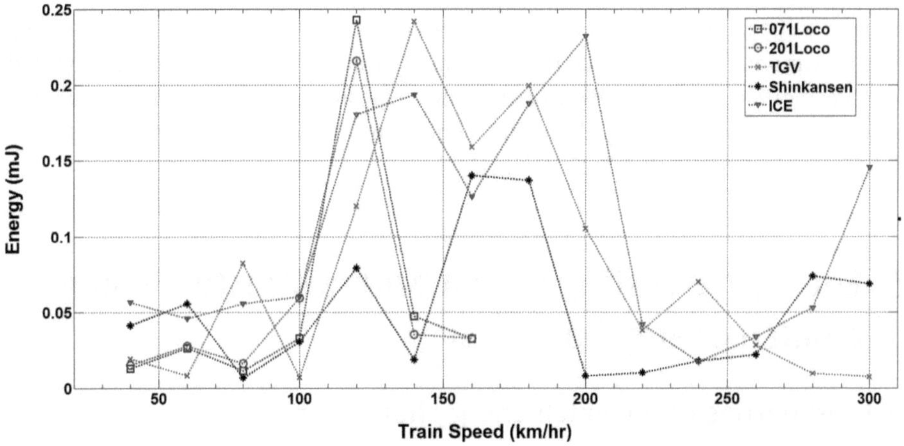

Fig. 7: Energy output from the cantilever energy harvester for single passages from the international fleet.

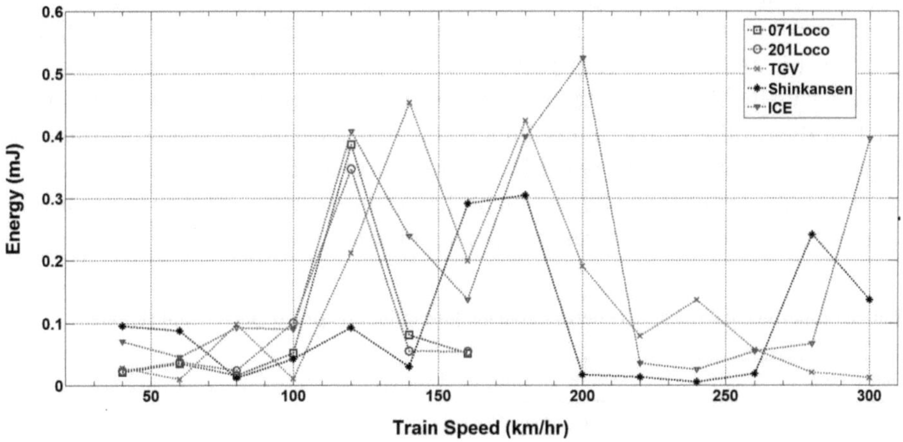

Fig. 8: Energy output from the cantilever energy harvester for double passages from the international fleet.

the 071Loco at a speed of 120 km/h, with a peak energy output of 0.243 mJ being obtained. The lowest energy being produced was during the passage of the Shinkansen at a speed of 80 km/h, with 0.006 mJ of energy being harvested from the model bridge. It was found that for the three trains capable of speeds in excess of 160 km/h, there exists a bandwidth for speeds during which the trains produced high levels of energy, ranging from 120 km/h to 200 km/h.

It was found for double train passages that the amount of energy harvested for most train speeds was higher in magnitude than their single passages counterpart (Figure 8). The maximum amount of energy that was harvested was from the ICE, traveling at a speed of 100 km/h, with an energy output of 0.526 mJ. The lowest energy output was once again obtained from the Shinkansen, with 0.005 mJ being harvested at a speed of 240 km/h. The differences in the single and double outputs illustrate the requirement for considering all loading conditions of civil infrastructure when considering energy harvesting deployments.

5 Energy harvesting from a bridge structure using a patch energy harvester

5.1 Properties of the patch energy harvester

To determine the energy harvesting potential from patch devices for integration with train bridges, a device formed from piezoelectric material of similar properties to that of the cantilever piezoelectric material, PZT, was considered. PZT exhibits strong piezoelectric properties, although being a ceramic, it can be brittle in nature [28]. The piezoelectric and physical properties of the PZT patch energy harvester are given in Table 3, with the size of the active piezoelectric material being larger than that considered for the cantilever device, resulting in a higher capacitance. Following from the strain response of the model bridge for the passage of the train fleet, as described previously, and using equation (5), the energy harvesting potential for each train passage using a PZT patch energy harvester was determined.

Tab. 3: Properties of PVDF patch energy harvesters for train–bridge interaction applications.

Parameter		Value	Unit
Capacitance	nF	224	nF
Piezoelectric constant	e_{31}	13.1	pC/N
Thickness	t_m	200	μm
Length	l	0.1	m
Width	w	0.05	m
Area	A	0.005	m^2
Permittivity constant	ϵ_{33}^S	8.96	nF/m
Resistance	R_L	1000	kΩ

5.2 Energy harvesting output from the patch energy harvester

Using the strain responses of the model bridge, the potential output of the patch energy harvesting devices can be determined. The location chosen was the same position as that used for the cantilever energy harvester, allowing for a direct comparison between the two. The voltage outputs of the patch device were found to be dependent on the train loading characteristics, with trains of greater axle loadings resulting in a higher voltage output and the train speeds and lengths determining the period energy harvesting. A typical voltage response for the five train vehicles all traveling at 100 km/h is illustrated in Figure 9. It was found that although the double train passages increased the voltage output from the patch harvester compared to the single passage counterparts, it did not lead to a doubling of the output.

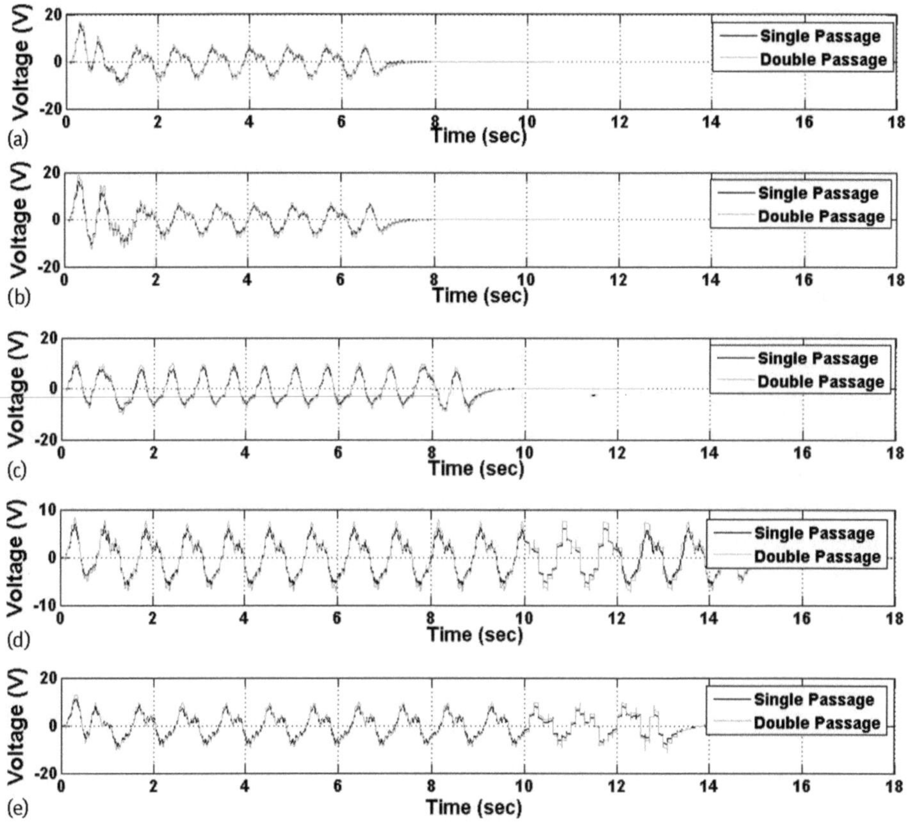

Fig. 9: Example voltage response for single and double train passages at 100 km/h for PVDF patch harvester from (a) 071Loco, (b) 201Loco, (c) TGV, (d) Shinkansen, (e) ICE.

For single passages of the five trains considered, it was found that with increasing train speed, the energy output from the PZT patch energy harvester decreased (Figure 10). This trend for patch harvesters is due to strain magnitudes being relatively constant for similar loadings, with the speed of the trains being inconsequent. There, with increasing speeds, the time duration for the energy decreases, resulting in a decrease in the energy outputs. It was found that the maximum energy output was the

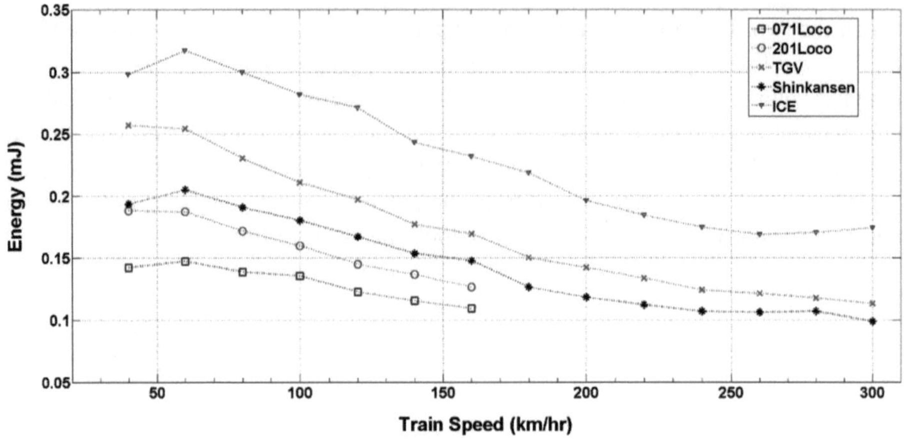

Fig. 10: PVDF energy harvester output for single passages of the international train fleet.

ICE traveling at a speed of 60 km/h, with a peak power output of 0.317 mJ, while lowest energy output was from the Shinkansen traveling at speed of 300 km/h, with 0.174 mJ harvested from the model bridge.

The energy outputs as a result of double passages were found to increase the total energy output for each train set and speed considered when compared to the power outputs from comparable single passage (Figure 11). For the double train passage, the highest output from the train fleet was, once again, the ICE traveling at a speed of 60 km/h, with an energy output of 0.429 mJ, with the energy harvesting outputs following a similar decrease with speed as found for the single passages.

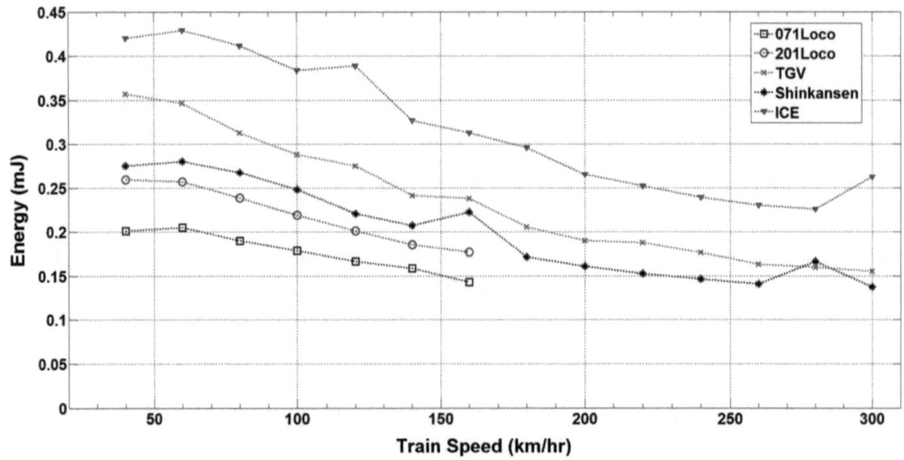

Fig. 11: PVDF energy harvester output for double passages of the international train fleet.

6 Conclusions

This chapter presents the feasibility of using train–bridge interaction for energy harvesting. Two piezoelectric energy harvester configurations were presented to capture different aspects of the dynamic response of the bridge, namely the strain and the acceleration. The bridge was created using finite element software and an international train fleet modeled to traverse the model bridge at a wide range of speeds. The voltage output for both device configurations was determined for the individual train passages and the energy output for the each of the trains over speeds ranging from 40 km/h to 300 km/h presented. This chapter further illustrates the immense potential for using piezoelectric energy harvesters for civil infrastructure applications.

Bibliography

[1] L. Zuo and X. Tang. Large-scale vibration energy harvesting. *Journal of Intelligent Material Systems and Structures*, 24(11):1405–1430, 2013.

[2] N. G. Elvin, N. Lajnef, and A. A. Elvin. Feasibility of structural monitoring with vibration powered sensors. *Smart Materials and Structures*, 15(4):977–986, 2006.

[3] E. Sazonov, D. Curry, and P. Pillay. Self-powered sensors for monitoring of highway bridges. *IEEE Sensors Journal*, 9(11):1422–1429, 2009.

[4] P. Cahill, N. A. N. Nuallain, N. Jackson, A. Mathewson, R. Karoumi, and V. Pakrashi. Energy harvesting from train-induced response in bridges. *Journal of Bridge Engineering*, 19(9):04014034, 2014.

[5] L. Frýba. A rough assessment of railway bridges for high speed trains. *Engineering Structures*, 23(5):548–556, 2001.

[6] Y.-B. Yang and J.-D. Yau. Vehicle-bridge interaction element for dynamic analysis. *Journal of Structural Engineering*, 123(11):1512–1518, 1997.

[7] K. Liu, D. Roeck, G., and G. Lombaert. The effect of dynamic train-bridge interaction on the bridge response during a train passage. *Journal of Sound and Vibration*, 325(1–2):240–251, 2009.

[8] M. Majka and M. Hartnett. Effects of speed, load and damping on the dynamic response of railway bridges and vehicles. *Computers and Structures*, 86(6):556–572, 2008.

[9] S. F. Ali, M. I. Friswell, and S. Adhikari. Analysis of energy harvesters for highway bridges. *Journal of Intelligent Material Systems and Structures*, 22(16):1929–1938, 2011.

[10] A. Erturk. Piezoelectric energy harvesting for civil infrastructure system applications: Moving loads and surface strain fluctuations. *Journal of Intelligent Material Systems and Structures*, 22(17):1959–1973, 2011.

[11] Y. Zhang, S. C. Cai, and L. Deng. Piezoelectric-based energy harvesting in bridge systems. *Journal of Intelligent Material Systems and Structures*, 25(12):1414–1428, 2014.

[12] P. Cahill, V. Jaksic, J. Keane, A. O'Sullivan, A. Mathewson, S. F. Ali, and V. Pakrashi. Effect of road surface, vehicle and device characteristics on energy harvesting from bridge–vehicle interactions. *Computer-Aided Civil and Infrastructure Engineering*, 31 (12), 921–935, 2016.

[13] M. Peigney and D. Siegert. Piezoelectric energy harvesting from traffic induced bridge vibrations. *Smart Materials and Structures*, 22(9):095019, 2013.

[14] P. Cahill, B. Hazra, R. Karoumi, A. Mathewson, and V. Pakrashi. Vibration energy harvesting based monitoring of an operational bridge undergoing forced vibration and train passage. *Mechanical Systems and Signal Processing*, 106:265–283, 2018.
[15] P. Cahill, B. Hazra, R. Karoumi, A. Mathewson, and V. Pakrashi. Data of piezoelectric vibration energy harvesting of a bridge undergoing vibration testing and train passage. *Data in Brief*, 17:261–266, 2018.
[16] N. Kaur and S. Bhalla. Combined energy harvesting and structural health monitoring potential of embedded piezo-concrete vibration sensors. *Journal of Energy Engineering*, 141(4):4014001, 2014.
[17] P. Cahill, R. O'Keeffe, N. Jackson, A. Mathewson, and V. Pakrashi. Structural health monitoring of reinforced concrete beam using piezoelectric energy harvesting system. In *EWSHM – 7th European Workshop on Structural Health Monitoring, Nantes, 8–11 July 2014*, 2014.
[18] S.-H. Kim, J.-H. Ahn, H.-M. Chung, and H.-W. Kang. Analysis of piezoelectric effects on various loading conditions for energy harvesting in a bridge system. *Sensors and Actuators A: Physical*, 167(2):468–483, 2011.
[19] T. V. Galchev, J. McCullagh, R. L. Peterson, and K. Najafi. Harvesting traffic-induced vibrations for structural health monitoring of bridges. *Journal of Micromechanics and Microengineering*, 21(10):104005, 2011.
[20] G. Gatti, M. Brennan, M. Tehrani, and D. Thompson. Harvesting energy from the vibration of a passing train using a single-degree-of-freedom oscillator. *Mechanical Systems and Signal Processing*, 66–67, 785–792, 2016.
[21] J. Wang, Z. Shi, H. Xiang, and G. Song. Modeling on energy harvesting from a railway system using piezoelectric transducers. *Smart Materials and Structures*, 24(10):105017, 2015.
[22] P. Kolakowski, J. Szelazek, K. Sekula, A. Swiercz, K. Mizerski, and P. Gutkiewicz. Structural health monitoring of a railway truss bridge using vibration-based and ultrasonic methods. *Smart Materials and Structures*, 20(3):035016, 2011.
[23] K. Sekula and P. Kołakowski. Piezo-based weigh-in-motion system for the railway transport. *Structural Control and Health Monitoring*, 19(2):199–215, 2012.
[24] C. R. Farrar and K. Worden. An introduction to structural health monitoring. *Philosophical transactions. Series A, Mathematical, physical, and engineering sciences*, 365(1851):303–315, 2007.
[25] N. Elvin and A. Erturk, editors. *Advances in energy harvesting methods*. Springer Science and Business Media, 2013.
[26] N. E. DuToit and B. L. Wardle. Experimental verification of models for microfabricated piezoelectric vibration energy harvesters. *AIAA Journal*, 45(5):1126–1137, 2007.
[27] J. Sirohi and I. Chopra. Fundamental understanding of piezoelectric strain sensors. *Journal of Intelligent Material Systems and Structures*, 11(4):246–257, 2000.
[28] J. F. Tressler, S. Alkoy, and R. E. Newnham. Piezoelectric sensors and sensor materials. *Journal of Electroceramics*, 2(4):257–272, 1998.

Feng Yang and Lin Du

Hybrid energy harvesting methodologies for energizing sensors towards power grid applications

Abstract: This chapter presents a hybrid magnetic, thermoelectric and vibrational energy harvesting (EH) system, which is oriented to power grid applications. The system can energize low-power wireless sensor networks (WSNs) that serve for condition-based monitoring of grid-connected power apparatus. Simulation and experimental results on the characteristics of the energy scavenging methodologies and the system are presented. Experimental evidence is provided to indicate the system's usability by observations on energy-autonomous Zigbee nodes, which are developed for proof-of-concept trials. This chapter is dedicated to giving an integrated case study for understanding the feasibilities and potentials of representative energy harvesting technologies, which can be promising as novel alternative power supplies in WSNs.

Keywords: Hybrid energy harvesting, WSN, power grid, magnetic, thermoelectric, vibration

1 Introduction

A power grid is an interconnected network for delivering electricity from suppliers to consumers. Unplanned power outages can result in a considerable impact on the network operation [1]. Therein, equipment failures represent a substantial proportion of the accidents. To have a reliable and sound operation of in-service power apparatus, it is necessary to identify their potential problems at an early stage before a catastrophic failure occurs [2]. In this regard, condition monitoring (CM) technology can facilitate online tracking of the equipment health and enabling utilities to respond to changing status with proactive arrangements. On the other hand, with recent interests and advancements worldwide to develop robust smart grids, at the bottom of the initiatives stands an enabling technique that is responsible for real-time monitoring of strategic power facilities and perceiving information, which is known as sensor technology.

Low-cost and miniaturized WSNs are preferable candidates to undertake massive and distributed monitoring tasks in an energy system. A number of demonstrative onsite trails to monitor the grid using WSN have been launched, such as the practice in Kentucky [3], where WSN has been utilized to acquire data from circuit breakers and transformers, as well as ambient temperature, etc. However, the limited lifetime as-

Feng Yang, Lin Du, State Key Laboratory of Power Transmission Equipment & System Security and New Technology Chongqing University, Chongqing 40044, China

https://doi.org/10.1515/9783110445053-021

sociated with the problematic power supplies of wireless sensors has become one of the challenges that need to be addressed before proliferation of WSNs in energy systems. The stated issue is seemingly very confusing to understand before a paradox is well recognized. Despite the vicinity of large electrical energy flows, there are no easy ways to directly wire a power source from high-voltage (HV) facilities by simply connecting the network [4], as the connection may compromise electrical insulation. On the other hand, batteries have a finite lifetime, and, therefore, regular recharging is needed. Motivated by recent interest in emerging energy harvesting (EH) technologies, in this chapter, we attempt to exploit potential energy sources that are present surrounding power grids to serve as power supplies for the unresolved issue.

2 Energy profiles and methodologies

2.1 Overview

Energy harvesting (EH) is the process of electronically capturing and accumulating energy from a variety of energy sources deemed wasted or otherwise said to be unusable for any practical purpose. Energy harvesting, is sometimes associated with capturing residual energy as a by-product of a natural environmental phenomenon or industrial process and is, therefore, considered free energy [5].

Implementation of EH under a power grid is justified by the reasonable ambient energy profiles associated with different power equipment. Undoubtedly, the grid is awash in the three forms of energy, which are essentially subjected to the three effects of current, i.e., calorific effect, electrodynamic effect and electromagnetic induction. For instance, an in-service transformer vibrates due to the electromagnetic force applied on its components. It was reported that the vibration level of an in-service 132/66 kV, 40 MVA transformer is above $1.0\,\mathrm{m/s^2}$ most of the time. A Perpetuum PMG 17 vibration energy harvester could produce 4.5 mW under such conditions [6]. Field measurements in a substation demonstrate that the maximum magnetic field intensity in a 500 kV substation reaches about 5 mT under a 35 kV dry-type air core reactor, indicating that the maximum magnetic energy density can be $9.9\,\mathrm{\mu W/cm^3}$ [7]. As a farmer must familiarize and prepare the land to harvest a predictable and useful crop, it is necessary to understand the potentials of an environment to harvest enough energy to accommodate a useful application. Therefore, a brief investigation of energy harvesting profiles under power grids follows.

2.2 Magnetic energy harvesting

A well-known fact is that the very essence of an energy system is electromagnetic induction, and to harness the power contained in the magnetic or electric field should always be the first choice. It is noteworthy that to harness the magnetic field energy can be more realistically termed as “draw out”, rather than “harvesting”, as electricity is transmitted via the very same medium which we are considering, the electromagnetic field encircling current-carrying conductors. Therefore, exploiting magnetic energy is usually not scavenging or recovering waste power, but extracting the energy from power flows.

Harvesting magnetic energy can be rewarding where intensive load currents are carried, e.g., medium and low-voltage busbars, cables and inductors. Before designing any dedicated energy harvesters, the profiles of the magnetic field should be clearly understood. We herein select a segment of a 35 kV busbar bridge (Figure 1) in a 220 kV substation for analysis.

Fig. 1: The 35 kV busbar bridge for analysis. There are three phase conductors on the bridge. The cross-section of the rectangular conductors is 125 mm × 10 mm. The conductors are equidistant with a distance of 0.4 m and loaded with symmetry currents of 1039 A.

The magnetic induction intensity amplitude B_m excited by the conductors of three phases can be calculated using the well-known Biot–Savart law. Figure 2 illustrates the distribution of B_m in the vicinity of a busbar bridge.

Figure 2 reveals that around a conductor, the magnetic field concentrates on the conductor surface but decays dramatically while moving away. With this understanding, the harvester needs to be stuck onto the surface in order to draw out the most intensive power with the highest efficiency. The current transformer (CT)-based structure is conventionally used [8, 9], however, in this manner, the apparatus needs to encircle overhead conductors. This structure can be problematic when live-line working is prohibitive or de-energizing the line is unscheduled; three-phase conductors are compact with insufficient remaining interspace to add any harvesters as phase insulation must be guaranteed; the iron in a CT gains weight that may compromise the structural bearing of the busbar bridge.

This chapter provides a type of one-side structure for the harvester. Theoretically, the harvester is applicable to enclosures of any equipment, as long as a magnetic field exists thereof. A special funnel-shaped iron core is used, which features a larger surface area at the two ends. Just as a funnel concentrates liquid, flux lines can be effectively intensified at the middle section of the core. Therefore, in this way, more magnetic energy can be harvested.

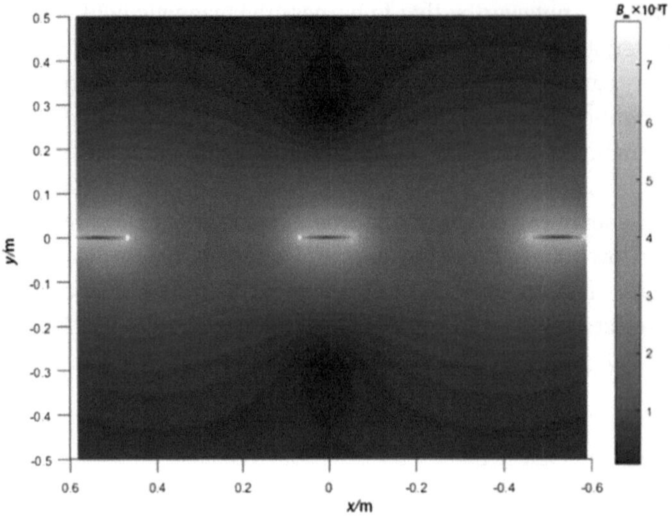

Fig. 2: Cross-sectional view of the distribution of B_m surrounding a busbar bridge.

The designed structure is then simulated using a finite element software. The present structure is compared with the conventional bar-type core and coreless winding, aiming to embody the changes of flux under different core shapes. In each case, the current in the busbar is set as 500 A and B values are inspected, see Figures 3–5. Comparing Figures 3 and 4, it is revealed that the funnel shape is proven to be capable of concentrating flux, as the maximum of B is increased by 71.9%.

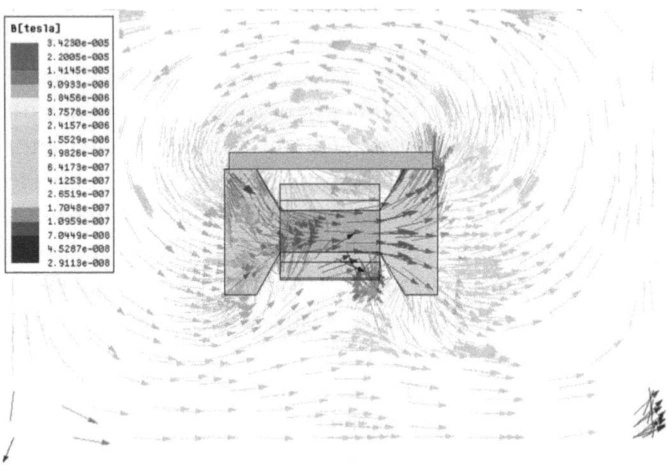

Fig. 3: Vector overlay of the magnetic induction intensity B in the case of using a funnel-shaped core (section view). Under these three cases of using three different core types, the harvester is attached on a rectangular busbar and a winding is wired on the intermediate part of the core.

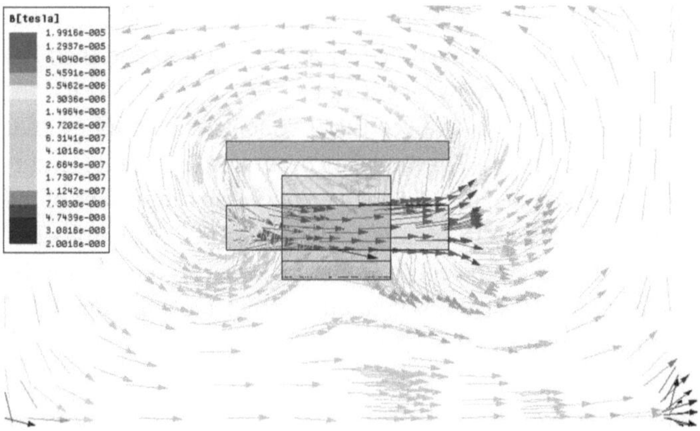

Fig. 4: Vector overlay of the magnetic induction intensity *B* in the case of using a bar-type core (section view).

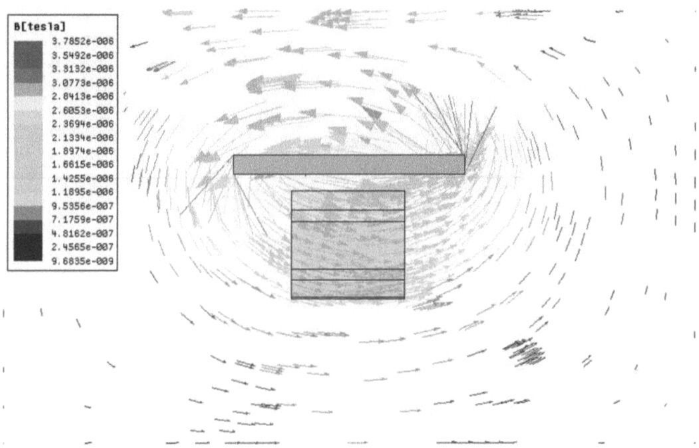

Fig. 5: Vector overlay of the magnetic induction intensity *B* in the case of using coreless winding (section view).

2.3 Thermoelectric energy harvesting

Given the statement that electromagnetic energy is the first choice, harvesting thermal energy can be the second choice. Heat is produced, whether desired or not, by the generation and propagation of most other energy forms. Hot spots on operational power assets can support thermoelectric generator (TEG) to convert dissipated heat into electricity as a result of the Seebeck effect. For instance, the literature [10] reports that on an investigated 63 MVA transformer, hot spots are located at the center of the tank and in the upper part of the radiator. Their temperatures range from 333 to 353 K. Assuming an ambient air temperature of 25 °C (298 K), this means there is a usable temperature gradient of approximately 55 K. With a given heat source and a harvesting asset housed onto it, heat flow can be absorbed by a concentrator, and then the residue heat is dissipated into the ambient by a heat sink.

A simulation model is established to analyze the thermoelectric coupling that achieves thermoelectric energy harvesting. The model incorporates a segment of a current-carrying busbar as the heat source. Four TEGs are arranged on the surface of the busbar, and they are connected in series electrically. Inside each TEG, 127 thermoelectric couples are arranged with their thermoelectric properties set as Bi_2Te_3. The model is solved using the finite element analysis (FEA) method. Figure 6 plots the simulated results of the potential distribution inside the TEGs.

Fig. 6: Potential distribution on the thermoelectric couples inside each TEG.

2.4 Vibrational energy harvesting

Untapped kinetic energy is almost everywhere in infrastructures, automobiles and industrial plants in the form of vibrations. The motion is plentiful in power plants and substations. For example, steam turbines and transformers vibrate periodically. The kinetic energy can be converted into electricity using piezoelectric transducers. An off-the-shelf PZT device named Volture™ V21BL (Figure 7(a)) has been employed. The transducer was forced by the sinusoidal motion of the bench (Figure 7(b)).

2.5 Hybrid energy management system

Power management circuits for EH applications should feature high conversion efficiency [11]. Therefore, ultra-low-power consumption circuitry is used to enable hybrid energy management. The structure of the management system is shown in Figure 8.

(a)

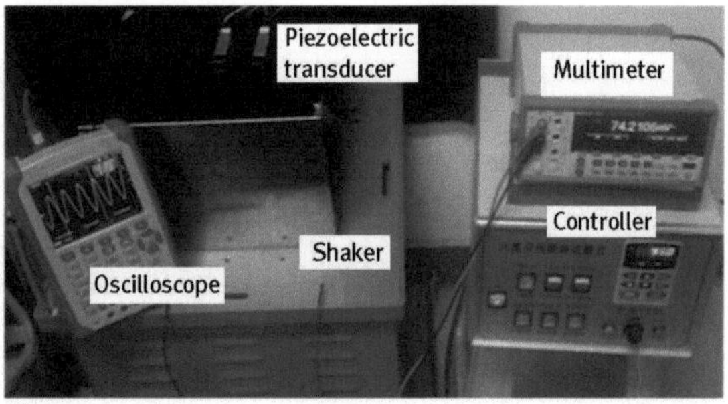

(b)

Fig. 7: (a) The piezoelectric transducer employed for experiments; (b) the vibration bench that consists of an oscilloscope, a shaker, a multimeter and a vibration controller.

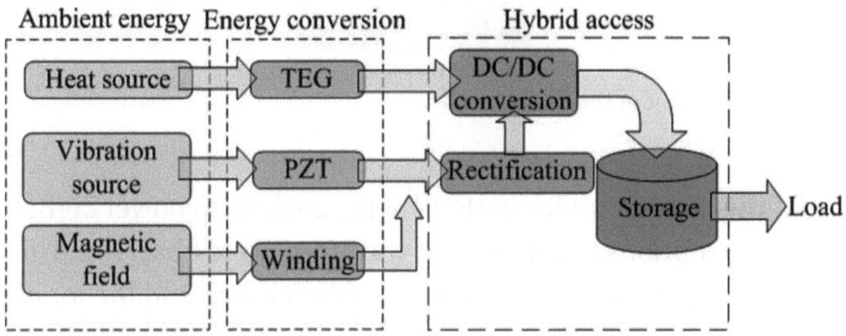

Fig. 8: Scheme of the hybrid energy management system.

Fig. 9: Setup of the tentative application.

A couple of Zigbee sensor nodes was adopted in this chapter to launch tentative application with genuine load. The sensor nodes are based on a system-on-chip solution for 2.4 GHz IEEE 802.15.4 for Zigbee applications. The gateway and sensor nodes have been programmed to process temperature measurement.

In the test, we observed that the sensor node reports data to the gateway node every 2 s continually as programmed without power interruptions. The received data of the ambient temperature and humidity are displayed on the interface in a host computer.

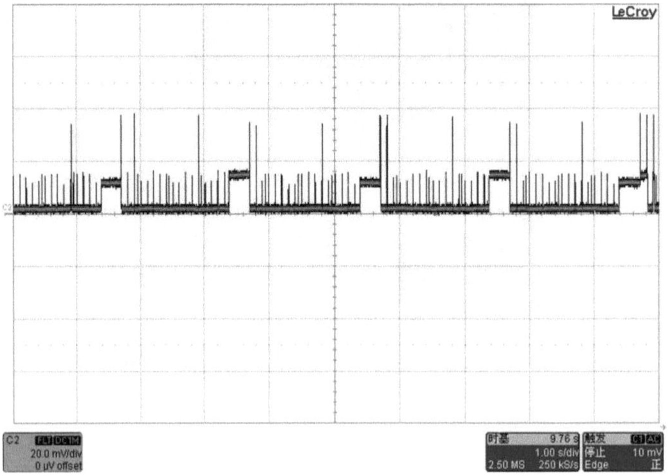

Fig. 10: Waveform of the operational current of the Zigbee node.

Figure 10 indicates that the quantity of harvested energy suffices to power ZigBee sensor nodes, and thus no interruptions in the current waveform were observed. This experiment indicates that the system can be fully energy-autonomous and the topology is capable of powering low-power sensors.

3 Envisioned prototype

Based on the existing work within the design scheme, the authors envision a miniaturized energy harvesting mote (Figure 11), as is planned as further work. The EH mote

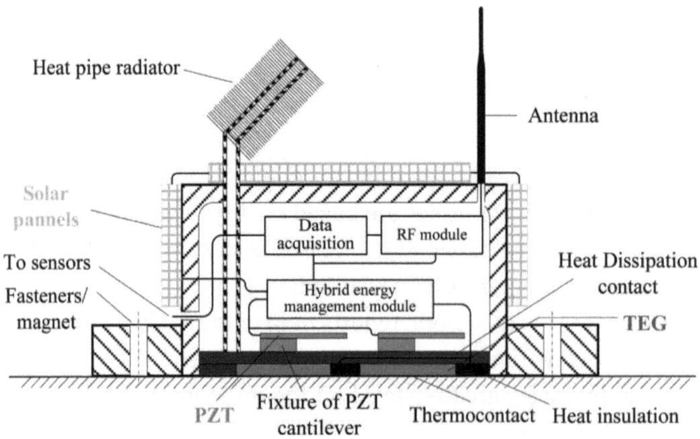

Fig. 11: Scheme of the envisioned prototype of an energy harvesting mote.

is mainly anticipated to be non-intrusive and easy to deploy. Assuming that the mote is attached handily to the enclosure of host equipment with a powerful magnet fixture if applicable (considering that most power apparatus have robust iron casing or components), the node can be powered by the discussed energy modalities, which may be locally available. The structure signifies the tailless and "deploy and forget" nature of a wireless sensor node in the future.

As the scheme demonstrates, solar, thermal and vibrational EH components are cooperative inside to constitute the power module. On-board solar panels cover all the outer surfaces to receive sunlight from all directions within a day. TEGs absorb heat from any radiating surfaces and channel the heat outside using an effective heat pipe radiator. PZTs are mounted tightly with the mote to obtain outer motions. Hybrid energy management, data acquisition and RF are the other modules involved in the sensing function.

Future work will incorporate trials on real-life apparatus for customized specifications and prototype fabrication to achieve maximized conversion efficiency.

4 Conclusions

Magnetic, thermoelectric and vibrational energy harvesting are discussed in this chapter to understand the feasibility of EH in scenarios that arise most commonly in a power grid. Tentative applications demonstrate that the EH approaches and the hybrid EH structures are able to energize low-power sensors. The research of this chapter is promising as solutions for the problematic power supplies of sensors for CM monitoring of power equipment.

Bibliography

[1] S. M. Kaplan. Electric power transmission: background and policy issues. *US Congressional Research Service*, 14:4–5, 2009.

[2] J. Singh et al. Condition monitoring of power transformers-bibliography survey. *IEEE Electrical Insulation Magazine*, 24(3):11, 2008.

[3] M. Erol-Kantarci and H. T. Mouftah. Suresense: sustainable wireless rechargeable sensor networks for the smart grid. *IEEE Wireless Communications*, 19(3), 2012.

[4] F. Yang et al. Hybrid energy harvesting for condition monitoring sensors in power grids. *Energy*, 118:435–445, 2017.

[5] A. Jain and S. Bhullar. Emerging Dimensions in the Energy Harvesting. *IOSR Journal of Electrical and Electronics Engineering*, pages 70–80, 2012.

[6] 2008. Perpetuum PMG17 datasheet, Perpetuum Ltd, Southampton, UK.

[7] L. Xu et al. *Analysis of power frequency magnetic field simulation of air core reactor based on the method of moments.* High Voltage Apparatus, 2015.

[8] Du, Lin, et al. A Novel Power Supply of Online Monitoring Systems for Power Transmission Lines. *IEEE Transactions on Industrial Electronics*, 57(8):2889–2895, 2010.
[9] N. M. Roscoe and M. D. Judd. Harvesting Energy From Magnetic Fields to Power Condition Monitoring Sensors. *IEEE Sensors Journal*, 13(6):2263–2270, 2013.
[10] K. Murakami, K. Sasaki, T. Shindo, et al. Development of Waste Heat Recovery System from Transformer with Bi-Te thermoelectric modules. In *2nd European Conference on Thermoelectrics, Poland*, 2004.
[11] M. H. Nehrir et al. A review of hybrid renewable/alternative energy systems for electric power generation: Configurations, control, and applications. *IEEE Transactions on Sustainable Energy*, 2(4):392–403, 2011.

Xinming Zhao, Thomas Keutel, and Olfa Kanoun

Energy harvesting for a wireless monitoring system of overhead high-voltage power lines

Abstract: For electric power transmission overhead power line monitoring plays an important role in the operating security for grid operators, as well as for the supply availability at optimized cost under consideration of renewable energies. In this chapter, an approach for harvesting energy from the electric field around the power line to supply wireless sensor nodes is presented. High requirements for functionality, reliability and limited accessibility result in the development of a specific energy management strategy. The prototypes of the harvester and the power module were realized and tested under high-voltage (HV) laboratory conditions. A field test was carried out under real operating conditions. The test results confirm that the power module can collect enough energy to power the sensor nodes. The chapter presents theoretical considerations of energy availability from the electric field, the design of the energy harvester and measurements with the prototype in the HV laboratory and in the field. The field test results demonstrate that the energy harvested is enough to supply the sensor nodes at 16 mW.

Keywords: Energy harvesting, wireless sensor network, online monitoring, HV transmission line

1 Introduction

The power distribution of newly installed renewable energy resources, for example, offshore wind farms, is becoming more challenging for grid operators. The amount of energy that can be transported on overhead power lines depends, the one hand, on the current state of the grid and, on the other hand, on conditions like the temperature of the power line and wind. Also, considering extreme weather events, especially in winter, involves a power line monitoring system. Therefore, several critical parameters must be monitored in time directly at the power line to guarantee their operability. In order to achieve low installation and maintenance effort, an autonomous wireless sensor network for online monitoring was developed [1, 2]. The wireless monitoring system incorporates sensors to measure temperature, the inclination and the current load of the power lines. The sensor nodes are directly mounted onto the conductor rope, and the data measured are transmitted wirelessly to a base station.

Xinming Zhao, Thomas Keutel, Olfa Kanoun, Chair for Measurement and Sensor Technology, Technische Universität Chemnitz, 09126 Chemnitz, Germany

https://doi.org/10.1515/9783110445053-022

Besides the electric field harvester, the potential resources for the energy harvesting include photovoltaic panels (PV), thermoelectric generation (TEG), kinetic energy (e.g., piezoelectric generation, wind turbines) and electromagnetic field harvesters. PV panels are generally not considered suitable for power supply because they need to be installed on the transmission tower, and the energy harvesting would be greatly impacted by the sunlight conditions. In the case of good light during the day, the harvester system can collect enough energy to supply the sensor nodes. However, in bad weather conditions, the fluctuant energy source can even threaten the normal operation of the monitoring system. In [3], a temperature monitoring system of an overhead transmission line is shown, which used two solar panels for the power supply. The two solar panels are separately fixed at the top of the tower. This method for energy harvesting not only results in shortcomings of restricted energy availability but also increases the installation and maintenance costs. TEG is also not suitable because of the limited and fluctuating temperature swing between the power line and the ambient environment. It is, therefore, difficult to achieve a power level of about 10 mW. Kinetic energy also depends on the ambient environment conditions and cannot deliver stable power to the sensor nodes.

Two promising methods for energy harvesting at a HV power line are the current transformer and energy harvesting based on the electric field. Compared to AC, AC voltage is a stable and reliable energy source to collect energy from the ambient electric field because of its independence from the conductor's current load. Energy harvesting from the electric field becomes a viable option. In [4], an approach for harvesting energy in substations is presented. It powers autonomous sensors from the ambient electric field, which gives the concept of energy conversion by switching at 100 Hz on the peaks of the electric field around the HV power line. From the simulation results, the power level harvested is in the range of hundreds microwatts, and there is no test module to verify the simulation results. Moreover, the switch needs to be synchronized with the phase of the voltage waveform. This control method is complex and susceptible to interference from the ambient electric field. Another approach for energy harvesting from the electric field is discussed in [5].

This paper presents experimental results for power harvesting based on the test module. However, few investigations have been carried out on the theoretical simulation, which aims to estimate the available power level from the harvester. Therefore, this paper focuses on a new approach to calculate the harvested power based on the physical model. Some theoretical fundamentals were discussed in [1, 2]. Furthermore, the first prototypes of the harvester and the power module will be introduced in this paper, and the total power system tested in the HV laboratory and in field tests, respectively. To summarize the different energy harvesting methods for the online monitoring system of the overhead transmission line, a comparison is shown in Table 1.

Tab. 1: Comparison of different energy harvesting methods for the online monitoring system of the overhead transmission line.

Method	Size	Complexity of system	Power rating	Energy availability
Solar energy (PV)	++	+	+++	--
TEG	+	+	--	--
Kinetic energy	--	--	+	+
Magnetic field	++	++	+++	++
Electric field	++	+++	++	+++

2 Theoretical consideration and simulation

2.1 Theoretical analysis

Figure 1(a) shows the principle structure of the energy harvester, which comprises a tube with the radius r_2, the axial length l, the dielectric material ϵ_r between the conductor and harvester electrode and the height d to the ground. Consequently, two capacitances are formed from the wire to the ground because of the cylindrical harvester electrode. The two-capacitance model is shown in Figure 1(b); C_{WE} (wire to the electrode) and C_{EG} (electrode to ground). Due to this physical model, there is the possibility to evaluate the energy obtained from the electric field around the conductor regarding all parameters of the harvester structure. The two capacitors, which are connected in series, can be seen as a capacitive voltage divider [6], as shown in Figure 1(b); u_0 is the high voltage on the conductor, C_{WE} is the capacitance of the cylindrical electrode around the conductor, C_{EG} is the capacitance of the cylindrical electrode to ground, Z_L

Fig. 1: (a) Structure of the cylinder electrode and (b) equivalent circuit of the cylindrical harvester.

is the load impedance and u is the voltage at the load. Equation (1) is derived based on the circuit theory:

$$u = \frac{\frac{Z_L}{1+j\omega C_{WE} Z_L}}{\frac{Z_L}{1+j\omega C_{WE} Z_L} + \frac{1}{1+j\omega C_{EG}}} u_0 \,. \tag{1}$$

The capacitances C_{WE} and C_{EG} are described by Equations (2) and (3)

$$C_{WE} = \frac{2\pi\epsilon_0\epsilon_r l}{\ln\left(\frac{r_2}{r_1}\right)} \,. \tag{2}$$

$$C_{EG} = \frac{2\pi\epsilon_0 l}{\cosh^{-1}\left(\frac{d}{r_2}\right)} \,. \tag{3}$$

The voltage (U) at the load is given by equation (4) through inserting Equations (2) and (3) into equation (1):

$$u = u_0 \cdot \frac{j\omega Z_L \cdot 2\pi\epsilon_0 l \ln\left(\frac{r_2}{r_1}\right)}{\ln\left(\frac{r_2}{r_1}\right) \cdot \cosh^{-1}\left(\frac{d}{r_2}\right) + j\omega Z_L \cdot 2\pi\epsilon_0 l \left[\ln\left(\frac{r_2}{r_1}\right) + \epsilon_r \cdot \cosh^{-1}\left(\frac{d}{r_2}\right)\right]} \tag{4}$$

By substituting following terms in equation (4):

$$\begin{cases} m = 2\pi\epsilon_0 l \cdot \ln\left(\dfrac{r_2}{r_1}\right) \\ n = \ln\left(\dfrac{r_2}{r_1}\right) \cdot \cosh^{-1}\left(\dfrac{d}{r_2}\right) \\ q = 2\pi\epsilon_0 l \cdot \left[\ln\left(\dfrac{r_2}{r_1}\right) + \epsilon_r \cdot \cosh^{-1}\left(\dfrac{d}{r_2}\right)\right] \end{cases}$$

the equation results in:

$$u = \frac{j\omega Z_L m u_0}{n + j\omega Z_L q} \,. \tag{5}$$

Equation (5) describes the voltage at the load, and the obtainable power from the harvester can also be calculated with respect to a load.

2.2 Analytical simulation

For a resistive load the power can be expressed as:

$$P = \frac{|U|^2}{Z_L} = \frac{(\omega m u_0)^2 Z_L}{n^2 + (\omega Z_L q)^2} \tag{6}$$

When, $\delta P/\delta Z_L = 0$, $Z_{L_{opt}}$ can be calculated by equation (7), and at the same time the maximal power P_{max} can be calculated at $Z_L = Z_{L_{opt}}$ by equation (8)

$$Z_{L_{opt}} = \frac{n}{q\omega} \tag{7}$$

$$P_{max} = P(Z_{L_{opt}}) = \frac{(\omega \cdot m u_0)^2 Z_{L_{opt}}}{n^2 + (\omega Z_{L_{opt}} q)^2} \tag{8}$$

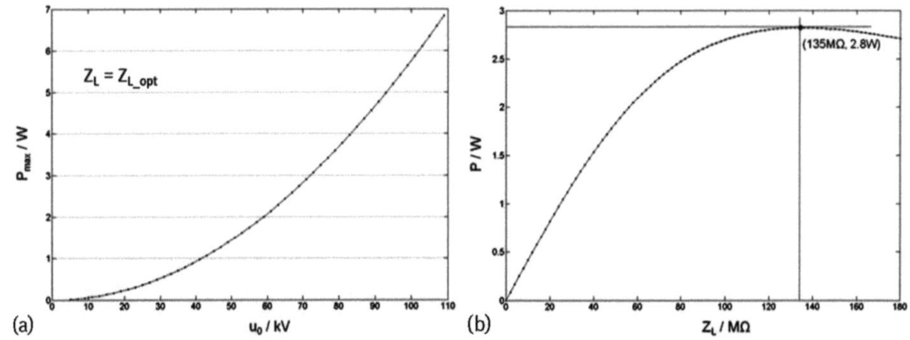

Fig. 2: Dependency of the maximal available power at the nominal voltage. The simulation parameters are $d = 90$ cm, $r_1 = 1$ cm, $r_2 = 7.75$ cm, $l = 52.5$ cm, $Z_L = Z_{L_{opt}}$. (b) Dependency of the available power on the load. The simulation conditions are $d = 90$ cm, $r_1 = 1$ cm, $r_2 = 7.75$ cm, $l = 52.5$ cm, $u_0 = 70$ kV.

In the ideal case, the maximal power can be obtained when the condition $Z_L = Z_{L_{opt}}$ is satisfied. Figure 2(a) shows the maximal available power (P_{max}) concerning the nominal voltage (u_0). The power continuously increases by an also increasing conductor voltage (u_0). Figure 2(b) shows the relation between the available power and the load. For the determined conditions given in Figure 2(b), the maximal available power can be achieved at the load resistance about 135 MΩ. From Figure 2(b), we can see that in order to achieve more power from the harvester, the impedance on the load must be increased as much as possible. Therefore, we propose one solution using a specific transformer configuration, because it acts as an impedance transformation. Besides this method for voltage conversion, other principles can also be considered, such as a switch transformer. The switching transformer needs a high resonant frequency. This leads to a higher complexity and increases the energy consumption to control the necessary components. Figure 3 shows the dependency of the available power on the nominal voltage (u_0) for the three different loads: 1 MΩ, 2 MΩ and 5 MΩ, which values are possible to implement with the method of impedance conversion. Except for the load impedance, the other simulation conditions are the same as in the example discussed above. The figure shows that more power can be obtained from a larger load impedance and a higher conductor voltage. As the results in this figure show, at the nominal voltage of 70 kV, the available power of 43 mW can be achieved from 1 MΩ, 84 mW can be achieved from 2 MΩ and 209 mW can be achieved from 5 MΩ. Because in real operational conditions the distance between the conductor and ground is about 15 m, which is superior to the height in the HV laboratory condition. The influence of the distance of the height between the conductor and the ground is simulated, and the result is shown in Figure 4. As expected, the power decreases with an increase of the height, but the decrease is nonlinear due to the nonlinear change of the capacitance. The available power decreases quickly with

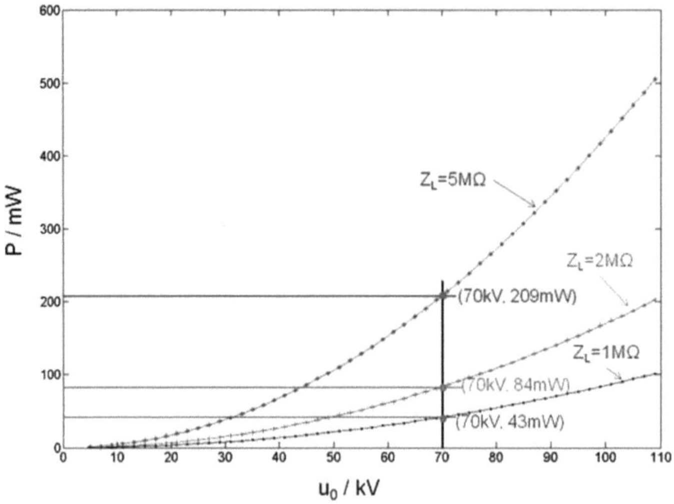

Fig. 3: Dependency of the available power on the nominal voltage under three different loads: 1 MΩ, 2 MΩ and 5 MΩ. The simulation parameters are $r_1 = 1$ cm, $r_2 = 7.75$ cm, $l = 52.5$ cm and $d = 90$ cm.

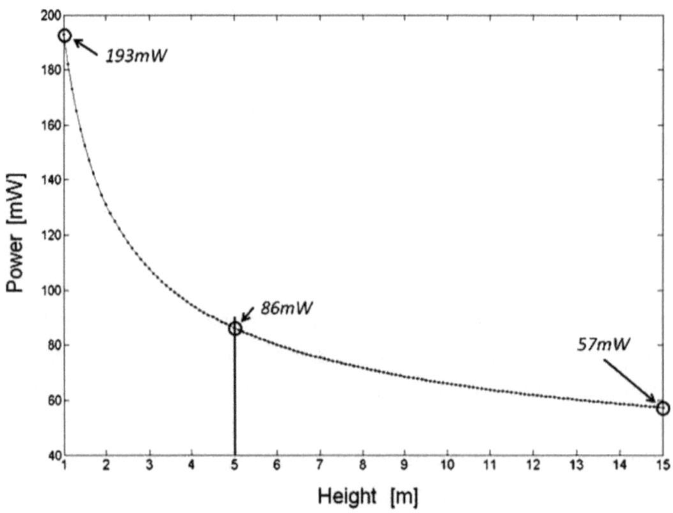

Fig. 4: Influence of height (d) on the available power. The simulation parameters are $u_0 = 110$ kV, $r_1 = 1$ cm, $r_2 = 7.75$ cm, $l = 52.5$ cm and $Z_L = 2$ MΩ.

the height at the initial distance from 1 to 5 m, and the decrease is rather slow after the height becomes greater than 5 m. An increase of the distance from about 1 to 5 m reduces the available power by about 55%, and the increase from about 5 m to approximately 15 m reduces the available power only by about 34%. Accordingly, the total decrease of the available power with an increase of the distance from 1 meter to 15 m is about 70%. Therefore, under laboratory conditions, with a lowconductor height, the harvester device can be tested at a relatively low nominal voltage to simulate similar power results at a nominally high voltage in the real operation.

2.3 Finite element simulation

For the normal operation of the electronic devices inside the harvester tube, it is mandatory to examine the electric field strength inside the tube. To verify the electric field strength and distribution inside the harvester tube, electric field analysis was carried out by finite element methods using the tool Comsol [7]. Figure 5 shows the FEM model for electric field analysis [8, 9]. The thickness of the tubular metal plate is 2 mm. Further parameters are chosen as stated in the caption of Figure 2. Electric field-based energy harvesting works as a capacitive voltage divider by using a metal tube mounted onto the transmission conductor to extract energy from electric field. As described above, the two equivalent capacitors are connected in series, where CWE stands for the equivalent capacitance between the HV transmission line and the tube, and C_{EG} represents the equivalent capacitance between the tube and the earth ground. The power conditioning and its load can be seen as an equivalent load, which are connected with CWE in parallel. If the power obtained from the harvester increases, the voltage stress on the tube to the power line will also increase. If there is no active load, potential differences across the tube could be as high as several thousand volts. The enormous field strength inside the harvester tube could easily destroy electronic devices, such as microprocessor, sensors and other electronic de-

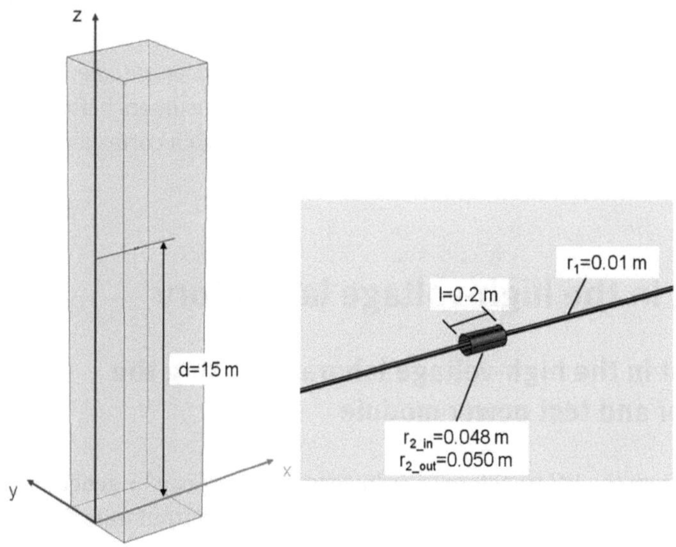

Fig. 5: FEM simulation setup for electric field analysis.

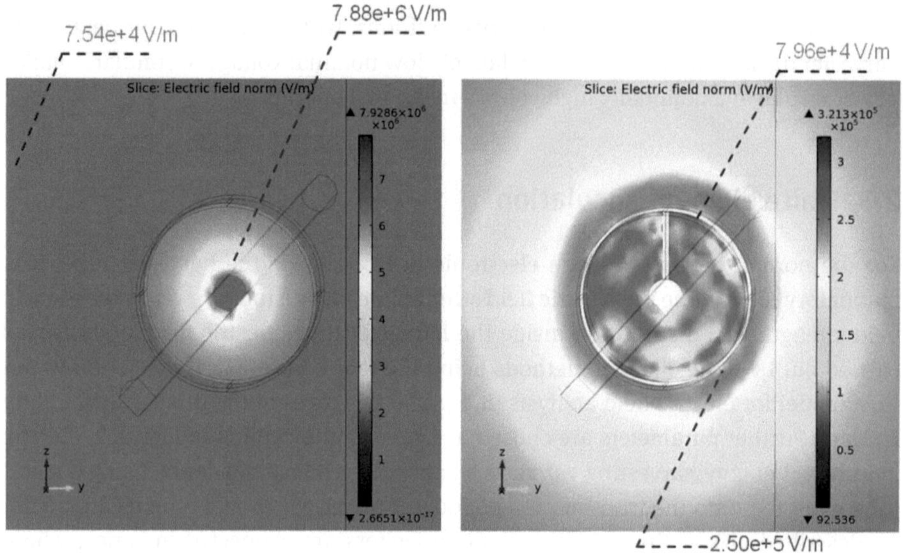

Fig. 6: Electric field strength inside and outside the tube (a) without active load and (b) with active load.

vices. Figure 6 shows the FEM simulation comparison of the electric field strength in the vicinity of a HV conductor for two cases: without active load and with active load. The tube is used as the energy harvester, but at the same time, it can also be used for electric field damping to reduce the field strength. The potential difference between the tube and the transmission conductor should be limited by a shunt regulator or active load. It is clearly shown that the electric field concentration without active load would be much higher than the case with active load. As the simulation in Figure 6(b) shows, the field strength inside the tube is low. Here, electronic devices are safe. Thus, for the circuit design, a dummy load should replace the energy consumers before the sensor nodes go into sleep mode and consume very low energy, which corresponds to the harvester tube without load.

3 Experiment in the high-voltage laboratory

3.1 Measurement in the high-voltage laboratory using the test harvester and test power module

In the HV laboratory, from the HV transformer, voltage up to 110 kV can be generated. The high voltage is adjustable in the range of 0 V to 110 kV. The HV output from the HV transformer is connected with the power line, on which the sensor node is mounted. On the other end of the power line, a ball terminal is installed in order to avoid un-

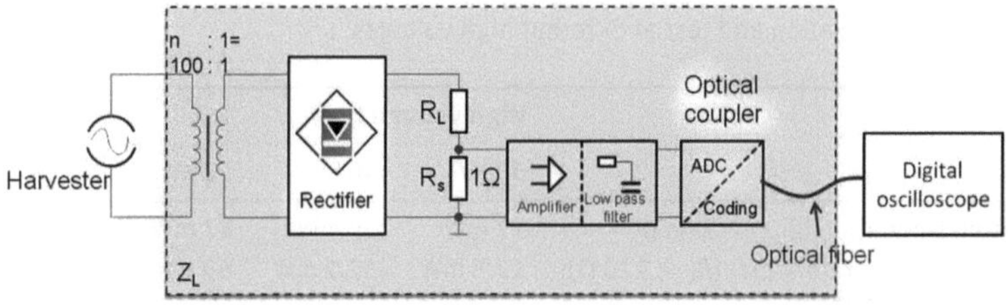

Fig. 7: Test module for power measurement inside the harvester tube.

desirable high-pressure discharge under the high voltage of the power line. Figure 7 shows the block diagram of the test module, which was placed inside the harvester tube for the power measurement.

Figure 8 shows the measurement results of the available power depending on the nominal high voltage for the different load resistances. For the first tests in the HV laboratory, the cylindrical harvester tubes were used to verify the previous theoretical investigation. Tube 2 has a length of 52.5 cm, and its diameter is 15.5 cm. The measurement results in Figure 8 show that the harvested power increases with an increasing conductor voltage. Meanwhile, the harvested power also increases with an increase of the load resistance. Therefore, more power can be obtained from a larger load resistance. Table 2 shows the comparison of the harvested power based on the simulation and actual measurement results at different nominal voltages (u_0) with a load resistance of 220 Ω. The two results between simulation and measurement in the HV laboratory are very close.

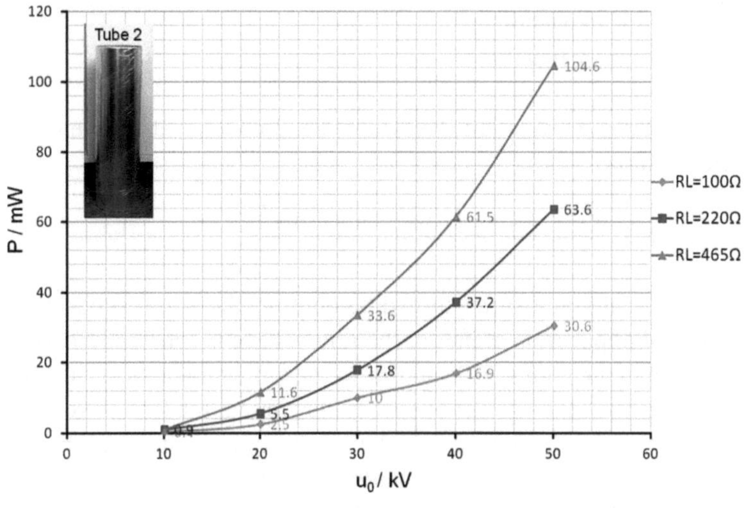

Fig. 8: Dependency of the available power at the nominal high voltage for different loads. Radius of tube 2: $r_2 = 77.5$ mm; length of tube 2: $l = 525$ mm and height of the power line to the ground: $d = 90$ cm.

Tab. 2: Harvested power of simulation and test at different high voltages.

	High voltage (u_0)		
	30 kV	**40 kV**	**50 kV**
Harvesting power from simulation (Z_L = 2.2 MΩ)	17 mW	30 mW	47 mW
Harvesting power from HV laboratory test (R_L = 220 Ω)	17.8 mW	37.2 mW	63.6 mW

3.2 Measurement in the high-voltage laboratory using the prototype of harvester housing and power module

According to the theoretical investigation for the design of the harvester, a cylindrical tube with a diameter of 10 cm and a length of 20 cm was chosen as the harvester housing. The first prototype of the power module is placed in the harvester housing. An optimized transformer with a turns ratio of 120 was proposed to increase the output power. The transformer is a special HV transformer with a maximum rating voltage of 3000 kV and a comparatively low rating power of 0.5 W. Three supercapacitor modules were used in the prototype of the power module with 3.6 F. The scavenged energy is stored in the supercapacitors and the harvested power can be calculated by equation (9),

$$W_e = \frac{1}{2} \cdot CU^2 \tag{9}$$

where

W_e is the energy stored in the supercapacitor, J;
C is the value of the supercapacitor, F;
U is the voltage of the supercapacitor, V;

The power can be calculated by equation (10), where the time interval between two measurement points is known:

$$P = \frac{W_{e_2} - W_{e_1}}{t_2 - t_1} = \frac{\frac{1}{2} \cdot C{U_2}^2 - \frac{1}{2} \cdot C{U_1}^2}{t_2 - t_1} \tag{10}$$

The test conditions in HV laboratory are a little different from the actual application conditions. In the actual application scenario, the grid voltage is constantly 110 kV. The height of power line to the ground is 15 m. Due to the former theoretical calculation and simulation, the higher the power line from the ground, the less power will be harvested. According to the theoretical calculation, it can be estimated that the power harvested at the 60 kV nominal voltage under test conditions in the HV laboratory is roughly equivalent to the power harvested in the actual application conditions (u_0 = 110 kV, d = 15 m). Figure 9 shows the photo of the prototype of energy

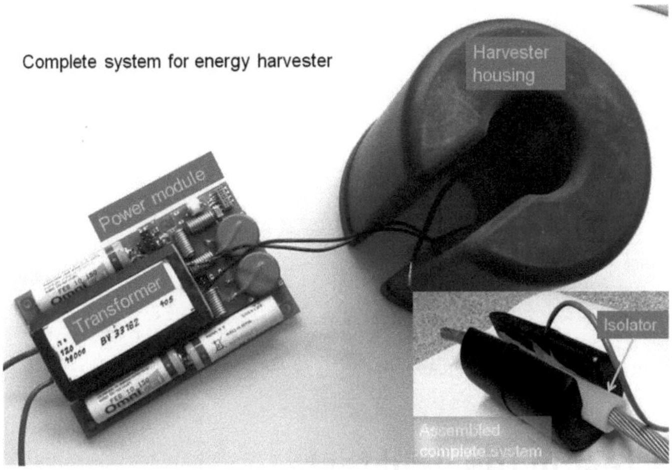

Fig. 9: Photo of the energy harvesting system.

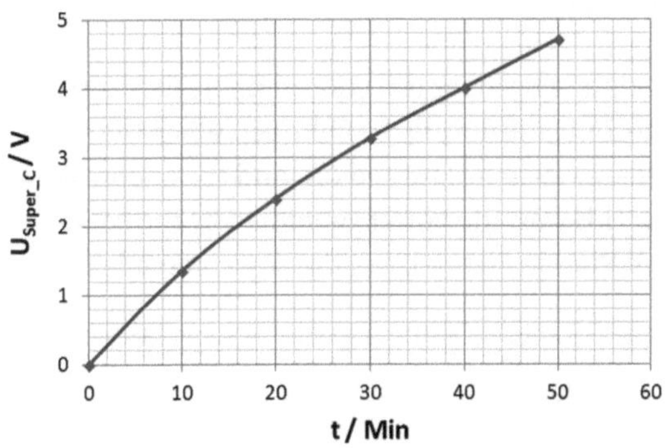

Fig. 10: Time-dependent voltage level of the supercapacitor.

harvester, which consists of the harvester housing and the power module. The transformer also belongs to the power module. The assembled system for energy harvesting is shown in the bottom right corner of this figure.

Figure 10 shows the time-dependent voltage level of the supercapacitor during a charging process with the nominal voltage (u_0) of 60 kV. Furthermore, the energy in the storage device can be calculated by equation (9). Its dependency on the charging time is displayed in Figure 11. The power in Figure 11 was calculated, and a power level of about 16.4 mW obtained at a nominal voltage of 60 kV. According to this obtained power and equation (6), the load impedance $Z_L \approx 4.8\,\text{M}\Omega$ can be calculated. For the field test, the power line is 15 m high and the nominal voltage is 110 kV. Inserting these parameters into equation (6), the available power level of about 17.3 mW can be calculated.

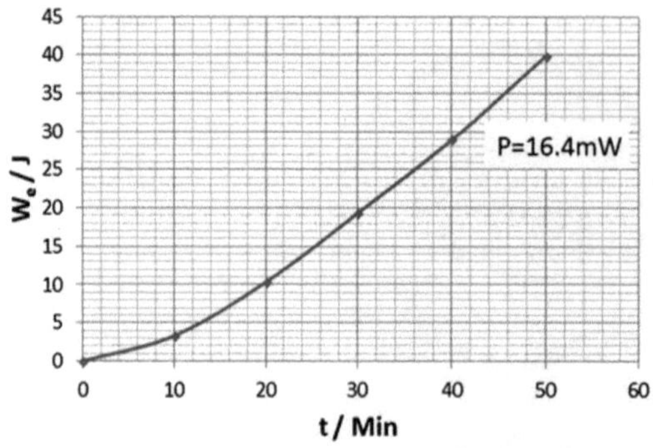

Fig. 11: Time-dependent energy stored in the supercapacitor.

4 Performance test in the field test

4.1 Test strategy for the field test

The test/debug block in the power module is used for test control in the field test, because the power module is difficult to reconfigure during the field test. When the supercapacitor is almost full, the load discharge switch is turned on, and the energy flows through the discharge resistor of 50 Ω. This way, the power generated by the harvester can be periodically measured. Through Pin 3 in Figure 12(a), the power module is con-

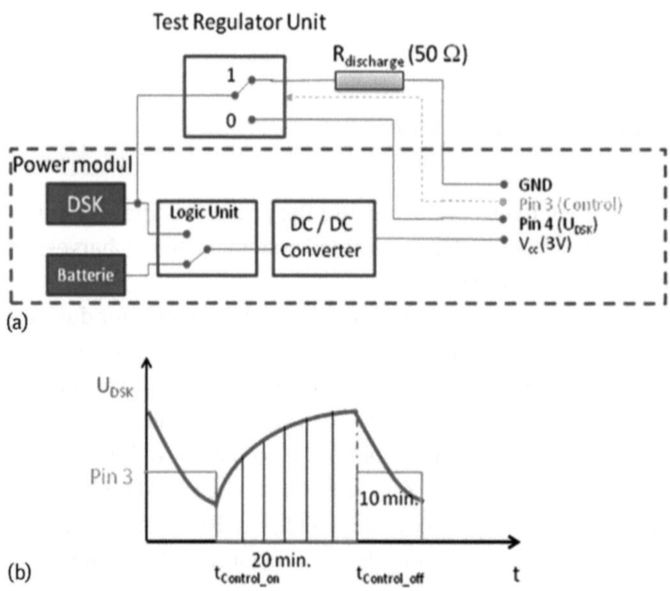

Fig. 12: (a) Principle of the test strategy, (b) timing diagram of the test unit.

Fig. 13: Experiment setup of the field test in Zwoenitz.

figurable to charge for 20 minutes and discharge for 10 minutes. Figure 12(a) sketches the base principle, and the timing diagram is shown in Figure 12(b). The experiment setup for the field test is shown in Figure 13.

4.2 Field test results

Figure 14 shows the results of the field test. Because of the discharge function, many periods can be obtained. According to the values in Figure 14, about 16.3 mW can be achieved from the ambient electric field, which is close to the theoretical estimation above. The power can be further increased by increasing the high voltage or enlarging the active harvester area (a larger diameter or length of the harvester) or optimizing the energy management by power matching between the source and the application.

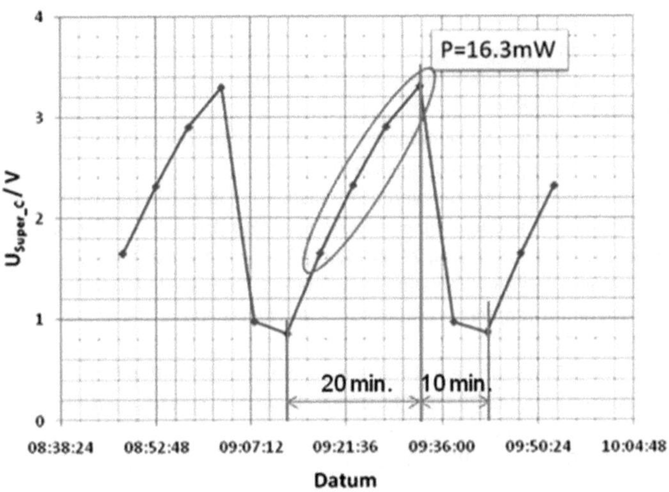

Fig. 14: Experiment results of the field test in Zwoenitz.

5 Conclusion

In this chapter, the energy supply for a sensor node applied in an online wireless monitoring system of overhead HV power lines is studied. The prototype of the harvester tube and the power module are developed and tested in a HV voltage laboratory and under field conditions. First, we presented theoretical and experimental investigations on energy scavenging from the electrical field of a HV power line. Simulations based on the model showed the feasibility of energy harvesting from the electric field around the HV conductor. Then, experimental results obtained from the field test proved that an energy harvester with a 20 cm long tube with a diameter of 10 cm can deliver an approximate power of 16 mW, which is enough to supply the wireless sensor nodes. In future, the influence of the bundle configuration in the power grid on the energy harvesting should also be considered. The bundle structure could be a two-bundle or a four-bundle conductor in the power grid. Furthermore, the influence of weather conditions should also be considered. For example, moisture in the rainy season results in problems like corona and partial discharge, which will disturb the wireless communication and may damage the electronic devices as well.

Acknowledgment: The authors thank the Professorship of Power Systems and High Voltage Engineering at the Chemnitz University of Technology for supporting the experiments in the HV laboratory. This work was funded by the German Federal Ministry of Education and Research (BMBF) within the project 16SV3771.

Bibliography

[1] X. Zhao, T. Keutel, M. Baldauf, and O. Kanoun. Power module for a wireless sensor node of a power line monitoring system. In *Eighth International Multi-Conference on Systems, Signals & Devices, Sousse, Tunisia*, pages 1–5, 2011.

[2] T. Keutel, X. Zhao, and O. Kanoun. Energy Scavenging for Monitoring of Overhead Power Line Networks. *SENSOR+TEST, Nuremberg, Germany*, pages 207–212, 2009.

[3] C. Bernauer et al. Temperature measurement on overhead transmission lines (OHTL) utilizing surface acoustic wave (SAW) sensors. In *Conference proceedings / 19th International Conference and Exhibition on Electricity Distribution: Vienna, 21–24 May 2007, Vienna, Austria*, 2007.

[4] M. Zhu, M. D. Judd, and P. J. Moore. Energy Harvesting in Substations for Powering Autonomous Sensors. In *2009 Third International Conference on Sensor Technologies and Applications, Athens, Greece*, pages 246–251, 2009.

[5] H. Zangl, T. Bretterklieber, and G. Brasseur. Energy Harvesting for Online Condition Monitoring of High Voltage Overhead Power Lines. In *2008 IEEE Instrumentation and Measurement Technology Conference, Victoria, BC, Canada*, pages 1364–1369, 2008.

[6] J.-L. Liu et al. Coaxial Capacitive Dividers for High-Voltage Pulse Measurements in Intense Electron Beam Accelerator With Water Pulse-Forming Line. *IEEE Trans. Instrum. Meas.*, 58(1):161–166, 2009.

[7] H. Ma and G. Wang. *COMSOL Multiphysics: User's Guide, Questions and Answers*. China Communications Press, Beijing, 2009.
[8] X. Yang and X. Li. *Engineering Electromagnetic Field*. China Electric Power Press, Beijing, 1st edition, 2007.
[9] W. B. J. Zimmerman. *COMSOL Multiphysics: Modeling and Analysis with Finite Element Methods*. China Communications Press, Beijing, 2007.

Series: Advances in Signals, Systems and Devices

Systems, Automation, and Control
F. Derbel, N. Derbel, O. Kanoun (Eds.), 2017
ISBN 978-3-11-046821-2, e-ISBN 978-3-11-047046-8,
e-ISBN (EPUB) 978-3-11-046850-2, Set-ISBN 978-3-11-047047-5

Sensors, Circuits and Instrumentation Systems
O. Kanoun, F. Derbel, N. Derbel, (Eds.), 2017
ISBN 978-3-11-044619-7, e-ISBN 978-3-11-044837-5,
e-ISBN (EPUB) 978-3-11-044624-1, Set-ISBN 978-3-11-044838-2

Power Systems & Smart Energies
F. Derbel, N. Derbel, O. Kanoun (Eds.), 2017
ISBN 978-3-11-046820-5, e-ISBN 978-3-11-047052-9,
e-ISBN (EPUB) 978-3-11-044628-9, Set-ISBN 978-3-11-047053-6

Communication, Signal Processing & Information Technology
F. Derbel, O. Kanoun, N. Derbel (Eds.), 2017
ISBN 978-3-11-046822-9, e-ISBN 978-3-11-047038-3,
e-ISBN (EPUB) 978-3-11-046841-0, Set-ISBN 978-3-11-047039-0

Bei Fragen zur Produktsicherheit wenden Sie sich bitte an:
If you have any questions regarding product safety,
please contact:

Walter de Gruyter GmbH
Genthiner Straße 13
10785 Berlin
productsafety@degruyterbrill.com